3rd International Conference on Energy and Environmental Protection 2018

E3S Web of Conferences Volume 46 (2018)

Krakow, Poland
13-14 September 2018

Editors:

M. Banas **T. Pajak**

ISBN: 978-1-5108-7294-3

Printed from e-media with permission by:

Curran Associates, Inc.
57 Morehouse Lane
Red Hook, NY 12571

Some format issues inherent in the e-media version may also appear in this print version.

Printed by Curran Associates, Inc. (2018)

For additional information, please contact EDP Sciences – Web of Conferences at the address below.

EDP Sciences – Web of Conferences
17, Avenue du Hoggar
Parc d'Activité de Courtabœuf
BP 112
F-91944 Les Ulis Cedex A
France

Phone: +33 (0) 1 69 18 75 75
Fax: +33 (0) 1 69 28 84 91

contact-edps@webofconferences.org

Additional copies of this publication are available from:

Curran Associates, Inc.
57 Morehouse Lane
Red Hook, NY 12571 USA
Phone: 845-758-0400
Fax: 845-758-2633
Email: curran@proceedings.com
Web: www.proceedings.com

TABLE OF CONTENTS

III INTERNATIONAL CONFERENCE

ENERGY AND ENVIRONMENTAL PROTECTION

LOCATION

AGH University of Science and Technology,
30 Mickiewicza Av., 30-059 Krakow
Conference Center, building B-2

DATE

13 - 14 September 2018

SCIENTIFIC EDITORS

- Marian Banaś – AGH UST Krakow (Poland),
- January Bień – Czestochowa University of Technology (Poland),
- Antoni Cieśla – AGH UST Krakow (Poland),
- Małgorzata Cimochowicz-Rybicka – Cracow University of Technology (Poland),
- Tadeusz Chmielniak – Silesian University of Technology, corresp. member of PAS (Poland),
- Ewa Czarnobilska – Medical College of Jagiellonian University (Poland),
- Valerio Cozzani – University of Bolonga (Italy),
- Georgij Drapak –Khmelnytskyi National University (Ukraine),
- Magdalena Dudek – AGH UST Krakow (Poland),
- Mariusz Filipowicz – AGH UST Krakow (Poland),
- Ján Gaduš – Slovak University of Agriculture in Nitra (Slovakia),
- Agata Godula-Jopek – Airbus Group Innovations (Germany),
- Andrzej Gołaś – AGH UST Krakow (Poland),
- Beata Gorczyca – University of Manitoba (Canada),
- Wacław Gudowski – KTH Royal Institute of Technology (Sweden),
- Stanisław Gumuła – AGH UST Krakow (Poland),
- Zbigniew Hanzelka – AGH UST Krakow (Poland),
- Krzysztof Jesionek – Wrocław University of Technology (Poland),
- Antoni Kalukiewicz – AGH UST Krakow (Poland),
- Bożenna Kawalec-Pietrenko – Gdansk University of Technology (Poland),
- Beata Kępińska – IGSMiE PAN Krakow (Poland),
- Jan Kiciński – IMP PAN Gdansk, corresp. member of PAS (Poland),
- Włodzimierz Kowalski – AGH UST Krakow (Poland),
- Władysław Kryłłowicz – Lodz University of Technology (Poland),
- Janusz Lewandowski – Warsaw University of Technology (Poland),
- Mychajło Łobur – Lviv Polytechnic National University (Ukraine),
- Janina Molenda – AGH UST Krakow (Poland),
- Miloš Nedeljković – University of Belgrade (Serbia),
- Wojciech Nowak – CoE AGH UST Krakow (Poland),
- Tadeusz Pająk – AGH UST Krakow (Poland),

- Lucjan Pawłowski – Lublin University of Technology, corresp. member of PAS (Poland),
- Vladimír Rataj – Slovak University of Agriculture in Nitra (Slovakia),
- Czesława Rosik-Dulewska – Institute of Environmental Engineering PAS (Poland),
- Stanisław M. Rybicki – Cracow University of Technology (Poland),
- Stanisław Rychlicki – AGH UST Krakow (Poland),
- Peter Quicker – RWTH Aachen University (Germany),
- Andrzej Salamon – AMK Iserlohn (Germany),
- Helmut Seifert – Karlsruhe Institute of Technology (Germany),
- Jakub Siemek – AGH UST Krakow, corresp. member of PAS (Poland),
- Tatyana Solomakhova – TsAGI Moskwa (Russia),
- Ulrich Stimming – Newcastle University (UK),
- Wojciech Suwała – AGH UST Krakow (Poland),
- Stefan Taczanowski AGH UST Krakow (Poland),
- Antoni Tajduś – AGH UST Krakow, Committee of Mining of PAS (Poland),
- Jan Taler – Cracow University of Technology (Poland),
- Rumen Tomov – University of Cambridge (UK),
- Witold Warzecha – MPEC Krakow (Poland),
- Roman Weber – TU Clausthal (Germany),
- Grzegorz Wielgosiński – Lodz University of Technology (Poland),
- Tadeusz Wójcik – AGH UST Krakow (Poland).

Statement of Peer review

In submitting conference proceedings to *Web of Conferences*, the editors of the proceedings certify to the Publisher that

1. They adhere to its **Policy on Publishing Integrity** in order to safeguard good scientific practice in publishing.
2. All articles have been subjected to peer review administered by the proceedings editors.
3. Reviews have been conducted by expert referees, who have been requested to provide unbiased and constructive comments aimed, whenever possible, at improving the work.
4. Proceedings editors have taken all reasonable steps to ensure the quality of the materials they publish and their decision to accept or reject a paper for publication has been based only on the merits of the work and the relevance to the journal.

Title, date and place of the conference

3rd International Conference on Energy and Environmental Protection
13 - 14 september 2018, Krakow, Poland

Proceedings editor(s):

Marian Banaś PhD DSc
Tadeusz Pająk PhD DSc

Date and editor's signature

17, avenue du Hoggar - PA de Courtabœuf – BP 112 - 91944 Les Ulis Cedex A (France)
Tél. : 33 (0)1 69 18 75 75 - Fax : 33(0)1 69 07 45 17 – www.edpsciences.org

E3S Web of Conferences **46**, 00001 (2018) https://doi.org/10.1051/e3sconf/20184600001
3rd International Conference on Energy and Environmental Protection

A research station with a two-disc cutting unit for studying the cutting process of energy

Henryk Rode[1,*]

[1]Warsaw University of Technology, Faculty of Civil Engineering, Mechanics and Petrochemistry, Institute of Mechanical Engineering, Łukasiewicza 17, 09-400 Płock, Poland

Abstract. The paper presents the construction of a new research station for studying the cutting process of energy plants. The station consists of 3 basic units: a truck with drive and cutting units, a frame which is a guide for the truck moving on it, and a base for placing the plant's stems. The station allows to determine the unitary energy of the cutting process for various rotary and travelling speeds of the cutting unit and for various sizes, spacing and angles of inclination of the cutting discs. It also allows to form freely the cut plants' stems' sockets.

1 Introduction

Depletion of energy from conventional sources and a continuous increase in energy consumption necessitates the search for new energy sources, preferably from renewable ones [1,5]. One of such sources is biomass [3].

Proper harvesting and fragmentation of energy crops have a great impact on the profitability of their production. Therefore, it is necessary to know about the processes occurring during cutting and fragmenting plants' stems, with particular emphasis on the factors that most affect the energy consumption of these processes [2,4,12].

A research on the cutting process of cereals and grasses has been carried out for several years at the Institute of Mechanical Engineering at Warsaw University of Technology in Płock [13], and recently also of energy plants. Laboratory stands of pendulum and rotational construction are used for the research, where the process of axial cutting and inertial cutting of plants takes place [7,8,9,10].

Due to the fact that the process of harvesting energy plants is most often carried out by a two-disc rotary unit [6], the latest research station for testing this process is equipped with such a cutting unit. It enables the measurement of the unitary energy of the cutting process of various plants for the variable construction and working parameters of the cutting unit. The unitary energy of the cutting process is the total energy needed to carry out the cutting process related to the cross-section area of the stems of the cut plants.

* Corresponding author: henryk.rode@pw.edu.pl

2 Research station - assumptions and criteria of its construction

The new research station must pursue the process of cutting plants in conditions similar to the ones on the field. The cutting unit is to consist of two interchangeable circular saw blades installed on the truck, which travels over the frame above the plants' stems placed in a special base with handles enabling the forming of plant sockets. The energy required to carry out the cutting process is recorded by the computer card.

The new research station should meet the following criteria:

- ensure safe work during testing,
- ensure durability of cutting elements,
- enable a wide range of rotary speed regulation of the cutting unit,
- enable a wide range of travelling speed regulation of the cutting unit,
- ensure the possibility to form freely the sockets of plants' stems',
- ensure the possibility of changing the inclination angle of the discs to the base surface,
- ensure the possibility of using discs of various diameters and different axle bases,
- ensure low construction costs.

3 Research station

The station (Fig. 1,2) consists of three basic units: a truck with drive and cutting units (Fig.3), a frame which is a guide for the truck moving on it and a base for placing the plant's stems (Fig.5).

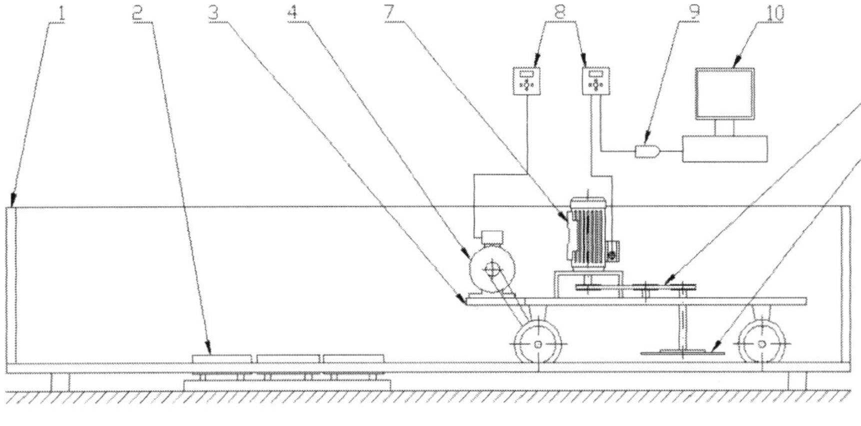

Fig. 1. Research station with a two-disc cutting unit for studying the energy consumption during the cutting process of energy plants [1]:1 – frame, 2 – a base for placing the samples, 3 – a truck with drive and cutting units, 4 – electric motor for the truck, 5 – disc, 6 – belt transmission, 7 – electric motor for cutting unit ,8 – frequency converters, 9 – signal converter, 10 – controlling computer with a software

The maximum rotary speed of the motor is 2890 rpm. The drive of the cutting unit is transmitted from the motor by means of a belt transmission with a V-belt, arranged in an eight figure, in order to obtain opposite rotation directions of the cutting discs. The tensioner attached to the truck's frame is responsible for the tension of the V-belt.

The other motor (Pro-Motor MS 80 1-4, three-phase with the power of 0,55 kW and maximum rotary speed of 1370 rpm) ensures the movement of the truck with the cutting unit (Fig.3). The feed motion is realized thanks to the angular and belt transmission. The turning moment is transferred to the shaft with wheels placed on the guide. The belt tension adjustment is made by changing the position of the motor.

Fig. 2. Research station – view without housing

Fig. 3. A truck with drive and cutting units

For the station drives a three-phase motor, Tamel 3SK6100L-2, with rated power of 3 kW was used (Fig.3).

The cutting unit, used for pilot research, consists of 2 discs with an external diameter of 400 mm and a thickness of 2 mm. The number of teeth of each disc is 80 (Fig.4). The axes of saw blades turn are spaced apart by 410 mm. The adjustment of the inclination angle of the discs was carried out using screws, which are also used to tension the belt of the transmission belt responsible for the movement of the truck with the cutting unit.

Placing the plant samples takes place on a specially prepared base. It is made of two sheets of metal, in which holes with diameters from 10 to 25 mm were drilled randomly (Fig.5). Thanks to such a construction, plant sockets, similar to those existing in reality on the plantation, can be easily formed. Immobilization of the samples in vertical position takes place by means of shifting the sheets of metal relative to each other using screws and nuts.

Fig. 4. Discs

Fig. 5. A base for placing the samples

Two frequency converters LS - iG5A series are responsible for controlling the rotary speed of electric motors (Fig.6). The supply of motors and frequency converters as well as their switching on and off takes place through a fuse strip. The control of travelling speed of the truck with the cutting unit is effected by changing the frequency of the current realized by means of the frequency converter located on the left side of the control board (Fig. 6). Changing the current frequency and switching the motor on and off is effected directly from the frequency converter with the use of a special keyboard.

Rotary speed control of the cutting unit is achieved by changing the current frequency in the second converter. The ADA - I9140 signal converter and a computer with dedicated DriveView7 control software are used to record the current consumption. Using a measurement and control card, the signal from the frequency converter is downloaded to the computer. The computer records the changes in electric power consumption of the motor driving the cutting unit in time. Autodesk AutoCAD determines the area under the graph corresponding to the energy collected for the cutting process of the energy plant. The total energy obtained refers to the cross-sectional area of the samples of the stems of the cut plants, obtaining the unit energy of the cutting process.

The station is protected by a housing that protects the cut material from falling out and at the same time allows to easily take out the cut samples and place the new ones.

Fig. 6. Frequency converters LS - iG5A series and fuse strip

4 Research material

The research material for tests on the presented station are energy plants. They are characterized by a very diverse morphological structure. Their common feature is that they can be used as energy material in the form of biomass. This fuel has one feature that cannot be overestimated. The combustion process does not affect the balance of CO2 in the natural environment. The amount of carbon dioxide emitted to the atmosphere during the combustion of biomass is balanced by the amount of CO2 absorbed by the plants during its reproduction in the process of photosynthesis. The use of biomass also allows to manage biological waste.

Energy plants used for biomass production are characterized by high resistance to diseases and pests, high calorific value, small amount of carbon dioxide emitted to the atmosphere in the combustion process, high increase in dry matter in the growing season, small soil requirements, the possibility of using wasteland for cultivation, high mass density, which is extremely important during transport and storage within the boiler room. The plants are also characterized by their susceptibility to mechanization of agro technical activities related to establishing a plantation, its maintenance and biomass harvest [3,5].
The group of these plants includes:

- basket willow,
- Jerusalem artichoke,
- poplars,
- Pennsylvania mallow,
- multiflorous rose,
- miskanthus giganteus,
- Sakhalin knotweed.

5 Pilot research

Basket willow was used as the research material in the pilot research [11]. The functioning of the station in the rotary speed range of the cutting unit from 1120 to 2400 rpm and the range of travelling speeds from 0.046 to 0.083 m / s were checked. Samples with diameters of plant stems from 7 to 22 mm were cut.

The study confirmed the validity of most of the assumptions. Drive units functioned correctly. The method of placing samples needs correction as the samples of plants turned out to be unstable. An additional third terminal plate should be used. The range of change of the inclination angle of the cutting discs to the base surface is worth to be increased in order to reduce the friction between the discs' surfaces and cross-cut sections of the plant stems.

6 Conclusion

The new research station is the most technically advanced of all stations for testing cutting process built at the Institute of Mechanical Engineering at Warsaw University of Technology in Płock. It gives the possibility to conduct a wide range of tests for variable construction and working parameters of the cutting unit as well as for variable parameters of the plant itself, e.g.: stem's diameters or moisture. The obtained results allow for a better understanding of the cutting process of energy plants carried out with a two-disc cutting unit. The research results will also enable the construction of more efficient and energy-saving cutting units for agricultural machines.

References

1. Dreszer K., Michałek R., Roszkowski A., *Energia odnawialna – możliwości jej pozyskiwania i wykorzystania w rolnictwie*, Wyd.. PTIR, Kraków – Lublin –Warszawa (2003)
2. Frączek J., Mudryk K., Metoda określenia oporów cięcia pędów wierzby energetycznej. Inżynieria Rolnicza, **8**, 83, pp. 91-98, (2006)
3. Gradzik P., Grzybek A., Kowalczyk K., Kościk B., *Biopaliwa*, Wyd. Wieś jutra, Warszawa, (2003)
4. Kowalski S., *Badania oporów cięcia wybranych roślin*. Zeszyt Prob. Post. Nauk Rol. 408, pp.297- 303, (1993)
5. Lewandowski W.M., *Proekologiczne odnawialne źródła energii*, Wyd. Naukowo-Techniczne, Warszawa (2007)
6. Lisowski A. i inni., *Technologie zbiorów roślin energetycznych*, Wyd. SGGW, Warszawa (2010)
7. Rode H., Witkowski P., The study of the rotary cutting process of chosen energy plants. TEKA Commission of motorization and energetics in agriculture. An international journal on motorization, vehicle operation, energy efficiency and mechanical engineering, **13**, 1, Lublin – Rzeszów, , pp. 139-144, (2013)
8. Rode H., Witkowski P., Stanowisko do badań procesu cięcia roślin energetycznych. MOTROL. Commission of motorization and energetics in agriculture, An international journal on operation of farm and agri – food industry machinery, **15,** 1, Lublin – Rzeszów, pp. 111-114, (2013)
9. Rode H., Witkowski P., Investigations Into the Process of Rotary Mowing of Selected Energy Crops. TEKA Commission of motorization and energetics in agriculture. An international quarterly journal on motorization, vehicle operation, energy efficiency and mechanical engineering, **14**, 3, Lublin – Rzeszów, pp. 95-98, (2014)
10. Rode H., Chęciński M., Research on Disc Cutting Unit for Cutting Energy Plants, TEKA Commission of motorization and energetics in agriculture, An international quarterly journal on motorization, vehicle operation, energy efficiency and mechanical engineering, **16**, 1, Lublin – Rzeszów, pp. 25-30, (2016)

11. Szczukowski S., Tworkowski J., Stolarski M.J., *Wierzba energetyczna*, Wyd. Plantpress Sp. z o.o., Kraków (2014)
12. Szymanek M., *Analysis of cutting process of plant material*, Teka Komisji Motoryzacji i Energetyki Rolnictwa – OL PAN, VIIA, Lublin, pp.107-113, (2007)
13. Żuk D. *Proces cięcia źdźbeł zbóż*, Prace Naukowe Politechniki Warszawskiej - Mechanika z. 95, Warszawa (1985)

Analysis of energy efficiency of photovoltaic installation in central Poland

Mariusz Sarniak[1,*]

[1]Warsaw University of Technology, Faculty of Civil Engineering, Mechanics and Petrochemistry, Institute of Mechanical Engineering, Łukasiewicza 17, 09-400 Płock, Poland

Abstract. The paper analyzes the electricity generation from photovoltaic micro-installation, which was not subjected to any maintenance operations such as cleaning or periodic change of the PV panel tilt angle, recorded in the years: 2015-2017. The PV micro-installation presented is a system called On-Grid, i.e. it is connected to a low voltage power network, and the PV generator is built from the two most commonly used types of PV modules: from mono- and polycrystalline silicon. Data sets for relative energy yields combined and broken down into PV generator manufacturing technologies were analyzed. The results of the measurements were also divided into percentage amounts of generated energy in two time intervals: "summer" months from April to September (on average over 80% of generated electricity) and others recognized as "winter". This division was also made for two technologies of making a PV generator.

1 Introduction

The development of photovoltaics in Poland is a derivative of the country's energy policy. Under the Directive of the European Parliament 2009/28/EC, Poland is obliged to achieve a minimum 15% share of energy from renewable sources in final gross energy consumption in 2020. The implementation plan for achieving this goal is the National Action Plan (NAP) in the field of energy from renewable sources. It assumes that the share of energy from renewable sources in the final energy consumption will be higher than the set minimum and will amount to 15.85%. Currently (until the end of 2017), 29197 solar systems with a total capacity of 281.4 MW have been installed in Poland. This number consists of 589 units of systems licensed by the Energy Regulatory Office (URE) with a total capacity of 107.7 MW and 28608 units of prosumer systems with a total capacity of 173.7 MW [1].

Peak power values, which are reported by the manufacturers of solar modules, are very rarely obtained in practice in Poland (Central and Eastern Europe). STC (Standard Test Conditions) in which new modules are tested in this location do not actually occur. Poland lies in the zone of large differences in the size of solar radiation in the winter and summer months. The genesis of the paper is the search for optimal solutions based on the

* Corresponding author: mariusz.sarniak@pw.edu.pl

observation of energy yields in PV systems that already exist in a given location. In the paper, the PV system connected to the utility network was tested.

At present, virtually all manufacturers of inverters used in PV systems provide adequate additional equipment that allows monitoring of their operation. PV system monitoring is not only an effective gadget, but above all, continuous registration of basic operating parameters, enabling analysis and detection of all kinds of irregularities. The preview of data from monitoring from mobile devices may also be a kind of security element protecting the external elements of the PV system against theft and vandalism.

2 Test stand description

The test stand is the demonstrative mini PV system, located in Płock (GPS coordinates: 52.546713N, 19.706945E), on the flat roof of a three-storey University building. The PV generator was built from modules of the two most popular types of mono- and polycrystalline silicon (Fig. 1). The PV modules are mounted in the azimuthal deviated by 27° from the direction of the South to the West and inclined at an angle of 45° to the horizontal.

Fig. 1. Photovoltaic PV generator: left side polycrystalline PV modules - **CL-130-12** model, and right side PV monocrystalline modules –**AEMF 130** model

The PV system uses inverters in the form of two Sunny Boy 240 micro inverters, which were connected to mono- and polycrystalline modules directly with the PV modules. An important element of the PV system is the Sunny Multigate device, which connects the PV mini system with the building's electrical network (connection to one phase) and the computer network router connected to the Internet. Fig. 2 presents a simplified wiring diagram of all elements of the PV system.

A maximum of 12 Sunny Boy 240 micro inverters can be connected to one Sunny Multigate network interface device. In order for the PV installation data to be sent to the Sunny Portal server, it is necessary to register the installation based on individual device numbers assigned by the manufacturer. After a properly carried out registration process, the system every 15 minutes sends data to the server separately for each micro inverter. There is also the possibility of local monitoring using the Sunny Explorer application, if the location of the system does not allow connection to the Internet. Fig. 3 presents a functional diagram of the PV system and ways registering of energy yields.

The maximum output power of Sunny Boy 240 micro inverters is 245W. PV modules with a total peak power of $260W_p$ were connected to each of them. This configuration is correct, because for the analysed geographical location the maximum allowable nominal power factor is NPR=1.2 (*Nominal Power Ratio*), and we define it as a ratio of peak power of the PV generator to the maximum power of the inverter. The NPR coefficient calculated according to the above definition is equal 1.06, which is acceptable due to the lack of real possibilities of obtaining peak power by the PV panel (see in Fig. 9). Such a load on the PV inverter would be wrong in the inter-tropic geographic zone, where the PV generator may even exceed its peak power value due to the higher than the nominal solar radiation intensity occurring in these areas.

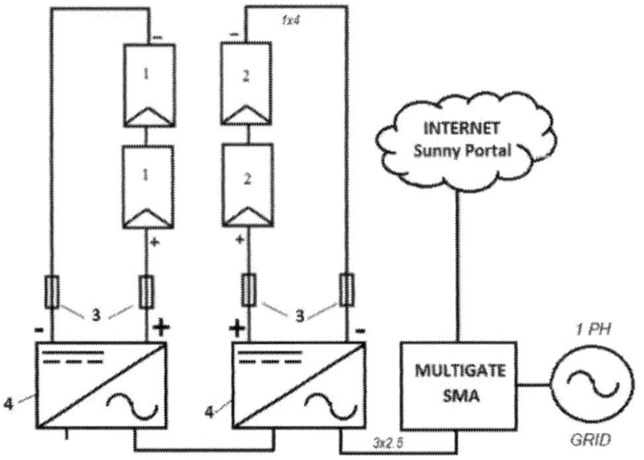

Fig. 2. Connection diagram for PV system components: 1-mono PV modules (AEMF130), 2-polycrystalline PV modules (CL130-12), 3-fuses, 4-micro inverters Sunny Boy 240

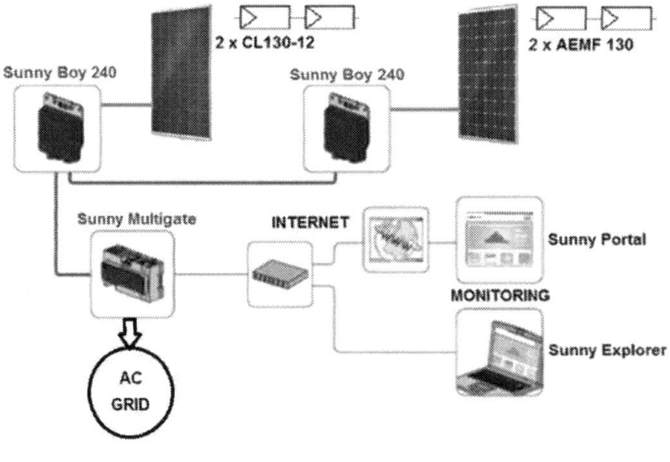

Fig. 3. Functional scheme of the PV mini system

3 Research methodology

No radiation measurements are made at the PV installation site. As comparative data was generated average monthly electricity production from the given system for four databases in PVGIS [2,3]. On this basis, it was estimated average monthly expected energy yield of PV system (Tab. 1).

The data obtained from the demonstration PV system in the years 2015-2017 were used for the analysis. The data was balanced on an annual basis and divided into individual months. The list was made for the whole system together and divided into outputs from mono- and polycrystalline PV modules. Analyzed values of energy yields are output data, energy coming from the micro inverters.

During the whole period of data collecting, the PV generator was not subjected to any operations. The results for 2017 are therefore subject to losses resulting from the accumulation of PV generator pollution. This is an important aspect because Płock is the seat of Poland's largest producer of liquid fuels, which affects significant air pollution. Near the PV generator, there is no position for measuring and recording the intensity of solar radiation and temperature.

Table 1. Average monthly electricity production from the given system (values for four databases in kWh) [2,3] and average expected energy yield in kWh/kW_p

Month	PVGIS-CMSAF	PVGIS-SARAH	PVGIS-ERA5	PVGIS-COSMO	Average expected energy yield of PV system
January	14.4	16.2	17.8	13.6	29.8
February	25.3	21.9	31.1	23.0	48.7
March	49.4	42.5	54.0	44.2	91.4
April	65.4	61.2	64.5	55.4	118.5
May	68.9	65.6	64.8	57.6	123.5
June	67.7	64.2	65.2	57.8	122.5
July	67.4	64.7	63.5	55.2	120.6
August	65.4	60.9	63.2	51.1	115.7
September	53.5	52.8	51.0	42.9	96.3
October	35.9	37.1	38.0	30.1	67.8
November	16.8	16.7	18.2	13.5	31.3
December	12.5	13.8	14.9	9.57	24.4
Annual sums of energy produced	542.6	517.6	546.2	453.97	990.5

4 PV system tests results

Figures 4-6 show the relative energy yields for subsystems with PV modules from mono- and polycrystalline silicon for each year, broken down by months. A thorough analysis of the measurement results shows a large difference in the generated energy between summer months: from April to September and winter months from October to March. The percentage distribution of generated energy and average values from three years are presented in Table 2.

Table 2. Percentage share of individual technologies in generating electricity

Generator type	Winter 2015	Summer 2015	Winter 2016	Summer 2017	Winter 2017	Summer 2017	Winter average	Summer average
Poly-crystalline	25%	75%	23.5%	76.5%	24.2%	75.8%	24.2%	75.4%
Mono-crystalline	10.9%	89.1%	10.6%	89.4%	14.6%	85.4%	12.0%	88.0%

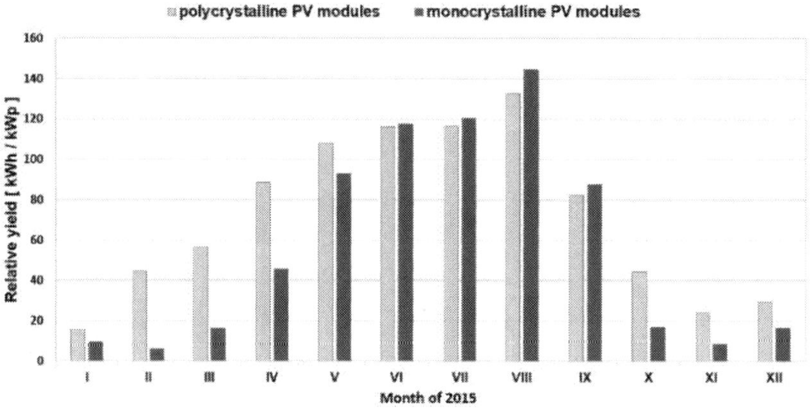

Fig. 4. Relative energy yields from the demonstrative PV system with division into months and generator manufacturing technologies in 2015

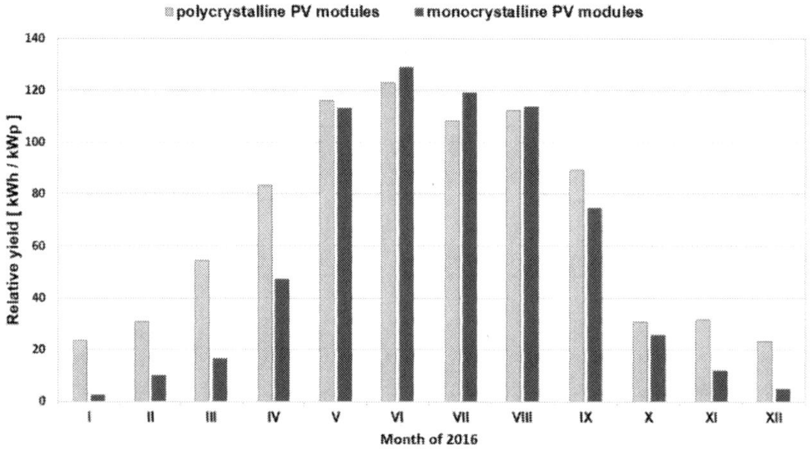

Fig. 5. Relative energy yields from the demonstrative PV system with division into months and generator manufacturing technologies in 2016

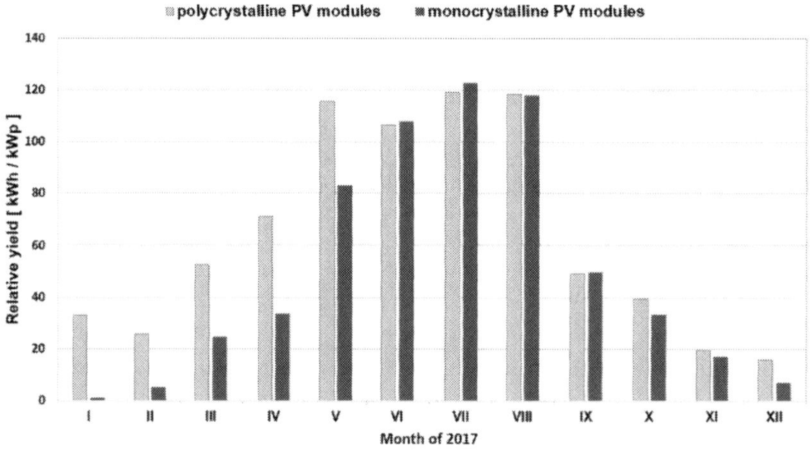

Fig. 6. Relative energy yields from the demonstrative PV system with division into months and generator manufacturing technologies in 2017

Figure 8 show balances for the entire PV system in total. The listings were made in the same way as before, but without a division into the technologies of the PV generator. Figure 7 shows the part of the balance relating to the summer months, which in this case accounts for over 80% of the relative energy yield for the PV generator in 2015-2017.

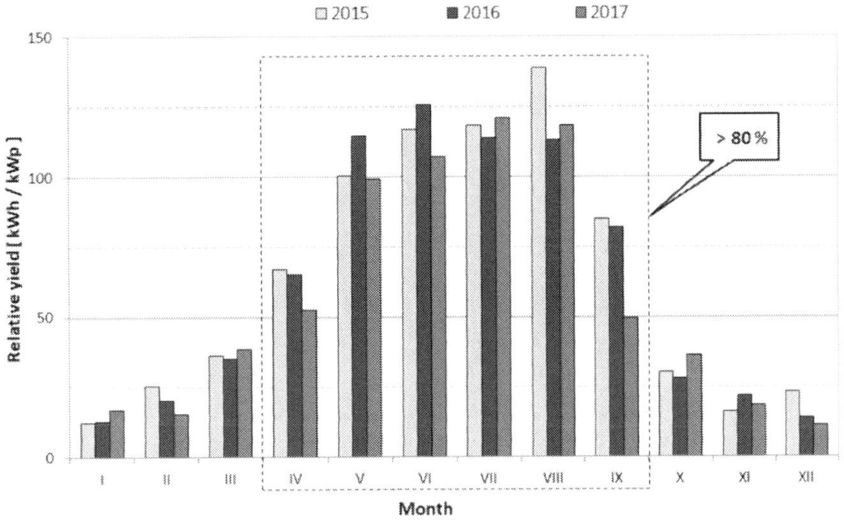

Fig. 7. Total balance of PV generator yields divided into months in 2015-2017

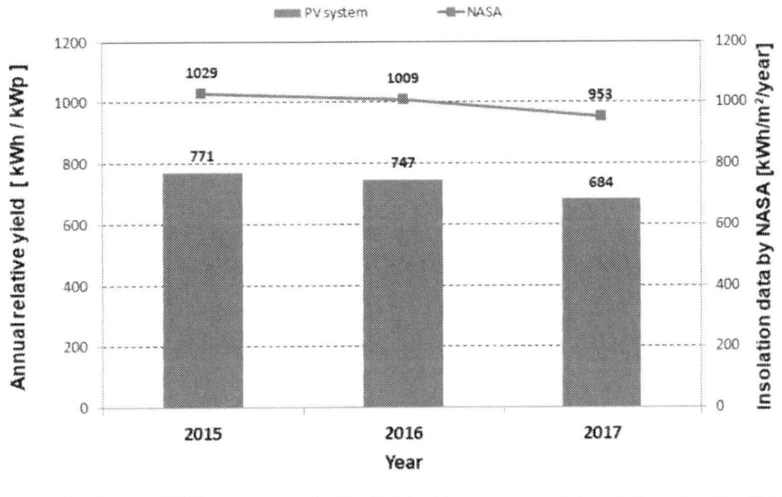

Fig. 8. The total balance of PV generator yields divided into years and insolation data by NASA [4]

Figure 9 shows the micro inverters power histogram for two selected days, characterized by the highest power values. This analysis show the real load of inverters throughout the considered period. This confirms the correctness of the PV generator selection for micro inverters at the NPR=1.06 level. Micro inverters did not work at maximum power in any time interval.

Figure 10 presents micro inverters power diagrams for an example of a "Winter Day" (2015-12-24) with the shading effect of a PV generator. The time of occurrence of the shading effect coincides with the direction of the location of two tall trees relative to the PV installation.

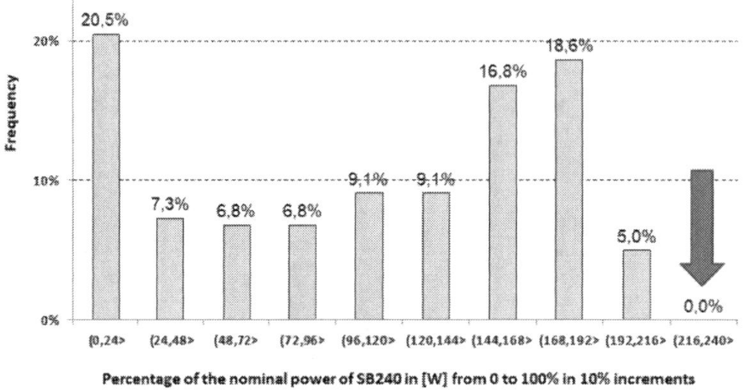

Fig. 9. Histogram of power distribution of micro inverters for two selected days: 2015-08-10 and 2015-08-11

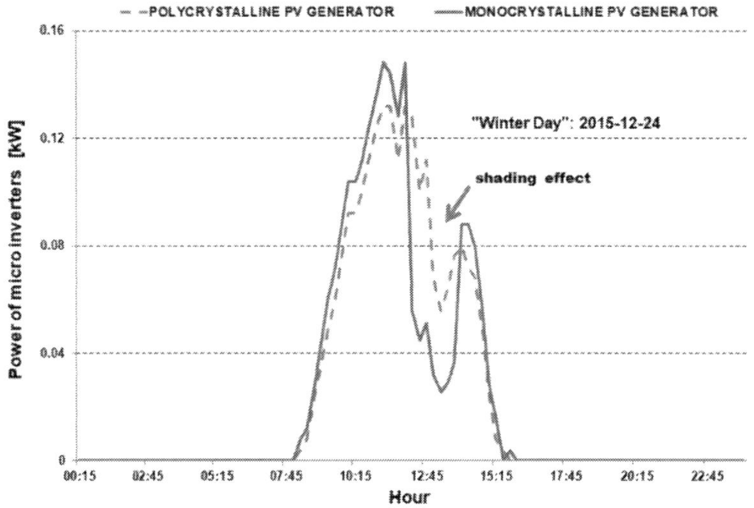

Fig. 10. Daily diagrams of power for micro inverters with an exemplary shading effect for the "Winter Day"

5 Conclusion

The analysis of energy yields for the described installation leads to the following conclusions:

- During the period of testing of the PV installation in 2015-2017, large differences in the annual values of generated electricity were observed (Fig. 8). In the summer months, i.e. from April to September, it was over 80% (Tab. 2, Fig. 7).
- For the considered location, energy yields from the PV system in the given location were smaller than theoretically estimated (Tab. 1) and were characterized by a clear downward trend (Fig. 8). The relative decrease in solar insolation (calculated on year to year basis) shown Figure 8 is respectively: 1.98% and 5.58%. The decrease in electricity produced (calculated in the same way) is 3.1% and 8.43%. This clearly indicates the impact of a other factor (temperature, soiling or shadingof PV module). It is also

recommended to install sensors for measurement of insolation and temperature at the PV installation location [5].

- One of the reasons for the lower generation of normalized energy is the shading effect visible in Fig. 10, which reduces the power of the generator made of monocrystalline PV modules. The exact of the shading effect on the PV generator's efficiency can be estimated on the basis of the 3D model of the shading elements relative to the location of the PV installation (e.g. in the PVSYST software) [6-8].
- Another recommendation resulting from the analysis of experimental results is to examine the influence of the of the tilt angle of PV panels on the degree of self-cleaning, while maintaining the criterion of maximizing energy yields, similar suggestions can be found in the literature, e.g. [9,10].
- The long time consumed by the research presented in the work (three-year observations) prompts to conduct such analyzes in a simulation manner on models. In the Matlab / SImulink package, two dedicated components were implemented: "Solar Cell" and "PV Array", which can be used to build a PV generator model [11-15].

References

1. Pietruszko S. Photovoltaic market in Poland - 2017, Magazyn Fotowoltaika, **1**, pp. 6-11, (1018)
2. Photovoltaic Geographical Information System (PVGIS 5 beta release) http://re.jrc.ec.europa.eu/pvg_tools/en/tools.html#PVP
3. Huld, R. Muller, Gambardella A., A new solar radiation database for estimating PV performance in Europe and Africa, Solar Energy, **86**, 6, pp. 1803–1815, (2012)
4. RETScreen - Clean Energy Management Software system for energy efficiency, renewable energy and cogeneration project feasibility analysis as well as ongoing energy performance analysis http://www.nrcan.gc.ca/energy/software-tools/7465
5. Rodziewicz T., Teneta J., Zaremba A., et al. Cheap Sensor Made of Multicrystalline Silicon for Insolation and Temperature Measurements, Ecological Chemistry and Engineering, **23**, 4, pp. 583-591, (2016)
6. Goss B., Cole I., Betts T., Gottschalg R. Irradiance modelling for individual cells of shaded solar photovoltaic arrays, Solar Energy, **110**, pp. 410–419, (2014)
7. Lua F., Guoa S., Walsha T.M.,Aberlea A.G. Improved PV Module Performance Under Partial Shading Conditions, Energy Procedia, **33**, pp. 248-255, (2013)
8. Fialhoa L. Melicioa R., Mendesa V.M.F., Figueiredoa J., Collares-Pereiraa M. Effect of Shading on Series Solar Modules: Simulation and Experimental Results, Procedia Technology, **17**, pp. 295-302, (2014)
9. Styszko K., Jaszczur M., Teneta J., et al., An analysis of the dust deposition on solar photovoltaic modules, Environmental Science and Pollution Research, (2018)
10. Jaszczur M., Teneta J., Styszko, Hassan Q., Burzyńska P., Marcinek E., Łopian N. The field experiments and model of the natural dust deposition effects on photovoltaic module efficiency, Environmental Science and Pollution Research, (2018)
11. Messenger R.A., Ventre J., *Photovoltaic Systems Engineering Third Edition*, CRC Press Taylor & Francis Group, (2010)
12. Sarniak M., *Budowa i eksploatacja systemów fotowoltaicznych*, Wydawnictwo Grupa Medium, Warszawa, (2015)
13. Sarniak M. Modeling of photovoltaic modules in Simulink and Simscape packages of Matlab software. Econtechmod. An International Quarterly Journal, **6**, 1, pp. 133-138, (2017)

14. Khatib T., Elmenreich W., *Modeling of Photovoltaic Systems Using MATLAB: Simplified Green Codes*, John Wiley & Sons, (2016)
15. Venkateswarlu G., Sangameswar Raju P.S. Simscape Model of Photovoltaic Cell, International Journal of Advanced Research in Electrical, Electronics and Instrumentation Engineering, **2**, 5, pp. 2320–3765, (2013)

E3S Web of Conferences **46**, 00003 (2018) https://doi.org/10.1051/e3sconf/20184600003
3rd International Conference on Energy and Environmental Protection

Analysis of selected aspects of collective water supply system operation in the selected municipality

Krzysztof Boryczko[1,*]

[1]Rzeszow University of Technology, The Faculty of Civil and Environmental Engineering and Architecture, Faculty of Management, Poznańska 2, 35-084 Rzeszów, Poland

Abstract. The purpose of the work was to analyze the exploitation of the collective water supply system (CWSS), on the example of Czudec, Podkarpackie CWSS. The water balance was prepared. The system operation was simulated using the Epanet software. The obtained data was presented in graphical form of pressure distribution on the network during the minimum and maximum demand. Water age simulations were also prepared, that is, the period of water residence in water distribution subsystem.

1 Introduction

Collective water supply systems (CWSS) existed already in antiquity, but in the 20th century that were introduced on a mass scale, which increased the hygiene of society's life. A significant increase in the length of the water supply network was observed at the end of the 20th century and in the 21st century, in particular in rural areas. In 1989, the length of the water supply network in Poland was 88.3 thousand km, while in 2014 it increased to 292.5 thousand km. The result of building new networks is increase people number who can use tap water. The percentage of urban and rural water supply has risen from 84.8% in 2002 to 91.6% in 2014, while 84% used it in rural areas in 2014, and 96% in urban areas. Acceleration of investments related to the build of water distribution subsystem (WDS) is related to the possibility of obtaining European funds by water companies and municipalities [4,5,6,8,9].

Water companies in Poland consume water primarily for industrial and household purposes. The highest water consumption took place in the second half of the 20th century. This was caused by low price of water. The increasingly water quality standards have forced companies to develop newer and more expensive water treatment technologies, which resulted in a drastic increase in prices [1-3]. The result was a reduction of water consumption per statistical inhabitant [7].

In the era of computerization, managers of water supply companies use computer programs that enable monitoring of water distribution subsystem, its correct operation, as well as the ability to quickly eliminate water pipes failures. The aim of the paper is analysis of the WDS Czudec commune operation.

* Corresponding author: kb@prz.edu.pl

2 Description of the analyzed system of collective water supply

The water distribution subsystem was created relatively recently. The first water pipes were built in 1994. Before that time, the inhabitants took water from their own individual wells. For the next 9 years, the water supply network did not increase. The breakthrough year was 2003, because the residents of the commune were running out of water. At that time, a decision was made to further expand WDS. It consists pipes with diameters: Ø50, Ø63, Ø90, Ø110, Ø125, Ø160, Ø225. The main material that was used to build the network is polyethylene (PE). Table 1 presents the characteristics of the WDS in 2009-2016.

Table 1. Characteristics of the WD in the Czudec commune in 2009-2016

year	Length	diameter	Number of residents	Number of residents connected to WDS	% residents connected to WDS
	[km]	[mm]	[-]	[-]	[%]
until 2009	86.569		11569	2606	22.5
2010	108.047	-	11588	3251	28.1
2011	1.250	Ø63	11749	3265	27.8
	0.750	Ø110			
	5.750	Ø160			
	3.080	Ø225			
Σ	118.877				
2012	3.765	Ø110	11777	3278	27.8
	5.835	Ø160			
Σ	128.477				
2013	4.139	Ø50	11721	3325	28.4
	4.832	Ø110			
Σ	137,448				
2014	0.024	Ø63	11723	3461	29.5
	1.662	Ø110			
	0.721	Ø160			
Σ	139.855				
2015	0.976	Ø50	11788	4139	35.1
	0.374	Ø63			
	0.081	Ø90			
	10.321	Ø110			
	1.345	Ø125			
Σ	152.952				
2016	1.351	Ø50	11798	4680	39.7
	0.134	Ø63			
	0.731	Ø90			
	10.504	Ø110			
	1.223	Ø125			
	0.336	Ø140			
	0.328	Ø160			
Σ	167.559				

At the end of 2016, the total length of WDS was 167.56 km, and the percentage of people connected to WDS reached 39.7%. The population of the Czudec commune in period 2009-2016 increased by 229 people and the number of water recipients at the end of 2016 reached 4680.

As shown in Figure 1, water intake increases. The exception is 2014, when it was 15.08 thousand m^3 of water less than in the previous year. A significant increase in water sales is also observed, which is associated with the WDS growth. In the period under analysis, a large drop in water losses can also be noticed.

Figure 2. present diagram of the WDS in the Czudec commune.

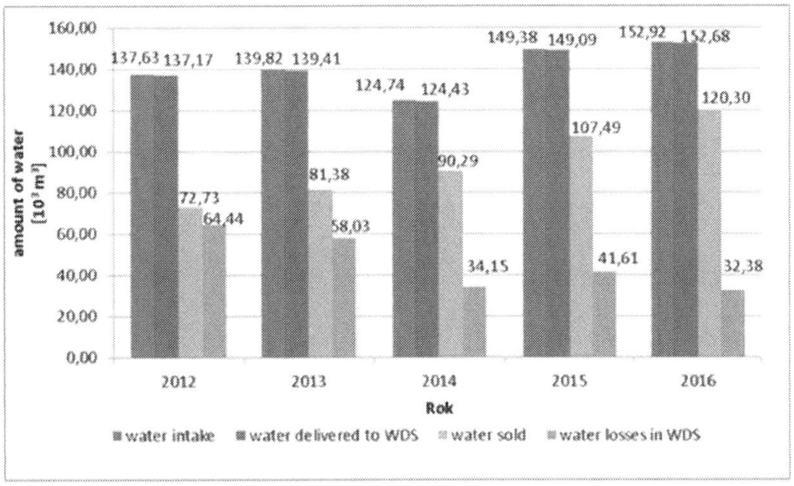

Fig. 1. Water balance in the municipality of Czudec

Fig. 2. Diagram of the WDS in the Czudec commune

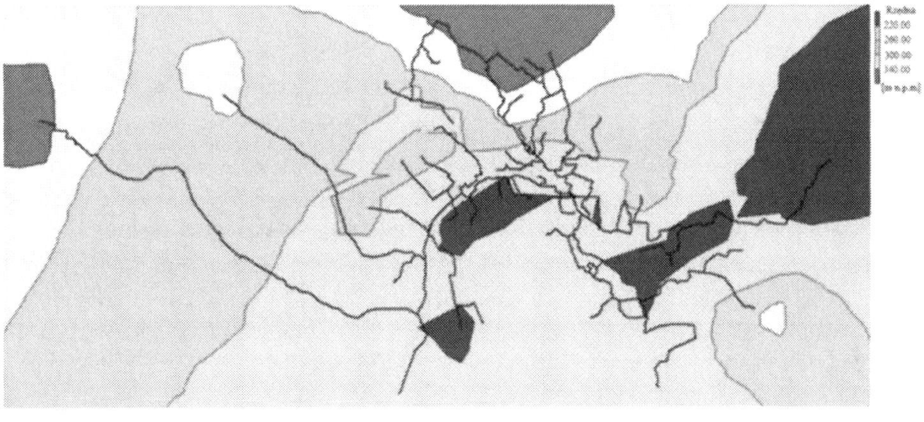

Fig. 3. The height layout of WDS in the Czudec commune

Water intakes and tanks are located in the southern and eastern part of the commune, and pumping stations are located on the entire surface of the supply area.

Czudec commune is located in a varied height. In figure 3 the height layout of WDS is presented.

3 Analysis of the pressure in WDS

For the purposes of the analysis, a hydraulic model of the water supply network was prepared. Figure 4 shows the distribution of pressure on the network during the minimum demand, while Figure 5 shows the frequency of pressure at the minimum demand.

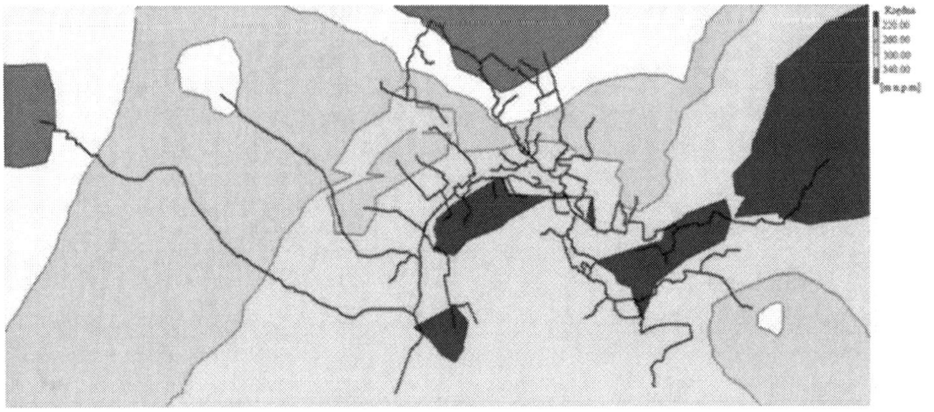

Fig. 4. Distribution of pressure on the network during the minimum demand

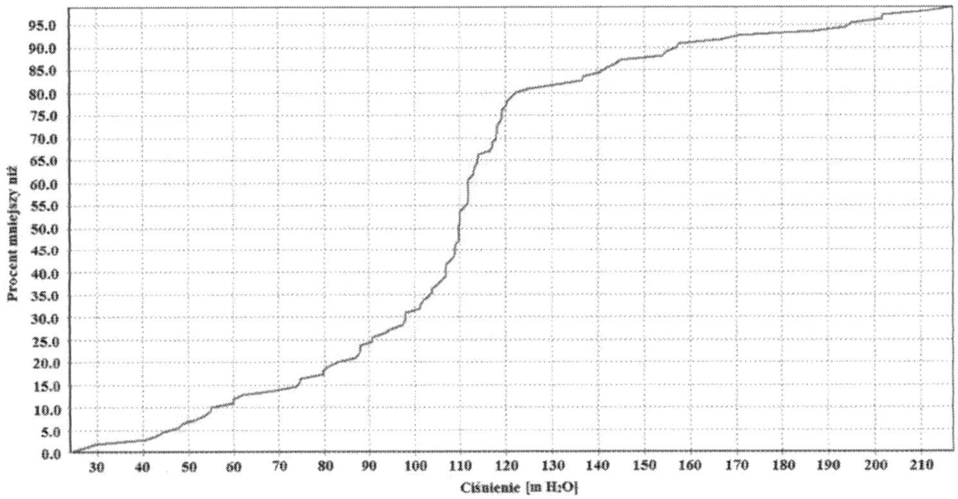

Fig. 5. Frequency of pressure occurrence during minimum demand (x-axis – pressure, y-axis - percent less than)

During the minimum demand, the pressure in the WDS ranges from 24 to 215 mH_2O, of which pressures up to 110 mH_2O account for 50%. The highest pressures occur in the northern part of the network and exceed the value of 120 mH_2O.

Figure 6 shows the distribution of pressure during the maximum water demand, while Figure 7 shows the frequency of pressure during this demand.

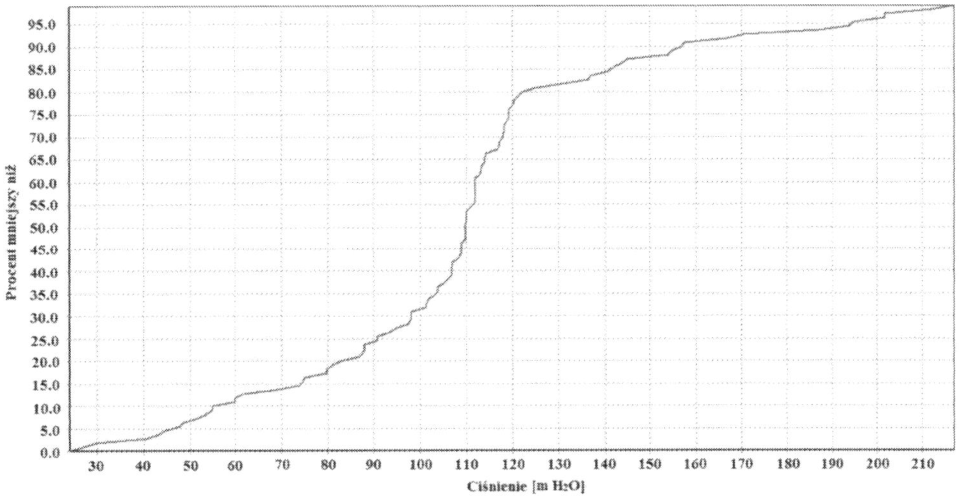

Fig. 6. Distribution of pressure on the network during the minimum demand

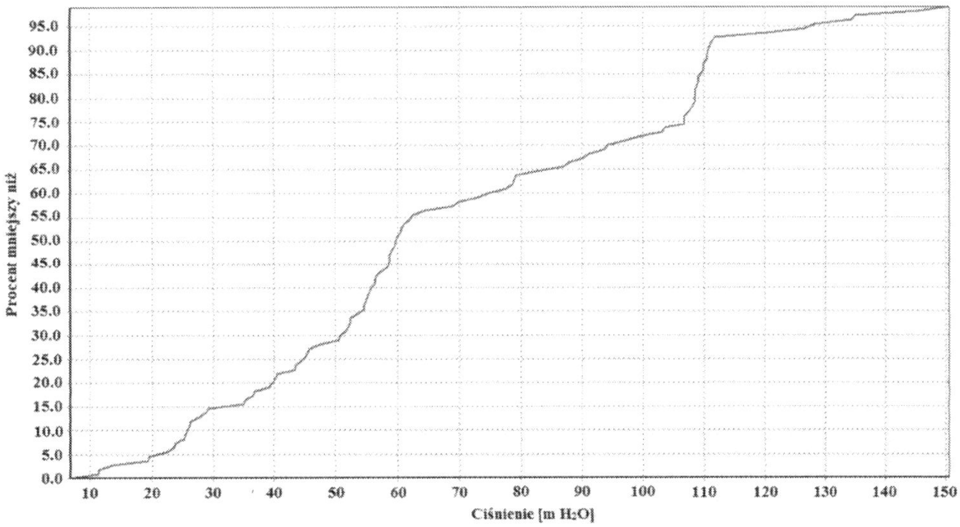

Fig. 7. Distribution of pressure on the network during the minimum demand (x-axis – pressure, y-axis - percent less than)

The highest pressure during the maximum water demand occurs in the north-western area of the WDS and exceeds 150 mH_2O. The pressure during this demand is in the range from 7 mH_2O to about 150 mH_2O, of which 50% are pressures below 60 mH_2O.

4 Water age analysis

Water age simulations were also prepared, that is, the period of water residence WDS. This simulation is shown in Figure 8.

The lowest water age below 24 hours occurs in the central, eastern and south-western parts of WDS. In the northwest and the western water supply area, the age exceeds 96 hours. These are mainly the ends of WDS in which the demand is negligible.

Fig. 8. Water age in the WDS

5 Conclusions

Proper operation of the CWSS allows to achieve continuous operation of the entire WDS, Thanks to hydraulic model it is possible to optimize the operation process.

During the minimum demand on a large part of the WDS, the pressure exceeds 60 m H_2O – recommended maximum for house connections. The reason for such high pressures is mostly the terrain. In places where pressure exceeds 60 mH_2O, pressure reducers are installed at house connections.

Due to the diversified terrain, there is no possibility of gravitational water supply entire WDS, therefore, where necessary, a pumping station is located (there are 11 pumping stations on the whole WDS).

The actions to be taken to reduce the pressure in WDS are primarily the installation of pressure reducers in regions where there is a gravitational supply of water. As a result, it is possible to reduce the risk of failure caused by the cracking of pipe due to high pressures. Another action that should be taken is the separation of more water pumping zones. This would reduce the lift height of the pump, and thus reduce the pressure, resulting in lower water losses.

References

1. Boryczko K. Water age in the water supply network as health risk factor associated with collective water supply, Ecol Chem Eng **23**, pp 33-43, (2016)
2. Boryczko K., Rak J. Assessment of water supply diversification using the Pielou index, ENVIRONMENTAL ENGINEERING V, Pawlowski L. Pawlowska M., Editor,: CRC PRESS-BALKEMA, pp 53-58, Leiden, Netherlands (2017)
3. Calzadilla A. , Rehdanz K.,. Tol R. S. J. The economic impact of more sustainable water use in agriculture: A computable general equilibrium analysis, J Hydrol **384**, pp 292-305, (2010)
4. Miszta-Kruk K., Kowalski D. Failure of water supply networks in selected Polish towns based on the field reliability tests, Eng Fail Anal **12**, pp 736-742, (2013)
5. Pietrucha-Urbanik K. Failure analysis and assessment on the exemplary water supply network, Eng Fail Anal, **57**, pp 137-142, (2015)
6. Pietrucha-Urbanik K., Tchórzewska–Cieślak B., Urbanik M. Analysis of the gas network failure and failure prediction using the Monte Carlo simulation method, Eksploat Niezawodn, **18**, pp 254-259, (2016)

7. Studziński A. Amount of labour of water conduit repair, Safety, Reliability and Risk Analysis: Beyond the Horizon, Van Gelder P.H.A.J.M. Steenbergen R.D.J.M., Miraglia S., Vrouwenvelder A.C.W.M., Editor,: Taylor & Francis Group, pp 2081-2084, London (2014)
8. Szpak D., Tchórzewska-Cieślak B. A Proposal of a Method for Water Supply Safety Analysis and Assessment, Ochrona Środowiska **37**, pp 43-47, (2015)
9. Tchorzewska-Cieslak B., Szpak D. Water producers risk analysis connected with collective water supply system functioning, Dependability Engineering and Complex Systems. Advances in Intelligent Systems and Computing, W; Mazurkiewicz Zamojski, J; Sugier, J; et al., Editor, Springer pp. 479-489, Switzerland (2015)

Analysis of the load capacity of the bolt joints of collecting electrodes with the rapper rod

Andrzej Nowak[1,*], and *Mirosław* Kurz[2]

[1]University of Bielsko-Biala, Willowa 2, 43-309 Bielsko-Biała, Poland
[2]Dedusting Installation Division, Rafako S.A., Górnośląska 3A, 42-200 Pszczyna, Poland

Abstract. This article presents an analysis of the load capacity of the bolt joints of collecting electrodes with the rapper rod, made in HUCK technology. A discrete model of the system of electrodes was worked out for the analysis of stresses in bolt joints, the validation of which was made on the basis of measurement data obtained on the test bench. Two variants of joining the plates of electrodes with the rapper rod were taken into account in calculations - the first one with the correct geometry in the rapper rod, the other one with an eccentric displacement of holes in the flat bars of the rapper rod. In both variants the modelled system was loaded with the impulse force corresponding to the impact of the hammer against the anvil, with the run obtained from measurements on the test bench. These models and the related calculations were performed in the ABAQUS environment. As a result of the analyses performed, recommendations were worked out intended to eliminate cases of the destruction of bolt joints closest to the anvil of the rapper rod.

1 Introduction

This article presents an analysis of the load capacity of the bolt joints of collecting electrodes with the rapper rod, made in HUCK technology. In the structure tested, the collecting electrodes of an electrostatic precipitator are periodically cleaned by inducing violent vibrations in them - the acceleration values definitely exceed 1000 ms^{-2}(1). These vibrations are caused by the impulse forces of drop-weight hammer impacts in the anvils of rapper rods [2]. Transfer of vibrations from the rapper rods to collecting electrodes is carried out by the bolt joints analysed in the study. The quality and stability of these joints is one of the important factors which indirectly affect the efficiency of cleaning exhaust gases from solid particles in electrostatic precipitators.

2 Models of the system

The models shown further on, and the related calculations were performed in the ABAQUS environment. Parts of the system were modelled using deformable discrete components described by the finite element method (FEM) [3], [4] - electrodes were

* Corresponding author: a.nowak@ath.bielsko.pl

modelled with coat components of the I order (shell, S4R), and the remaining parts of the system with continuous components also of the I order (solid, C3D8R) [5]. The upper, articulated suspension of the electrodes in the beam was performed as a mobile, rotary joint of hinge type. In the bolt joints friction was taken into account - accepted coefficient of friction μ = 0.4 - according to the idea of stressed or rivet joints, where the load capacity of the joint is obtained by the effect of friction induced by compression, and not by shearing and pressure as in conventional joints. Validation of models was based on the measurement data obtained on the test bench, as in earlier works of the co-authors [1], [6].

2.1 FEM model of the collecting electrodes

Two discrete models of the system of 9 collecting electrodes were used in the analysis of stresses in bolt joints consisting of 2 reinforced electrodes and 7 standard electrodes of the type SIGMA VI, length 16400 mm, made of sheet with a thickness of 1.5 mm, suspended in articulate form on the supporting beam and joined inseparably with the rapper rod (Fig. 1), different in the way of making bolt joints of HUCKBOLT C50L type (Fig. 2).

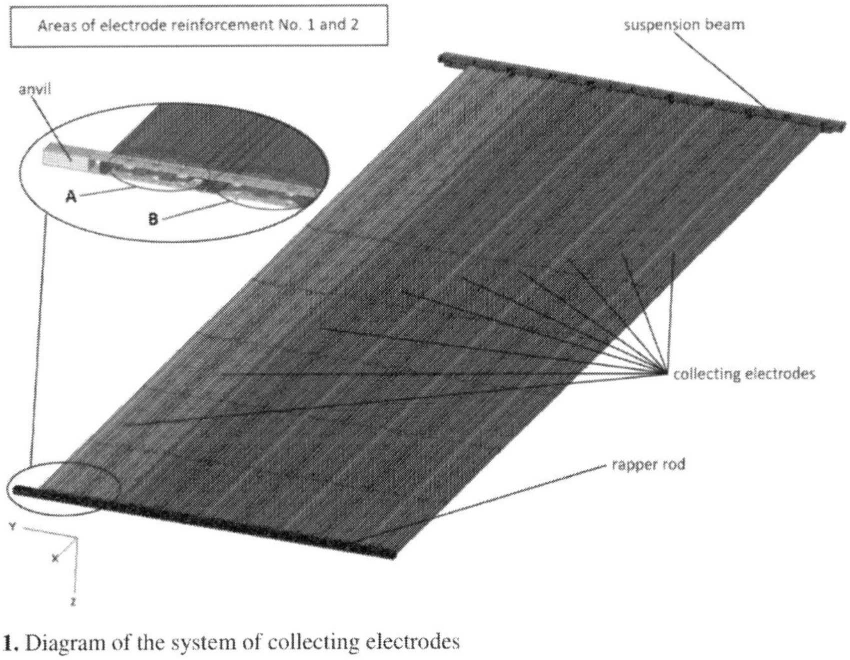

Fig. 1. Diagram of the system of collecting electrodes

2.2 HUCK type connection models

For the system of figure 1 two variants of bolt joints were worked out. In the first variant (v.1) the correct geometry of the holes in the rapper rod was modelled (Fig. 2a), i.e. longitudinal axes of the holes in the flat bars of the rapper rod overlap. The other variant (v.2) maps the skew position of the bolt and ring in relation to the holes in the rod - the axes of the bolt and ring are tilted by about 1°30' from the longitudinal axes of the holes in the flat bars "a" and "b" (Fig. 2b). This is the consequence of the displacement of the axis of holes by 3 mm in the flat bar "a" in relation to the relevant axes of holes in the flat bar "b" in the plane **xz** (system of coordinates **xyz**, as in figures 1 and 3), in the towards axis **x**.

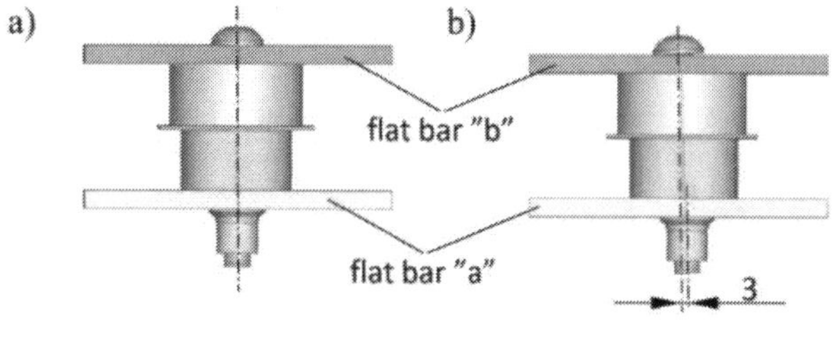

Fig. 2. Variants of bolt joints: a) v.1 - joint technologically correct, b) v.2 - skew joint

3 Calculations

As mentioned in the introduction, discrete models of the collecting electrode system and bolt joints were used in numeric calculations to designate stress reduced in bolt joints, paying particular attention to the zone of the contact of the bolt with the ring. The method of numbering bolts and the coordinate system **XYZ** was accepted as shown in Figure 3.

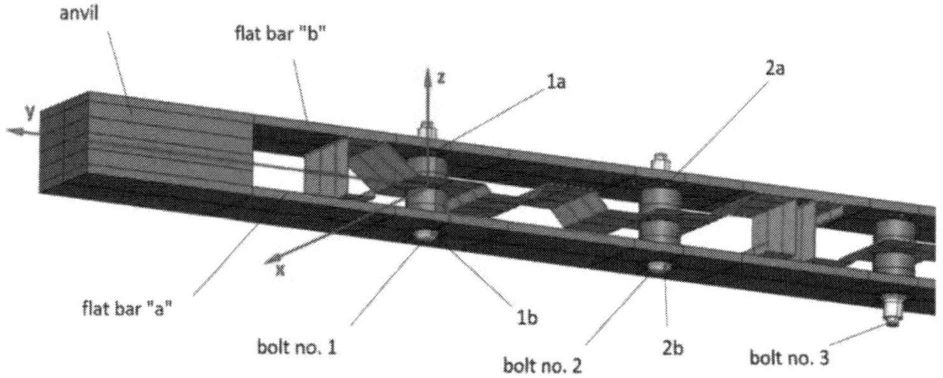

Fig. 3. The numbering of bolt joints and check points of the rapper rod

In both variants the system modelled was loaded with the impulse force corresponding to the impact of a hammer with a weight of 8 kg against the anvil, with the run obtained from measurements on the test bench. In addition, in bolt joints the force of pre-stress equal to N= 53.5 kN or N_{nom}= 62 kN was taken into account. The time of the analysis in each of the variants was 0.01 s. Calculations were made with a variable integration step, and the results were recorded in each iteration. In the calculations shown in the further part of study the following physical and geometric parameters were accepted: steel with density 7850 kgm^{-3}, Young's modulus E =206 GPa, Poisson's ratio ν = 0.3, coefficient of friction in joints μ= 0.4.

3.1 Acceleration in the rapper rod

Acceleration in the rapper rod induced by the impulse of impact was determined in the model for variant v. 1. As a result of the simulation performed, runs of acceleration at check points of the rod were obtained. For these runs the maximum values of the constituents of total acceleration were determined (Fig. 4).

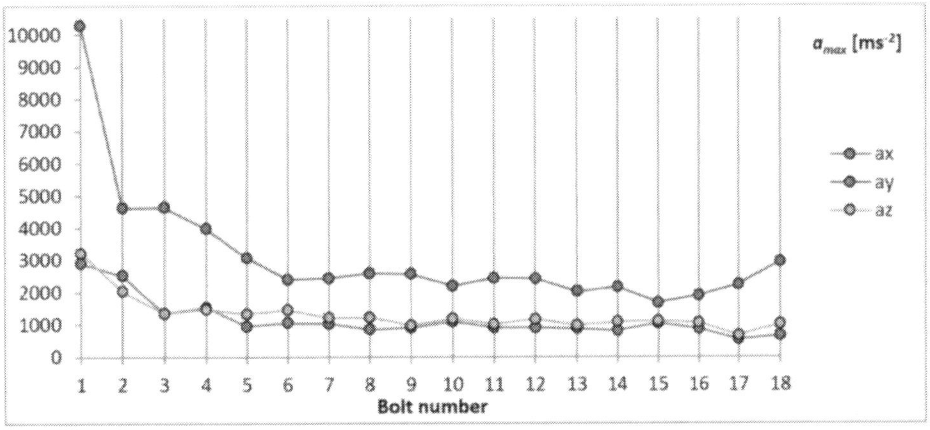

Fig. 4. The maximum total acceleration at checkpoints of the rod (variant v.1)

The analysis of these values shows that the greatest acceleration is at the points closest to the anvil, which is also in line with the experimental data and is one of the elements of model validation [1], [6].

3.2 Stress analysis in the bolts

In Figures 5 and 6 the distribution of reduced stresses σ_{red} typical for the variants analysed are shown, found in the elements being part of the joint (fragments of flat bars of the rod, EZ fragment, spacer sleeves, bolt and ring of the bolt). The status of stresses after loading the joint with only minimal force of pre-stress- N_{min}= 53.5 kN is shown on them. And since in the analysed joint tensile stresses are predominant, according to the Mises hypothesis there was reference to the condition: $\sigma_l \cong \sigma_n \leq \sigma_{lim}$.

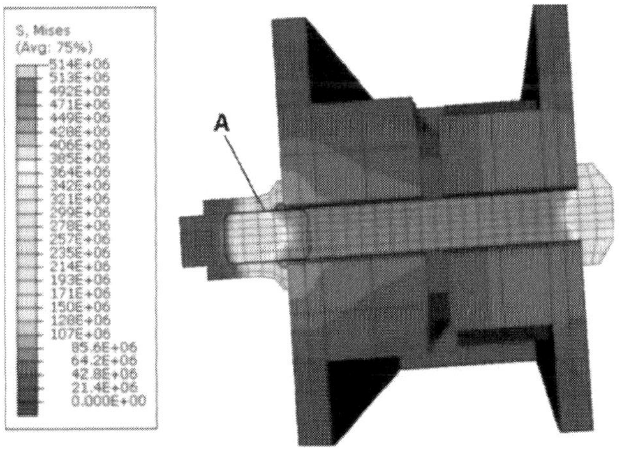

Fig. 5. Distribution of stresses reduced in the bolt joint of joint No 1 under the action of force of pre-stress equal to 53.5 kN (variant. 1)

Area A (Fig. 5) constitutes a sensitive zone of the bolt because of the largest stresses - this is the area of necking in the real bolt, i.e. change of the diameter of the bolt and the area where grooves appear at the entrance to the clamping ring. It should be noted that the A area includes the same zone of the bolt where in the actual objects its cracking was found. This fact confirms the correctness of the calculation model accepted and its conformity with the real object.

The model of the bolt joint in variant 2 mapped the skew position of the bolt and ring relative to the holes in the rod. As a result of shifting the axis of holes in the flat bars of the rod in areas B, C, D concentrations of stresses appear induced by the non-concentric arrangement of the ring relative to the bolt during pressure (areas B and C). The cause of stresses in region D is, inter alia, the lack of adherence of the frontal area of the head of the bolt to the side area of the rod (E).

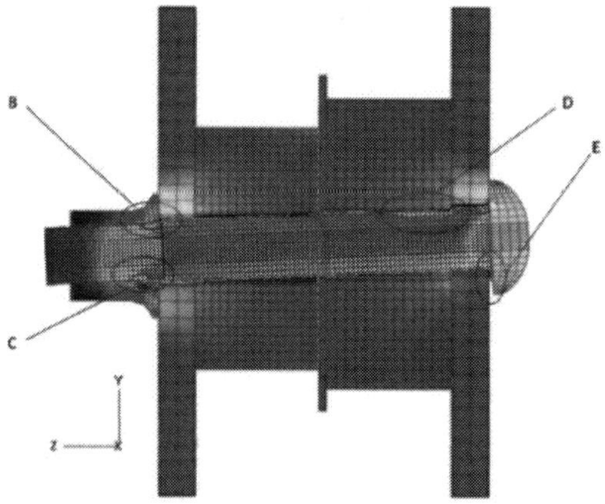

Fig. 6. Distribution of stresses reduced in the bolt joint of joint No 1 under the action of force of pre-stress equal to 53.5 kN (variant. 1)

Under the load of pre-stress (from the minimum stress force of the rivet) during clenching, the maximum value of reduced stresses does not exceed the variations in v.1 and v.2, in any bolts respectively: 530 MPa, which is approx. 76% of the permissible tensile stress, which is σ_{lim} = 699 MPa and 602 MPA for the cross-section of the bolt tested, i.e. approx. 86% of the permissible tensile stress σ_{lim}. The value of the maximum permissible tensile stress σ_{lim} was determined taking into account the minimum diameter of the bolt $d_{s,min} = (11.9 \pm 0.15 \text{ mm}) = 11.75$ mm appearing in area A, and the acceptance of permissible tensile strength, which according to the manufacturer of the bolts is N_{max} = 75.8 kN.

In the case of the impulse force, reduced stresses occurring in the system are variable in time. In particular, in the bolts they oscillate around the "fixed" value coming from the force of pre-stress - figure 7 shows, for example, oscillations of reduced stress values in the selected finite element in area A of bolt Nos. 1, 5, 14 and 18 for variant v. 1. The oscillations of the values of reduced stresses in variant v.2. are of the same nature.

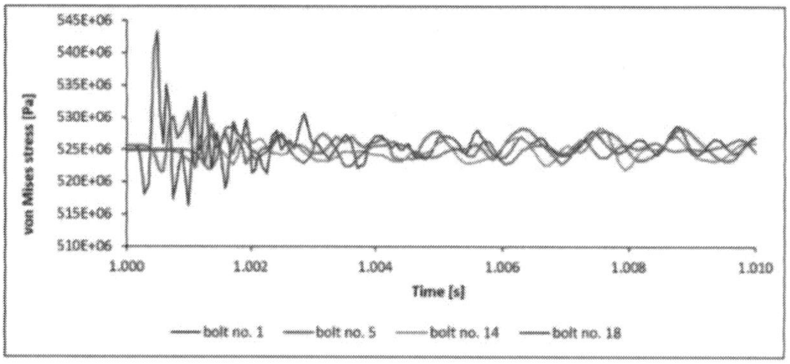

Fig. 7. Oscillations of reduced stress values in one of the finite elements of area A in bolt Nos. 1, 5, 14 and 18 (variant v. 1, the static load of bolts from the pre-stress in combination, and the load of the whole system with impulse force when the hammer hits the anvil)

Figure 8 shows the comparison of the values of oscillation amplitudes of reduced stresses in all bolt joints in both variants.

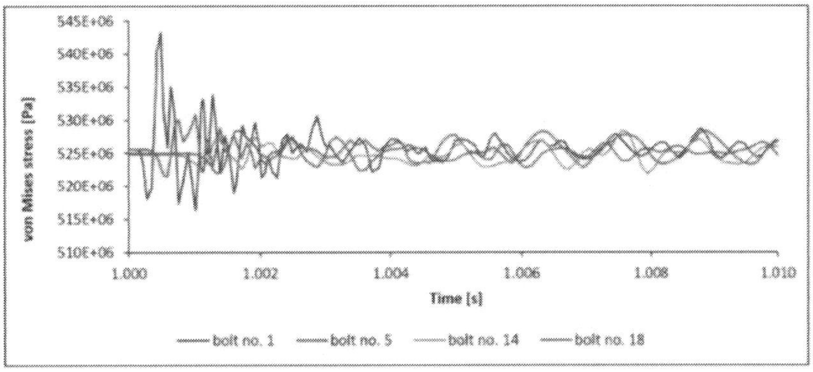

Fig. 8. Amplitudes of reduced stresses in the bolt joint of joint No 1 under the action of force of pre-stress equal to 53.5 kN (variant. 1 and variant 2)

These analyses were performed for the minimum calculation stress force stressing the bolt N_{min}= 53.5 kN specified by the manufacturer. According to the manufacturer's data, the actual stressing force can reach in some rivets higher values, of N_{nom} = 62.0 kN. This is due to the technology of clenching the rivet - by breaking the mandrel in the necking, during the operation of its stressing. For this increased force of stressing, on the basis of the assumptions specified earlier, Figure 9 shows the distribution of the maximum reduced stresses in the joints of bolts.

In conclusion, in variant v.1, with the bolts joints made in the right way, the maximum values of total reduced stresses - from the loads of bolt stress and the impulse of the impact of the hammer against the anvil, do not exceed the 634 MPa. In each of the bolts they are much lower than the stresses permissible for this type of bolt, which according to the data of the manufacturer are σ_{lim}= 699 MPa and significantly below the average tensile strength of the material of the bolt specified for $R_m \approx$ 851,5 MPa. The area of maximum stresses in the bolt is located in region A (Fig. 5), i.e. in the section of clenching the bolt with the sleeve. As can be seen in each of the bolts, stresses are well below the permissible tensile stress. In the case of variant v.2 in the bolts of joint Nos. 1 and 2 the value σ_{lim}= 699 MPa was exceeded, and in bolt No. 3 the value of reduced stresses received is close to σ_{lim}.

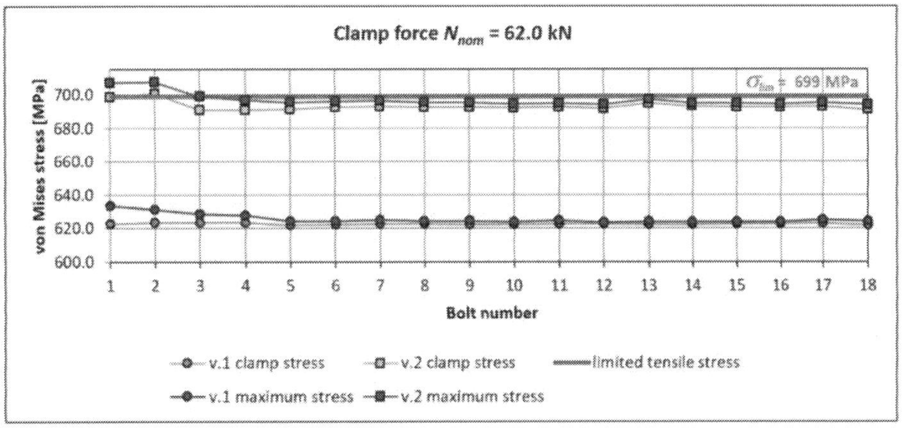

Fig. 9. Distribution of the maximum reduced stresses in the joints of bolt Nos. 1 ÷ 18 (variants v.1 and v.2) under the action of only the pre-load with the stress of force N_{nom}=62.0 kN (green) or pre-load from pre-stress with force N_{nom} and the impulse force of impact (red)

For the quantitative evaluation of dynamic stresses caused by the impulse force of the impact of the hammer against the anvil, the comparison of the effective values of the runs of these stresses was applied. After isolating the dynamic runs of reduced stresses, the effective values of reduced stresses were determined. It was accepted that the total effective value of reduced stresses in all joints is equal to 100%. The comparison of the relative effective values (i.e. related to the total values) allows, indirectly, the quantitative evaluation of the share of individual bolt joints in the transfer of energy - from the impulse force - from the rod to electrodes. Such a comparison for variant v.1 is shown in Figure 10.

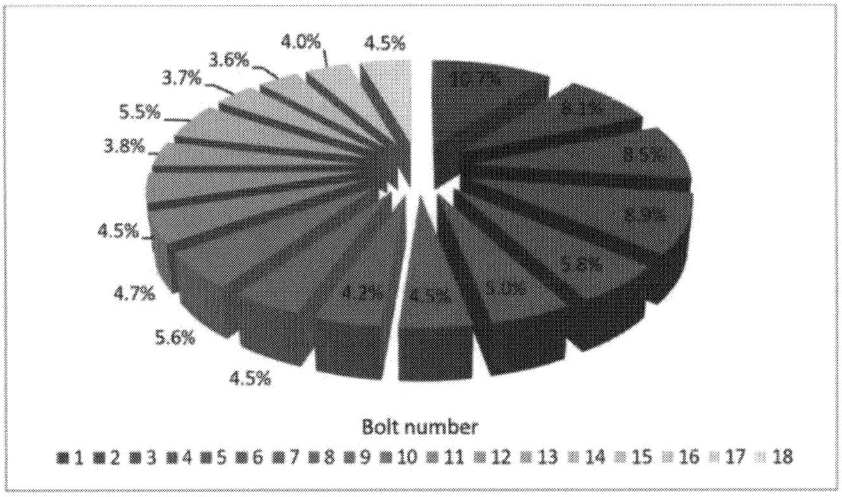

Fig. 10. The share of bolt joints in the transfer of energy from the impact of the hammer against the anvil to electrodes (variant v.1)

It can be noted that the first four joints from the side of the anvil (bolts no. 1÷4) transfer more than 36% of the energy of impact, including bolt joint No 1, almost 11%. This indicates that the most vulnerable to the impact of dynamic loads are bolts in joint Nos. 1 to No 4 from the side of the anvil. In other rivet joints this impact is significantly smaller. In variant v.2 similar dependencies take place.

4 Conclusion

The analyses shown in this study make it possible to conclude that with the bolt joints made properly (variant v.1) the maximum values of total reduced stresses - from the loads of the pre-stress of the bolt and the impulse force of the impact of hammer against the anvil in each of the bolts are significantly lower than the values of permissible stresses. However, the incorrect workshop execution of bolt joints, involving the non-alignment of the pairs of holes on the rivets, results in the skew position of the bolt in the holes during the installation of electrodes. The calculation analysis of such joints (variant v.2) showed a significant increase of pre-stresses in the bolts as a result of additional bending during stress. In special cases of the simultaneous occurrence of the skew execution of the joint and pre-stressing with the maximum force N_{nom} bolt joints which are on the electrodes close to anvils can be damaged, which in turn will result in the decrease of the effectiveness of the rapping of these electrodes, resulting in a decrease in the efficiency of cleaning exhaust gases in the chamber of the electrostatic precipitator.

References

1. Nowak A. Vibration of collecting electrodes in electrostatic precipitators – modelling, measurements and simulation tests, Journal of Electrostatics, **70**, 3, pp. 327-332, (2012)
2. Nowak A., Nowak P., Awrejcewicz J. Influence of impulse force loading on vibrations of the collecting electrodes, International Journal of Structural Stability and Dynamics, **17**, 5, pp. 1740011-1-1740011-15, (2017)
3. Zienkiewicz O.C., Taylor R.L.: The Finite Element Method, Vol. 2 Solid Mechanics, Fifth ed., Butterworth-Heinemann, (2000)
4. Zienkiewicz O.C., Taylor R.L., Zhu J.Z.: The Finite Element Method, Its Basis&Fundamentals, Sixth ed., Butterworth-Heinemann, (2005)
5. Rakowski G., Kacprzyk Z.: Metoda elementów skończonych w mechanice konstrukcji, Oficyna Wydawnicza Politechniki Warszawskiej, Warszawa (2005)
6. Nowak A., Nowak P., Kurz M., Ryguła C. Study of the effect of the stiffness of the anvil beam for vibration propagation in the system of collecting electrodes, Przegląd Elektrotechniczny, pp. 232-236, (2017)

Analysis of the structure of the frame of a trailer for transport of bales of compressed straw in the aspect of minimization of materials and energy consumption

Dariusz Lodwik[1,*], and *Jerzy* Pietrzyk[1]

[1]Warsaw University of Technology, Faculty of Civil Engineering, Mechanics and Petrochemistry, Institute of Mechanical Engineering, Łukasiewicza 17, 09-400 Płock, Poland

Abstract. The article presents the development of a structure model and numerical analysis of the trailer frame assembly intended for transporting bales of compressed straw, which may be a potential source of cheap and ecological energy. The aim of the analysis was to minimize the use of construction materials (reduction of the total weight), simplification of manufacturing technology, and thus energy savings in the production and operation. The aforementioned aspects also play a key role in environmental protection issues. The digital models of frame assemblies were made and subjected to static strength analysis. The structure was evaluated in terms of strength, based on the calculations made using the finite element method. The distribution of simulation parameters in the area of the tested structure made it possible to partially optimize the load-bearing system of the trailer, due to the adopted decision criterion in the form of reduction of the total weight and simplification of the structure, while meeting the limitations resulting from the values of stresses, displacements, safety factor in the admissible range. The material savings obtained for the optimal variant will result in simplification of manufacturing technology, energy savings and reduction of costs during production and operation.

1 Introduction

The use of modern technologies in plant and animal production results in a significant surplus of straw, which can be used, inter alia, as an ecological fuel for energy purposes. Excess straw in Poland is estimated at around 9÷12 million tons, which is equivalent to 5÷7 million tons of hard coal. An inseparable element of the logistic chain of using straw as an ecological fuel is transport. In transporting bales of compressed straw, trailers with an increased loading space are used. The advantages of such a solution is the simplicity of use and the convenience of loading, which makes it possible to reduce transport costs in all working conditions.

* Corresponding author: dariusz.lodwik@pw.edu.pl

The subject of deliberations in this paper is the structure model and static numerical analysis of the transport system frame assembly, in order to obtain material savings and to simplify the manufacturing technology, and thus energy savings. For the proposed structural solution, an attempt was made to assess the strength of the structure, based on stress analysis as large as the material properties and the required safety considerations allow. The model was made using a three-dimensional computer image in Autodesk Inventor [4, 9]. Strength calculations were carried out using the finite element method, for the adopted range of decision variables, geometrical and material features of the parts used to build the frame model. In the area of the tested structure, partial optimization of the load-bearing system of the trailer was performed, due to the accepted criterion of reduction of the total weight, while meeting the limitations resulting from the values of stresses, displacements, safety factor in the admissible range.

2 Research methodology

The concept of trailer structure has been developed on the basis of a frame made of rectangular steel profiles, supplemented by the structure of a loading platform and supporting walls, limiting the movement of the transported load. The model was supplemented with components offered on the market, creating a full-sized and compact structure of the trailer for transporting bales. The visualization of the developed model of the trailer, in a view with a limited level of detail together with the designation of the basic assemblies subjected to simulation tests in the FEM analysis environment, is shown in the Figure 1.

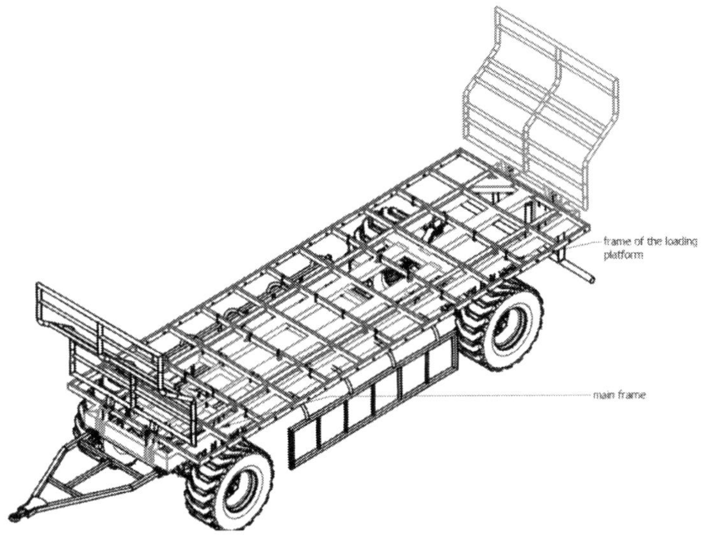

Fig. 1. Digital model of the trailer structure with marked components subjected to FEM analysis

Geometrical models of frames (Fig. 2), developed as spatial parametric 3D structures, in which standard steel profiles were used for construction with closed profile, square and rectangular sections, and sheet for the construction of: nodes and elements fastening the subassemblies of the wheel set, floor covering. The elements shaping the structure of the frames in the notation of structural features have been given appropriate geometric features based on databases of structural shapes, as well as material features. E360 structural steel was adopted in the construction of the frame model. The shaping elements were selected based on the database of standardized Content Center library profiles.

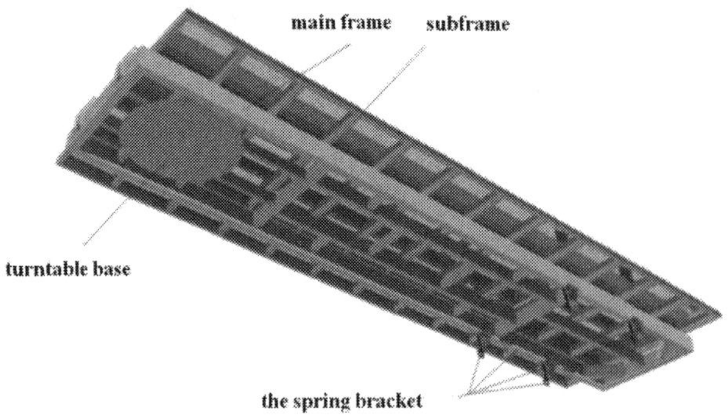

Fig. 2. Geometric model of the load bearing part of the trailer

The structure was written as N set of numbers and functions representing structural features and shaping the elements of the model. Frame models in the construction space were described by a vector [2, 3 ,8, 9]:

$$X = \left(X_1, \ldots, X_N\right) \in E_k \tag{1}$$

The group of vector coordinates (1) and frame models are set parameters which values have not changed during the numerical analysis and decision variables, which were the quantities selected in the construction process. The set of decision variables was adopted as a point in the Euclidean space of solutions E_x (2).

$$y = \left(y_1, \ldots, y_N\right) \in E_x \tag{2}$$

The assembly was performed using static assembly constraints of the model, taking the degrees of freedom of the frame structure profiles. In the geometric three-dimensional record of the trailer assembly model, a subassembly of the main frame, the frame of the loading platform were separated, which were marked in Figure 1. For these structures, a decision variable belonging to the E_x space was adopted [1, 2, 8]:

$$y = \left(y_1\right) \in E_x \tag{3}$$

where :

y_1 – thickness of the walls of the frame model sections.

For the decision variable y_1,a discrete range of variability was determined taking into account the market availability of rectangular profiles and pipes (standard types of profiles):

$$y_{1\min} \leq \left(y_1\right) \leq y_{1\max} \tag{4}$$

An attempt was made to partially optimize the discussed components, due to the adopted decision criterion in the form of reduction of the total weight, while maintaining the external dimensions of steel profiles and the value of safety factor at the required level, and the permissible value of displacement and stresses generated in the analyzed structure.
A model was used for numerical static analysis of the assembly, in which the suspension construction was excluded from the simulation. The model does not take into account the rigidity of the suspension. The bonding of the contacting surfaces made by screw connections has been replaced by contact models that simulate pressures and friction in the

joints. For contact pairs between the construction elements of the frame subassemblies, the surface-to-surface contact type with the "bound" attribute (no possibility of relative displacement of components) was used. The structure model was written as a vector of the decision variable y=(y$_1$), for which a set of acceptable solutions Ø(y) was obtained. The optimal solution was those for which the optimization criterion in the form of reduction of the total weight will reach the minimum value m(y).

$$y = (y_1) \in E_D$$
$$\varnothing = \varnothing(y) \subset E_D \tag{5}$$
$$\varnothing(y) = (m(y))$$

where :

m(y)	–	minimum weight.
ρ	–	air density [kg m^{-3}],
A	–	frontal area of the vehicle [m^{-2}],
v		relative fluid velocity [m s^{-1}].

In order to assess the structure on the basis of the adopted criterion, the restrictions which fulfillment was considered necessary were introduced:

$$w_{\max}(x, y, z) \le w_{dop}$$
$$\sigma_{S\max} \le k_r \tag{6}$$

where :

$w_{max}(x,y,z)$	–	maximum displacement
σ_{Smax}	–	maximum normal stress.

For the analyzed systems made of plasticity materials, the criterion of safety was the lack of the possibility of permanent deformation. It was considered that the condition for allowing the structure to work is to obtain stresses $\sigma_{Smax} \le k_r$ (k_r- allowable tensile stress), [2, 8].

Discretization of the analysis area was carried out for assembling a model from finite element types: beam and shell. For the analyzed case, a grid with the following parameters was generated: number of elements - 1138709, number of nodes - 2061117, average size of the element (fraction of the diameter of the model) - 0,1, minimal element size (fraction of medium size) - 0,2, average size of the element in the shells - 0,05, gradation coefficient - 1,5, the maximum angle of the grid triangle - 60 deg. In order to map the load locations in the form of bales to the surface of the trailer loading platform (Fig. 3), the floor plane was divided and the contact area between the log and the floor was determined.

The tested structure was subjected to forces, derived from the weight of the transported load, taking into account the force of gravity. The structure was designed for transporting up to 26 bales with a unit weight of m=600 kg each. The ideological distribution of the load from the weight of the load, in relation to the designated strips of the platform floor plane, is shown in Fig. 3.

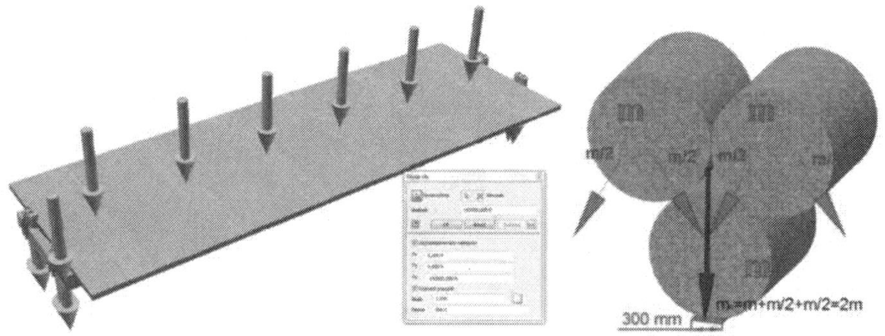

Fig. 3. Loading the transport platform with forces simulating the load weight

In strength calculations it was assumed that for each bale of mass m in the lower layer, there is an additional load in the form m (m/2+m/2), coming from two bales of the upper layer (Fig. 3). The remaining load in the form m (m/2+m/2) coming from the upper bales, decomposes into neighboring lower bales. The entire load acts on the contact plane between the lower bale and the floor, for each bale from the lower layer. The calculation includes the total maximum load (26 bales with a mass m=600 kg each), with a distribution of 92% (24 bales) of the maximum load in the area of contact between the load and the floor, and the remaining load was directed to the support wall brackets (weight of 2 bales). Maximum strength F=142000 N, distributed on 6 levels of the loading platform (interaction of 24 bales, stacked one above the other). The load with the value F1=14000 N, was applied to the support walls, imaging the interaction of 2 bales. In the construction simulation model, non-displaceable articulated supports were used to simulate a bolt connection between the bracket and the suspension components, a fixed support that reflects the geometry of the loading platform's support by the turntable.

3 Research results

In the calculations, the criterion of optimal frame construction was adopted as a function dependent on the value of the decision variable. The criterion was to minimize the weight of the frame for the proposed structural solution while meeting the strength limitations (6), which is in line with the main objective of the analysis, which is to minimize material consumption and reduce energy during production and operation. The structure made of assemblies was subjected to preliminary simulation tests. The selected research result in the form of a contour map of displacements are shown in the figure 4.

Maximum stresses in the place of their highest concentration did not exceed 148 MPa. Fig. 4 shows the local values of the displacement parameter within the examined unit. The value of the maximum displacement in the tested object was 3.7 mm, which was recorded in the floor plating (sheet with a thickness of 2.5 mm). The reason is the significant value of the load affecting the plane with a small thickness, in the place of maximum deflection of the main frame (central part of the structure).

There was a small value (1.5 mm) deflection of the main and auxiliary frames, which indicates a significant rigidity of the structure and the possibility of introducing partial optimization of the structure.

The minimum value of the safety coefficient (2.03) of a local nature was recorded at the points where the floor meets the beams of the subframe. Large values of the safety coefficient were observed in the load-bearing elements of the structure (7÷15), which indicates the possibility of attempting to partially optimize the discussed components. The load bearing system of the trailer was transferred to the framework analysis environment.

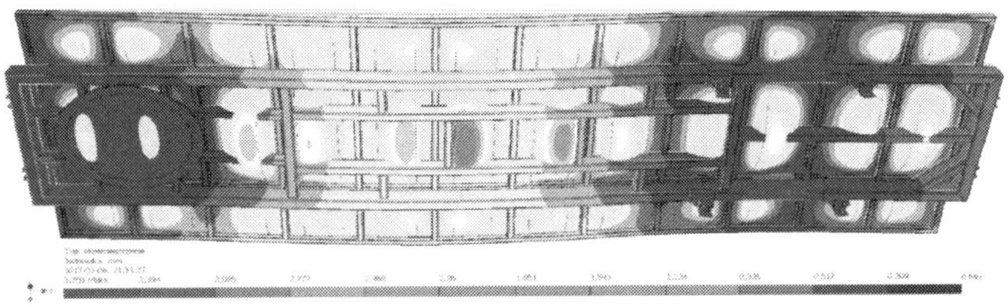

Fig. 4. Distribution of structure displacement - bottom isometric view

As a result of the transformation of the system, 40 beams, 261 nodes, and 105 rigid connections were obtained. At the place where the main frame was connected to the turntable base, a system of 4 non-displaceable supports was used, and 4 fixed pivot supports were installed in the suspension mounting points and the load distribution in the area of the tested frame system was mapped in the form of continuous load (10 N/mm), where the value of previously assumed loads was taken into account, and the length of the beams directly supporting the bales.

In the main part of simulation tests, calculations were made for various design variants. The simulation procedure was carried out for varying wall thicknesses of the profiles used in the structure, in accordance with their standard type series. The adopted assessment criterion was to obtain a structure with the lowest possible mass, while meeting the strength limitations (6). The selected research results, a set of permissible solutions Ø(y), meeting the strength criteria are summarized in the Table 1.

The effect of partial optimization was to reduce the total weight of the tested frame system by about 187 kg (variant 5). It was considered that the construction meets all the necessary conditions, ($\sigma_{Smax} \leq 200$ MPa, $w_{max}(x,y,z) \leq 5$ mm, $x_{min} \geq 1{,}5$). Figure 5presents the distribution of normal stresses S_{max}, for variant 5 of simulation calculations.

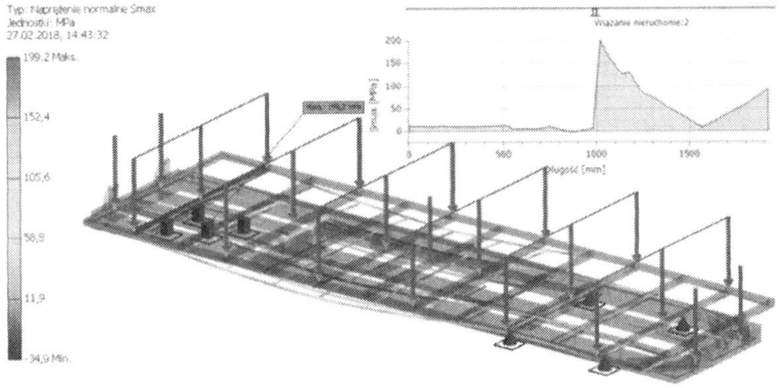

Fig. 5. Distribution of normal stresses Smax in the area of the tested model - variant 5

Maximum valueσ_{Smax}=199,2 MPa of normal stresses, occurs only in the area of non-sliding fixed support, which may be caused by the local stiffening of the longitudinal beam structure, for which a displacement of w = 3.6 mm has been recorded in the central part of the main frame structure (Fig. 6). In the remaining part of the structure, stresses not exceeding the values σ_{Smax}=110 MPa were recorded. Maximum deflection with a value w_{max} =4,17 mm (Fig. 6), occurs only in the central part of the auxiliary structure.

Table 1. Results of simulation tests for selected construction variants

Steel E360		Re=365 MPa		Rm=690 MPa	Young's mod. G=210 GPa	
		variant 1	variant 2	variant 3	variant 4	variant 5
No. of pieces	Length [mm]	Beams markings				
Main frame						
2	7400	200 x 100 x 5	200 x 100 x 4	200 x 100 x 4	200 x 100 x 4	200 x 100 x 4
2	1400	200 x 100 x 5	200 x 100 x 4	200 x 100 x 4	200 x 100 x 4	200 x 100 x 4
2	3320	160 x 80 x 5	160 x 80 x 4	160 x 80 x 4	160 x 80 x 4	160 x 80 x 4
2	1860	160 x 80 x 5	160 x 80 x 4	160 x 80 x 4	160 x 80 x 4	160 x 80 x 4
2	1200	160 x 80 x 5	160 x 80 x 4	160 x 80 x 4	160 x 80 x 4	160 x 80 x 4
2	504	160 x 80 x 5	160 x 80 x 4	160 x 80 x 4	160 x 80 x 4	160 x 80 x 4
1	1860	120 x 60 x 3	120 x 60 x 3	120 x 60 x 4	120 x 60 x 4	120 x 60 x 2,5
2	1200	120 x 60 x 3	120 x 60 x 3	120 x 60 x 4	120 x 60 x 4	120 x 60 x 2,5
2	520	100 x 60 x 3	100 x 60 x 3	100 x 60 x 3	100 x 60 x 3	100 x 60 x 2,5
4	260	100 x 60 x 3	100 x 60 x 3	100 x 60 x 3	100 x 60 x 3	100 x 60 x 2,5
Auxiliary frame						
4	7200	80 x 40 x 3	80 x 40 x 3	80 x 40 x 3	80 x 40 x 2,5	80 x 40 x 2
2	2400	80 x 40 x 3	80 x 40 x 3	80 x 40 x 3	80 x 40 x 2,5	80 x 40 x 2
11	2300	80 x 40 x 3	80 x 40 x 3	80 x 40 x 3	80 x 40 x 2,5	80 x 40 x 2
Total mass		**1046,5 kg**	**971 kg**	**986 kg**	**934 kg**	**859 kg**
Displacement W_{max} [mm]	Main frame	1,9	3,3	2,5	3,2	3,6
	Auxiliary frame	2,79	3,8	3,65	3,73	4,17
Normal stresses	Smax	102,8 MPa	179 MPa	153 MPa	153,9 MPa	199 MPa
	Smin	24,5 MPa	27,4 MPa	26,9 MPa	29,5 MPa	32,4 MPa
	Smax(Mx)	92,6 MPa	180,4 MPa	157,8 MPa	159,8 MPa	200 MPa
	Smax(My)	59,9 MPa	113,9 MPa	93,9 MPa	94,4 MPa	133,3 MPa
Transverse stresses	Tx	88,9 MPa	117,9 MPa	116,2 MPa	128,3 MPa	142 MPa
	Ty	17,5 MPa	41,1 MPa	32,2 MPa	32,4 MPa	48,5 MPa
Torsional stresses	T	14,7 MPa	22,8 MPa	19,9 MPa	20,2 MPa	26,3 MPa
Minimum safety factor x_{min}		3,5	2	2,4	2,4	1,8

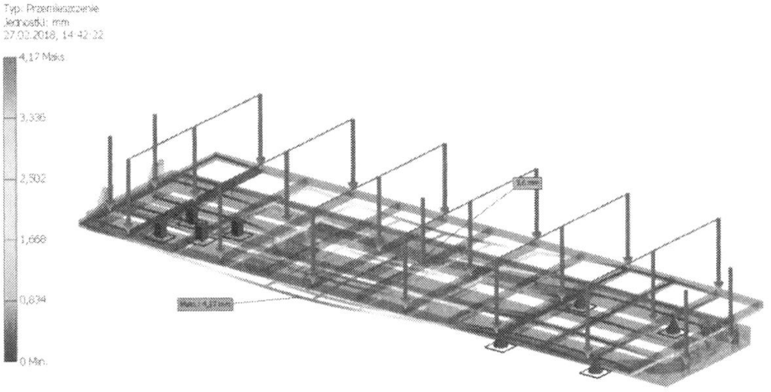

Fig. 6. Distribution of the displacement of the tested object - variant 55

4 Conclusion

Based on the conducted research, the following conclusions can be formulated:

- the presented variants of the trailer frame structure are included in the group of acceptable solutions. In all cases, the safety criterion expressed in the accepted strength conditions is met ($\sigma_{Smax} \leq 200$ MPa, $w_{max}(x,y,z) \leq 5$ mm, $x_{min} \geq 1{,}5$),
- variant 5 was considered optimal, for the assumed frame structure, the objective function has reached value m=859 kg, ($\sigma_{Smax}=$ 199,2 MPa, $x_{min}=1{,}7$, $w_{max}(x,y,z)=4{,}17$ mm),
- the results of the calculations showed that the maximum permissible load weight should not exceed 15.6 tonnes,
- obtaining lower values ($\sigma_{Smax}=153{,}9$ MPa, $w_{max}(x,y,z)=3{,}73$ mm, $x_{min}=2{,}2$), with a slight increase in the value of m = 75 kg of frame weight, was obtained for t variant 4,
- calculations indicate the possibility of further optimization of the structure by changing the material features or changing the model structure at the place of attachment of the rear axle of the wheel set,
- in the next stage of the work, the selection of the best solution among the conducted experiments can be supplemented by carrying out the task of the multi-criteria optimization process,
- considerable material savings obtained for the optimal variant 5 (about 190 kg) will result in simplification of manufacturing technology, energy savings and reduction of costs during production and operation.

References

1. Branowski B., *Wprowadzenie do projektowania.* Wydawnictwo Naukowe PWN, Polska, (1998)
2. Ditrich M., *Podstawy konstrukcji maszyn, t. 1*, wyd. III. Wydawnictwa Naukowe-Techniczne, Polska, (1999)

3. Dziama A., *Metodyka konstruowania maszyn*, Wydawnictwo Naukowe PWN, Polska, (1985)
4. Jaskulski A., *Autodesk Inventor metodyka projektowania*, Wydawnictwo Naukowe PWN, Polska, (2011)
5. Lodwik D., Malesa W., An analysis of selected container structures with built-in multi-layer composite sheets with the use of fem. *TEKA Kom. Mot. I Eng. Roln.*, **14**, 3, (2014)
6. Lodwik D., Jakliński L., Malesa W., [i in.], The test stand calculations of the research station of sub-atmospheric pressure press with implementation of the finite element method. TEKA Kom. Mot. I Eng. Roln., **12**, 2, (2012)
7. Mac Donald B., *Practical Stress Analysis with Finite Elements*, Glasnevin Publishing, Ireland, (2011)
8. Osiński Z., *Podstawy konstrukcji maszyn, wyd. II*, Wydawnictwo Naukowe PWN, Polska, (2012)
9. Tickoo S., *Autodesk Simulation Mechanical 2014 for Designes*, CadCim Technologies, USA (2014)

Analysis of the suitability of using the selected wind turbine blades for wind power appliances based on numerical analyses

Krzysztof Pytel[1,*], *Szymon* Szpin[1], *Wiktor* Hudy[2], *Małgorzata* Piaskowska–Silarska[2], and *Stanisław* Gumuła[1]

[1]AGH University of Science and Technology, Department of Power Systems and Environmental Protection Facilities, Mickiewicza 30 Av. 30-065 Cracow, Poland
[2]Pedagogical University of Cracow, Podchorążych 2, 30-084 Cracow, Poland

Abstract. The aim of this study was a comparative analysis of the suitability of the use of airfoil for wind power. Based on numerical analyses and analytical methods, information on the power factor was obtained. The analyses were carried out for the wind turbine blades and rotors of a vertical axis wind turbine. The tests were performed for the constructed profile and compared with the profile DU 06-W-200 used in the construction of a wind turbine rotors. A vertical axis wind turbine model equipped with designed blade profiles was prepared. The main predicted purpose of the device is to supply electricity to the household. The blade profile models were prepared and then a numerical analysis was performed using the CFD application. The obtained results for given wind speeds and types of profiles were compared with each other. The conducted research allowed to determine the sense of applying the unaudited profile based on the determined value of wind turbine power coefficient. Studies have shown that the accurate preparation of the optimal rotor blade with respect to flow of air stream strongly influences the characteristics of the wind turbine.

1 Introduction

One of the most important issues appearing in the power energy sector is development and examining of modern technologies enabling maximization of obtaining energy from renewable sources. Pollution resulting from burning fossil fuels in conventional power plants have become a serious ecological and economic problems in the whole world [1 – 3]. The development of appropriate technology to intensify the use of renewable energy sources in this context is very important [4].

Wind power plants, are most often associated with large high masts and turbines with blades of an enormous span. This is not the only solution for wind turbines. The result is the construction and continuous development of vertical axis wind turbines. Technology of wind turbines with a vertical axis of rotation is a technology that tries to convince investors to back to that idea. Technical parameters are promising. A turbine of this type can work with very

* Corresponding author: kpytel@agh.edu.pl

low wind speed. The design allows the investor to achieve a sufficiency high efficiency without the need to build high masts and allows the facility to be assembled from functional segments [5–7]. An additional advantage may be the long working period of the device. A vertical axis wind turbine (VAWT) works well enough in urban areas. A classic horizontal axis wind turbine (HAWT) are sensitive to an air turbulence and can be a source of noise. The VAWT can be used as a energy source for powering single-family houses, campsites and summerhouses, yachts, measuring and signalling devices, advertising banners, street lamps. VAWT can be used wherever access to a standard power grid is difficult or impossible. The main element of the wind power plant is the rotor converting wind energy into mechanical energy. The rotor is mounted on a low-speed shaft and rotates mostly at a speed of 15 to 180 rpm. This speed is multiplied by the transmission and transferred to the generator using a high-speed shaft. The generator produces electricity from mechanical energy [8,9].

Design requirements for modern turbines are high. New, advanced technologies, light and durable materials and the ability to perform complex aerodynamic simulations give designers great opportunities to create modern airfoils and efficient turbines. The operation of the wind turbine is based on the use of lift and drag forces. Wind power generators designed to start up at low wind speeds and high functionality are just a few of the advantages of the Darrieus type wind turbine. Despite not very complicated construction, such a power plant allows you to react to wind from any direction. Despite the disadvantage of this turbine associated with a very small starting torque, such a power plant with a low rotational rotor speed is characterized by a lower noise propagation during machine operation and the possibility of mounting closer to the population centres. Wind turbines constructed today are based on the achievements of aviation technology (Fig. 1a) and the production of wind turbine components has a lot in common with the aviation industry. These similarities mainly concern operating conditions, which in both cases take place at variable wind speeds and directions [10, 12]. Both in an aviation as well as in a wind power engineering, the geometry of the blade has a decisive impact on characteristics of the wind turbine or an aircraft performance. The changes in a surface smoothness and a shape of the airfoil are of great importance and can cause significant differences in the performance of the power plant with the same dimensions of the blades. Even slight differences in a shape can significantly change the characteristics. Therefore, choosing the right rotor blade is very important, and properly profiled blades guarantee a higher efficiency of the conversion of kinetic energy of wind into mechanical energy of the rotor [13 – 15].

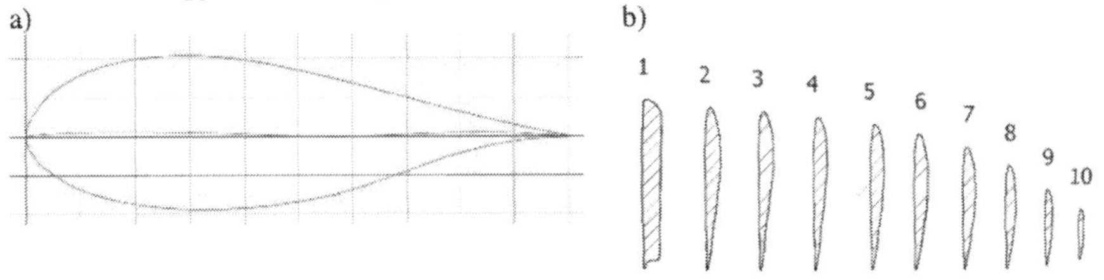

Fig. 1. Airfoils; (a) the DU 06-W-200 rotor blade profile used in the construction of wind turbines; (b) the designed rotor blade profile

Airfoils are characterized by an efficient work even with very large Reynolds numbers of 150.000 and 700.000. The increased blade width allows power plants to maintain the assumed capacity in a greater range of wind speed. Due to the thickening of the profiles, it is possible to obtain a higher strength of the structure.

A new wind turbine blade has been modelled (Fig.1b). Assuming that each cross-section of the blade behaves like a profile in a two-dimensional flow field, the rotor has been analysed in a simplified manner (Tab.1). In fact, it is impossible to use the total wind power. In practice, it turns out that only part of the wind power is actually used to set the turbine in motion. The mechanical power of a wind turbine is obtained by reducing the wind power by the power coefficient C_p.

Table 1. Assumptions for calculations

Airfoil	New blade	DU 06-W-200
c – chord length	105 mm	100 mm
H – max thickness	440 mm	445 mm
N_b – number of blades	3	3
R_z – radius of rotor	250 mm	250 mm
D – diameter of rotor	500 mm	500 mm
v – wind speed	$5-25\frac{m}{s}$	$5-25\frac{m}{s}$
rated wind speed	$v_{rated}=12\frac{m}{s}$	$v_{rated}=12\frac{m}{s}$
start-up wind speed	$v_{min}=5\frac{m}{s}$	$v_{min}=5\frac{m}{s}$
max wind speed	$v_{max}=25\frac{m}{s}$	$v_{max}=25\frac{m}{s}$
shape	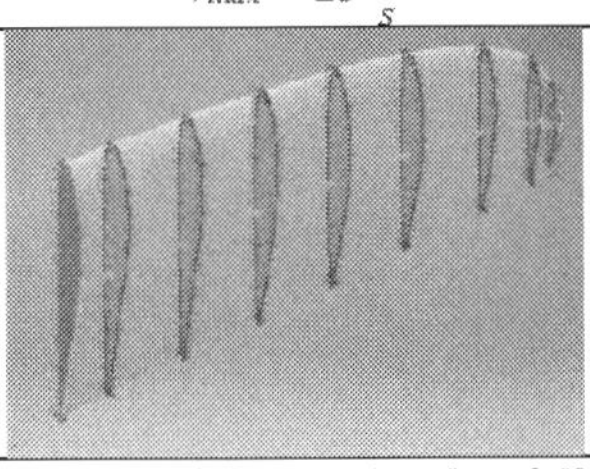	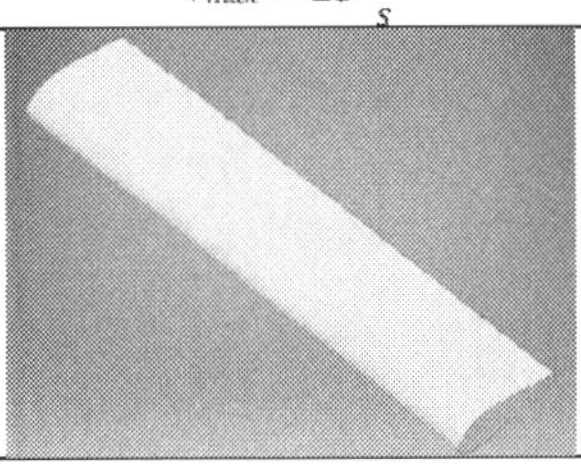

Thus, the energy on the rotor shaft is calculated as follows:

$$E=\frac{1}{2}v^3\cdot A\cdot C_p \tag{1}$$

where :

- C_p – power coefficient,
- P – power, W,
- ρ – density of air, kg/m^3,
- v – wind speed, m/s,
- A – area of a rotor, m2, given by: $A=\pi R^2$.

a) b)

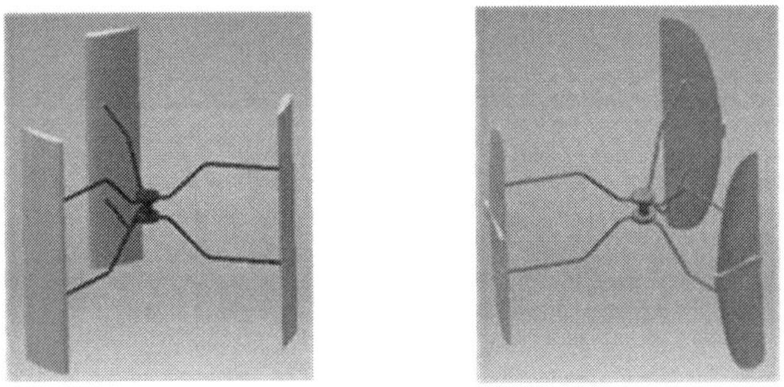

Fig. 2. Wind turbine model for numerical analysis; (a) with DU-06-W-200 profiles; (b) with designed airfoils

The profile DU-06-W-200 based on the data [16] and the new designed airfoil were modelled. The results obtained during the numerical analysis were referenced to these profiles. The modelled blade made it possible to obtain the final version of the turbine, which was used during the numerical analysis (Fig. 2a, 2b).

2 Power coefficient calculations

The value of the power coefficient C_p depends on the speed of the air before and after the rotor. The velocity of the air stream entering the rotor and the air velocity after the rotor is expressed as v_1 and v_2, respectively. Three analytical methods of calculating the power coefficient C_p are presented.

The Betz limit is a hypothetical boundary on the total amount of power that could be extracted by a wind turbine from a wind movement [17 – 19]. In the first method power coefficient C_p using the Betz limit is calculated as follows:

$$C_p = \frac{1}{2}\left[1 - \left(\frac{V_2}{V_1}\right)^2\right] \cdot \left[1 + \frac{V_2}{V_1}\right] \tag{2}$$

where :

v_1 – speed of the air stream entering the rotor, m/s,

v_2 – air speed behind the rotor, m/s.

Based on the above formula, a plot of C_p coefficient depending on the air speed was prepared (Fig. 3a). Calculated power coefficient values reach in reality, in fact, lower values due to the occurrence of various types of aerodynamic losses.

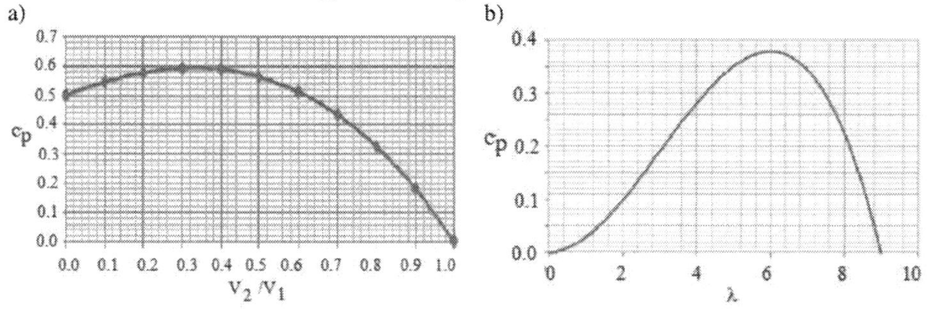

Fig. 3. Power coefficient C_p of wind turbine based on analytical analyses; (a) as a function of wind speed V_2/V_1 (Betz limit); (b) as a function of TSR λ;

The second method enabling the determination of the power coefficient C_p is based on the tip speed ratio (TSR) λ, which makes the rotational angular velocity dependent on the wind speed [15, 21, 22]:

$$\lambda = \frac{R_z \Omega}{V_1} \tag{3}$$

where :

Ω – angular velocity of the rotor wheel, rad/s,

v_1 – wind speed, m/s,

R_z – external radius of the impeller, m.

The rotational speed of the rotor can be express as $\Omega=(\pi\ n)/30$, where n is a rotor speed, rpm.

It is possible to find a relationship between C_p and TSR:

$$C_p = \left(\frac{3C_{p,max}}{\lambda_{max}{}^2}\right) \cdot \lambda^2 - \left(\frac{2C_{p,max}}{\lambda_{max}{}^3}\right) \cdot \lambda^3 \tag{4}$$

where :

λ – TSR,

λ_{max} – maximum TSR,

$C_{p,\,max}$ – maximum power coefficient.

Based on the above formula, a plot of C_p depending on the TSR was prepared (Fig. 3b).

The third method of determining the power coefficient is based on specific aerodynamic coefficients of the wind turbine rotor blades. This method is especially useful for comparison of a turbine performance with different number of blades [15, 23] and it can be presented as:

$$C_p = \frac{16}{27}\left(1 - \frac{C_1}{N \cdot \lambda}\right)^2 \cdot \left(e^{\lambda^{\frac{-C_2}{1.29}}} - \frac{C_d}{C_l} \cdot \lambda\right) \tag{5}$$

where :

- N – a number of wind turbine blades,
- λ – a TSR,
- C_1, C_2 – coefficients, in the literature they appear in a simplified form C_1=0.416 and C_2=0.35,
- C_d – a drag coefficient,
- C_l – a lift-coefficient.

Based on the above analyses, the power coefficient (C_p) can be the measure of the wind turbine rotor efficiency at a given wind speed, and it provides an approximation of the actual power produced by the rotor of the wind turbine.

3 Experimental procedure

The aim of the research is to determine the impact of the setting of the rotor axis on the power coefficient. Moreover, the role of optimal shape of the blades of the wind power plant was specified. Based on the computational algorithm and following the similar solutions found in the technique, the rotational speeds of the wind turbine rotor have been determined [15, 21, 23].

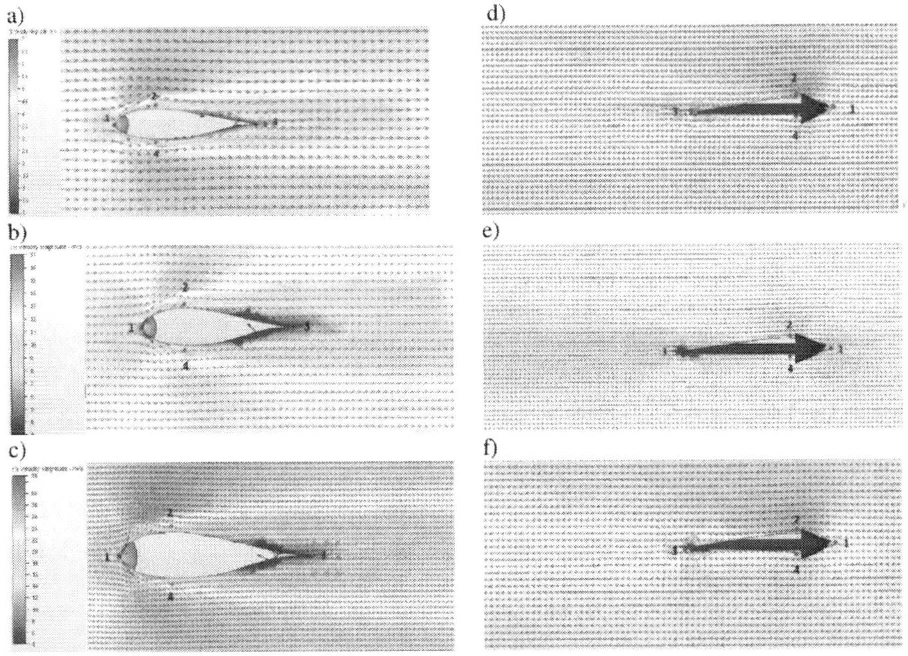

Fig. 4. The distribution of air speed for DU 06-W-200 and designed blade profile at different speeds of the airflow: (a) DU 06-W-200 at air speed 5 m/s; (b) DU 06-W-200 at air speed 12 m/s; (c) DU 06-W-200 at air speed 25 m/s; (d) designed blade at air speed 5 m/s; (e) designed blade at air speed 12 m/s; (f) designed blade at air speed 25 m/s.

The analysis was carried out at three rotational speeds of rotor depending on three wind speeds: minimum, rated and maximum wind speed. Minimum speed of the turbine describes a rotational speed of the rotor at which the turbine can start automatically. Rated rotational speed describes the rotor speed at which the wind turbine should achieve the nominal power.

Maximum speed describes the rotor speed at which the turbine can operate safely, without exposing the rotor structure to the destructive effect of centrifugal forces, vibrations and other factors that can damage the turbine rotor.

Considering that researches conducted on similar types of the wind turbines showed that constant-geometry Darrieus turbines can start-up during wind gusts and work at wind speeds of even 2- 5 m/s, the minimum wind velocity has been assumed as v_{min}=5 m/s.

Considering similar technical applications presented on the market by turbine manufacturers, taking into account the strength aspects and the correct and safe operation of the turbine structure, the maximum wind speed was assumed at the level of v_{max}=25 m/s, and the rated wind speed at the level v_{znam}=12 m/s.

Analyses were prepared for chosen rotor blades (Fig. 4). The speed of the moving air is shown using graphical vectors illustrating the distribution of airflow velocity and measured at four characteristic measuring points. The air speed values read from the measurement points at given air speeds are listed in the table (Tab.2, 3).

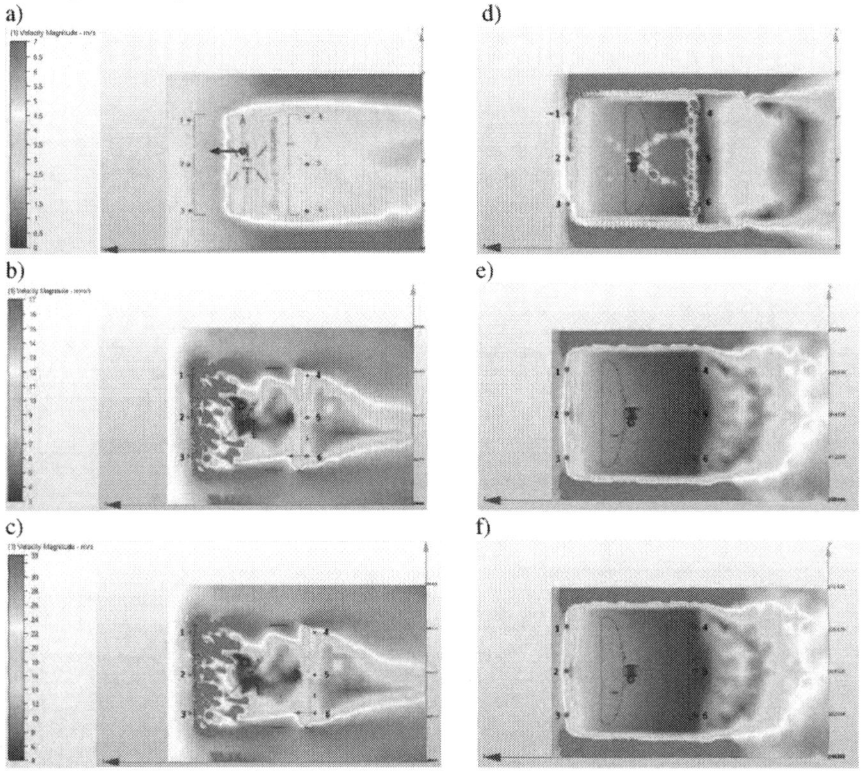

Fig. 5. The distribution of air speed for DU 06-W-200 and designed blade profile at different speeds of the airflow: (a) DU 06-W-200 at air speed 5 m/s; (b) DU 06-W-200 at air speed 12 m/s; (c) DU 06-W-200 at air speed 25 m/s; (d) designed blade at air speed 5 m/s; (e) designed blade at air speed 12 m/s; (f) designed blade at air speed 25 m/s.

In the same way as in the case of the tested blades profiles, a volume was generated, materials were assigned for the generated volume and turbine models, and the boundary conditions of the analyzes were determined. As for the blade profiles, the following: 5 m/s, 12 m/ s and 25 m/s air velocities were assumed: The rotor has been given a motion with a constant angular velocity. The appropriate angular velocities, determined on the basis of the air flow value, were imposed on the specified axis and the center of rotation (the turbine hub).

The following angular velocities were adopted: 63 1/s, 145 1/s and 310 1/s. The visible color difference between the upper and lower side of the profile (Fig. 5) results from the difference in speed and pressures. The lift force acting on the rotor blades and setting it in motion is formed as a result of these differences. Analyses were conducted for DU 06-W-200 and new designed blade profile.

The speed of the moving air is shown using graphical vectors illustrating the distribution of airflow velocity and measured at four characteristic measuring points. The air speed values read from the measurement points at given air speeds are listed in the table (Tab.4, 5).

Table 2. Airflow velocities for the DU 06-W-200 profile at four characteristic points

point	air speed before blade, m/s	measured air speed, m/s	air speed before blade, m/s	measured air speed, m/s	air speed before blade, m/s	measured air speed, m/s
1	5	0.83	12	2.09	25	4.56
2	5	3.37	12	8.97	25	21.95
3	5	0.89	12	2.00	25	4.38
4	5	3.25	12	8.65	25	17.01

Table 3. Airflow velocities for the designed blade profile at four characteristic points

point	air speed before blade, m/s	measured air speed, m/s	air speed before blade, m/s	measured air speed, m/s	air speed before blade, m/s	measured air speed, m/s
1	5	3.24	12	9.23	25	21.34
2	5	5.95	12	14.45	25	30.24
3	5	1.67	12	4.21	25	8.09
4	5	4.86	12	12.06	25	25.39

Table 4. Airflow velocities for the rotor made of DU 06-W-200 profile at six characteristic points

point	air speed before blade, m/s	measured air speed, m/s	air speed before blade, m/s	measured air speed, m/s	air speed before blade, m/s	measured air speed, m/s
1	5	4.75	12	17.45	25	26.54
2	5	4.86	12	16.85	25	27.52
3	5	4.69	12	14.93	25	27.14
4	5	1.84	12	11.84	25	17.92
5	5	1.75	12	7.78	25	15.68
6	5	1.83	12	8.93	25	20.41

Table 5. Airflow velocities for the rotor made of designed blade profile at six characteristic points

point	air speed before blade, m/s	measured air speed, m/s	air speed before blade, m/s	measured air speed, m/s	air speed before blade, m/s	measured air speed, m/s
1	5	6.81	12	12.5	25	27.32
2	5	7.23	12	10.9	25	25.11
3	5	5.82	12	13.4	25	26.53
4	5	1.25	12	8.3	25	8.32
5	5	2.74	12	7.2	25	6.45
6	5	3.52	12	7.5	25	11.21

4 Determination of values of power coefficient by analytical methods

For averaged measurements of air flow velocities collected at the measuring points before and after the turbine, the value of the power coefficient C_p was determined. The value of C_p

coefficients is calculated taking into accout Betz limit in succession for v_{min}=5[$\frac{m}{s}$], v_{rated}=12[$\frac{m}{s}$] and v_{max}=25[$\frac{m}{s}$] (Tab. 6).

Table 6. Values of C_p coefficients at three different flow rates

blades	C_p at V_{min}=5 m/s	C_p at V_{rated}=12 m/s	C_p at V_{max}=25 m/s
DU-06-W-200	0.242872	0.265886	0.196005
new designed	0.246465	0.251948	0.299166

On the basis of the range of the power coefficient can be concluded that results are possible to obtain.

While determining the value of the Cp coefficient with the use of the tip speed ratio, for the tested type of turbines, the maximum value of the power coefficient $C_{p,\ max}$=0.35 was assumed. The maximum TSR λ_{max}=6.4 and TSR λ=3.15 were assumed. In this method, identical results were obtained for all types of analyzed turbines. The distinguishing feature of TSR is dependent on the rotational speed, which has been adopted for all types of turbines as the same. In addition, in all types of rotors, the same radius of rotor was assumed, and in addition, this analysis was carried out with identical air flow conditions.

$$C_p = \left(\frac{3\cdot 0.35}{6.4^2}\right)\cdot 3.15^2 - \left(\frac{2\cdot 0.35}{6.4^3}\right)\cdot 3.15^3 = 0.171 \tag{6}$$

The value of the power coefficient calculated for this type of wind turbines, taking into account the adopted dimensions and parameters, was equal to : $C_p = 0.171$.

While determining the value of the Cp coefficient with the use of the aerodynamic coefficients, for the tested type of turbines, the values of the lift force was determined by determining the value of the lift and drag force coefficients. Based on the assumptions: the number of wind turbine blades N=3, the TSR as λ=3.15 and coefficients C_1=0.416 and C_2=0.35, for the new profile we calculate:

- for the air speed $v_{min} = c_{\infty 1} = 5\left[\frac{m}{s}\right]$

$$C_{p1} = \frac{16}{27}\left(1-\frac{0.416}{3\cdot 3.15}\right)^2\cdot\left(e^{3.15\frac{-0.35}{1.29}} - \frac{0.2309}{0.1027}\cdot 3.15\right) = 0.1547 \tag{7}$$

- for the air speed $v_{rated} = c_{\infty 2} = 12\left[\frac{m}{s}\right]$

$$C_{p1} = \frac{16}{27}\left(1-\frac{0.416}{3\cdot 3.15}\right)^2\cdot\left(e^{3.15\frac{-0.35}{1.29}} - \frac{0.1796}{0.1876}\cdot 3.15\right) = 0.2965 \tag{8}$$

- for the air speed $v_{max} = c_{\infty 3} = 25\left[\frac{m}{s}\right]$

$$C_{p1} = \frac{16}{27}\left(1-\frac{0.416}{3\cdot 3.15}\right)^2\cdot\left(e^{3.15\frac{-0.35}{1.29}} - \frac{0.1701}{0.2213}\cdot 3.15\right) = 0.3451 \tag{9}$$

The values of the power coefficient for the turbine with the DU 06-W-200 profiles were calculated. The Cp coefficients for the analyzed profiles, determined by three methods, are compared (Tab. 7).

Table 7. Values of C_p coefficients at three different calculation

blades	C_p at Betz limit	C_p at TSR	C_p at aero. coefficients	C_p average
DU-06-W-200	0.2659	0.1710	0.2240	0.2203
new designed	0.2519	0.1710	0.2965	0.2398

According to the first calculation method, the Cp coefficients for the turbine with the new designed blade profile is slightly worse to the comparative rotor.

The second method gave the same result for all types of analyzed rotors. This method depends on the type of turbine and its dimensions (diameter of the rotor wheel, the number of blades in the rotor), and do not depend on the parameters of the rotor's motion or the type of profiles used.

In the third method, better results ware obtained for the turbine with a new designed blade profile, and also the arithmetic average of all methods, calculated for each type of profile, indicate the new designed profile as better sollution.

5 Conclusion

A wind farm with a vertical axis of rotation is a simple construction that gives the possibility of obtaining electricity in places where access to it is limited or impossible. A model of a wind turbine with a vertical axis of rotation was developed with the use of new profiles. In order to interpret the obtained results and reference to reality, for comparative analysis the comparative profile DU 06-W-200 and the turbine with their use were used.

Based on the numerical analysis, it can be concluded that the new profile achieves comparable or even better results than the comparative profile in relation to the aerodynamics itself and the value of wind power used by the turbine.

The CFD analysis of the flow around the profiles showed a greater difference in speed between the air speeds at the bottom and the back of the blade for the designed profile, which suggests the creation of more lift force.

Also based on the power coefficient C_p determined by three methods, it is possible to suggest to use the new profile in the technology. The average arithmetic value of the C_p coefficient calculated on the basis of the above methods gives a better result for the designed blade profile than currently use one.

However, in addition to the properly engineered power plant blade, there are still many elements affecting the efficiency of the turbine: the structural parameters of the wind power plant, drives and control systems, materials from which the power plant was built, elements of the climate [24–27]. In order for the investment to bring profit to investors, each of these factors must have a positive impact on the venture.

References

1. S. Gumula, W. Hudy, M. Piaskowska–Silarska, K. Pytel, Chemical Industry, **94,** pp. 1000–1003, (2015)
2. H. Noga, Information technologies and technical education, **1,** pp. 165–169, (2009)
3. I. Turekova, J. Depesova, A. Haskova, T. Bagalova, EDULEARN 15 (2015) 1913–1921.
4. A. Bianchini, G. Ferrara, L. Ferrari, Energy Conversion and Management, **89**, pp. 690-707, (2015)
5. T. Prauzner, Electrotechnical overview, **88,** pp. 205–208, (2012)
6. J. Suchanicz, G. Stopa, J. Kusz, M. Zubko, W. Hofmeister, M. Antonova, A. Kalvane, M. Dambekalne, A. Sternberg, I. Jankowska-Sumara, B. Garbarz-Glos, D Wcisło, K. Konieczny, K. Pytel, A. Finder, Journal of Materials Science, **45,** pp. 1453–1458, (2010)
7. P. Dulian, B. Garbarz-Glos, W. Bąk, M. Antonova, C. Kajtoch, K. Wieczorek-Ciurowa, H. Noga, Ferroelectrics, **497**, pp. 62–68, (2016)
8. S. Gumula, K. Pytel, M. Piaskowska-Silarska, Polish Journal of Environmental Studies, **23**, pp. 2315–2320, (2014)
9. W. Hudy, K. Jaracz. K. Pytel, *International Carpathian Control Conference*, **16,** pp. 81-185, (2015)
10. M. L. Niculescu, M. G. Cojocaru, M. V. Pricop, D. Pepelea, A. Dumitrache, D. E. Crunteanu, *AIP Conference Proceedings*, **1863,** 420003, (2017)
11. L. Borowik, P. Ptak, Przeglad Elektrotechniczny, **88**, pp. 142–145, (2012)
12. G. Ronsten, Journal of Wind Engineering and Industrial Aerodynamics, **39**, 1, pp. 105-118, (1992)
13. Chen, Jin, et al. Applied Mathematical Modelling, **40**, 3, pp. 2002–2014, (2016)
14. Spinato, F., et al. "Reliability of wind turbine subassemblies." IET Renewable Power Generation, **3**, 4, pp. 387–401, (2009)
15. D. Mazur, Strength calculations and vibrations of the Darrieus type turbine, Rzeszow, PR, (2012)

16. http://airfoiltools.com, access June 1st, (2018)
17. S. Gumula, K. Pytel, M. Piaskowska–Silarska, Polish Journal of Environmental Studies, **23**, pp. 2321–2325, (2014)
18. K. Pytel, W. Hudy, H. Noga, W. Kulinowski, *International Carpathian Control Conference*, **17**, pp. 401–406, (2016)
19. K. Pytel, W. Hudy, H. Noga, W. Kulinowski, *International Carpathian Control Conference*, **17**, pp. 407–412, (2016)
20. S. Gumuła, T. Knap, Determination of the optimal geometry and movement parameters of the wind power plants depending on the average annual wind speed. *National Forum "Small Energy - '97."* Zakopane (1997)
21. S.Gumuła, Wind Energy, Kraków, AGH, (2006)
22. R. Jablonski, M. Turkowski, R. Szewczyk (Eds.). *Recent Advances in Mechatronics. Proceedings of 7th International Symposium Mechatronics held Sept. 19–21 in Warsaw*, Poland. (2007)
23. F.D. Bianchi, H. de Battista, R.J. Hernán, Wind Turbine Control Systems. Principles, Modelling and Gain Scheduling Design, Springer, (2007)
24. B. Garbarz-Glos, W. Bąk, H. Noga, M. Antonova, A. Kalvane, W. Śmiga, Integrated Ferroelectrics, **173**, pp. 12–18, (2016)
25. W. Śmiga, B. Garbarz-Glos, W. Piekarczyk, H. Noga, D. Sitko, M. Karpierz, M. Livinsh, Integrated Ferroelectrics, **173**, pp. 46–52, (2016)
26. W. Hudy, M. Piaskowska-Silarska, K. Pytel, S. Gumuła, Archives of Mining Sciences, **62**, pp. 579-596, (2017)
27. D. Olszewska, T. Prauzner, P. Ptak, H. Noga, Przegląd Elektrotechniczny, **12**, pp. 191-193, (2016)

Analysis of water flow velocity criteria for the selection of copper pipes on the example of production buildings

Alina Żabnieńska–Góra[1,*], and *Edyta* Dudkiewicz[1]

[1]Wrocław University of Science and Technology, Faculty of Environmental Engineering, Wybrzeże Wyspiańskiego 27, 50–370 Wroclaw, Poland

Abstract. The article presents different aspects of application of copper pipes in water supply system, which have importance on properly work of installation and people health safety. The article analyses the water flow velocity criteria and their influence on the operation of the installation. It is particularly dangerous for copper pipes to exceed the permissible water flow velocity, as this results in corrosive processes. It is also not recommended that the flow velocity be too low, as it affects the development of biofilm and multiplication of bacteria.

1 Introduction

According to the European Copper Institute 80% of copper extracted so far is still in use or is intended for reprocessing. The possibility of obtaining copper from secondary raw materials, without any loss in its quality, is a big advantage. Today 40% of Europe's demand for copper is met by copper obtained from secondary raw materials, which is used even in modern equipment. In generally copper pipes are more expensive than steel and plastic ones [1]. However owing to resistance to corrosion, their appropriate influence on water quality and ease of installation, copper and copper alloys are used in water installation [1, 2]. The use of copper pipes in drinking water installations reduces the occurrence of biofilm and bacterial proliferation. However, in order for the copper water supply system to work properly, it is necessary to ensure proper working conditions [3].

On the properly working of copper water installation have impact the factors:

- quality of water delivered to water supply systems [2],
- temperature of water [4],
- kind of pipe insulation materials [1],
- hydraulic of conditions of work [2, 5],
- conditions of exploitation of water installation [2].

* Corresponding author: alina.zabnienska@pwr.edu.pl

2 Water corrosivity criteria for copper

According to the Regulation of the Minister of Infrastructure and Development [6] "products used in water supply systems should be selected taking into account the corrosivity of water, so that its quality and durability do not deteriorate and such effects do not cause mutual influence materials from which these products were made". The Regulation of the Minister of Health [7] in Appendix 3 specifies the indicators (chlorides, sulphates, conductivity, pH) that should be taken into account when assessing the corrosive properties of water. This is particularly important when designing copper pipe systems. Water with inadequate chemical parameters in the copper installation may cause corrosion. It is precisely for reasons of water quality that the use of copper pipes for water installations is not recommended in some regions of Poland. The literature, e.g. [2], states that 10% of tap water has a composition that is not conducive to the formation of protective layers on the internal surfaces of copper pipes.

Limiting values determining the corrosive risk of metal are specified in the standard PN-EN 12502:2006 [8]. The standard [8] refers to the guidelines for testing the possibility of corrosion in water distribution and storage installations [10]. The requirements for tap water for copper feed systems refer to pH, nitrate ion content ($C_{NO_3^-}$), sulphate ion ($C_{SO_4^{2-}}$) and general alkalinity ($Z_{ogólna}$). The numerical values of the indicators of the above parameters are presented below:

$$\text{pH} > 7 \tag{1}$$

$$C_{NO_3^-} < 30mg/l \tag{2}$$

$$\frac{Z_{ogólna}}{C_{SO_4^{2-}}} > 2 \tag{3}$$

General corrosion can occur in copper water installations if they are not installed correctly. This corrosion may be caused by low pH value of water (below 6.5), presence of sulphate, chloride, nitrate ions, high content of carbon dioxide. In addition, e.g. in water treatment processes, depriving water of organic compounds, which are inhibitors of copper corrosion, also leads to corrosion of pipes. Soft water does not have carbonate hardness (which regulates the level of acidity and alkalinity of the water). This leaves carbon dioxide in the water, which transforms into unstable carbonic acid. When water flows through the water supply system, copper pipes corrode and copper flows into drinking water, which, in too high a concentration, is considered a toxic element. Such water may acquire a specific smell and taste unpleasant for the users [2]. According to the Regulation of the Minister of Health [7] the content of copper per liter of drinking water should not exceed 2 mg.

Pitting corrosion is when perforation of the pipe wall occurs as a result of particles and deposits on the inside surfaces of the pipes or, for example, as a result of leakage of soft soldering flux on the inside surface of the pipe (especially when the flux contains chloride ions that are corrosively hazardous). The corrosion of copper pitting takes place in different ways for cold and hot water. In cold water, corrosive risks are only present in unfavorable operating conditions and assembly installation errors. In hot water, the components of water influence the corrosion process of copper in various ways [4]. The phenomenon of selective corrosion may occur when the installation includes elements (e.g. valves) made of copper alloys such as brass (brass resistant to dezincification must be used).

The simultaneous action of corrosion and a turbulent stream of water, which locally destroys the oxide layer (the passive layer, which is the protective layer of the metal) causes erosion corrosion. In this case, the turbulent flow of water results not only from exceeding the permissible speeds specified for copper pipes, but also from any flow disturbances where the speed increases locally and turbulence occurs. Such conditions arise in the area of pipe cross-sectional changes, around soldering overhangs, and even in the areas of water flow direction changes (elbows, long radius elbow).

In order to prevent erosive corrosion of copper pipes, it is important to follow the recommended maximum flow rates for copper pipe installations at the design stage.

3 Water flow velocity criteria

One of the basic conditions of properly working is to ensure that the water flow velocity is adequate. This parameter is relevant for dimensioning the water supply system including the diameter selection of the water pipes. Criterion of the flow velocity of the water varies depending on the manufacturers guidelines, the material used and the law.

Ensuring the water velocity in the installation in accordance with the recommendations for pipes is important. Because the speed of water affects:

- the degree of corrosion of the pipe,
- the noise level of the system,
- hydraulic impacts,
- linear and local pressure losses in the system,
- the development of biofilms, which, in addition to the temperature and stagnation of the water, are essential to prevent and combat the proliferation of Legionella bacteria [9].

The common relations stated in the standard PN-92/B-01706 [10] provide the basis of designing water supply system in buildings. However, although according to Polish Committee for Standardization this standard bears the status of "withdrawn without replacement", it is cited in the amendment of the Regulation of the Minister of Infrastructure and Development [6] as concerns, among other things, the general rules of sizing the water supply conduits and determining the design water flow rates in residential buildings, office buildings, hotels and department stores.

Admissible maximum velocities of water which are decisive criterion in selecting the diameters of piping are, according to PN [10], as follows:

- up to 1.5 m/s in connections from riser to draw-off points and in risers of water supply system,
- up to 1 m/s in main distribution lines and home service pipes.

Manufacturers of plastic pipes specify the following admissible velocities:

- for risers and service lines to draw-off points – up to 2.5 m/s,
- for main distribution lines – up to 2 m/s,
- for water supply service pipes – up to 1.8 m/s.

In accordance with PN-EN 806-3:2006 [11], the maximum speed in a single connections from riser to draw-off points can be 4 m/s, while in other installation pipes it can be up to 2 m/s. Then the requirements on admissible velocities in water system pipes acc. to DIN [12], depending on the duration of maximum flow rate and local resistances of the fixtures on the plot, are specified in Table 1.

For the reason of erosion-corrosion in the copper water installations the water flow velocity should not exceed 1 m/s in horizontal conduits and 2 m/s in connections [2, 3, 5]. It is recommended to use long radius elbow instead of bends, which are less sensitive to scratches and destruction of the oxide layer due to pollution carried in tap water. At the beginning of the copper installation (behind the water meter set), a mechanical filter with a retention capacity of particles larger than 80 μm should be installed [3].

For the reason of corrosion steel pipes are using primarily in installation of cold water or to deliver water to tank water. In the steel water installations the water flow velocity should not exceed 1 m/s in horizontal conduits and 1,5 m/s in connections and in risers [10, 13].

Table 1. Water flow velocity acc. DIN 1988-300:2012-05 [12]

Type of water-pipe	**Water velocity [m/s]**	
	Duration of calculated flow water < 15 min	Duration of calculated flow water ≥ 15 min
Water supply connection	2.0	2.0
Sections in the installation with elements with resistance coefficient $\zeta<2.5$	5.0	2.0
Sections in the installation with elements with resistance coefficient $\zeta\geq2.5$	2.5	2.0

Most design guidelines and standard [10] specify the criteria for water velocity by distribution line and riser, while DIN [12] takes into account the flow time and local losses of the installation. In single-storey buildings without traditional water risers, the water supply system has main distribution lines and connections from main distribution lines to draw-off points. In this section, the speed criteria should be used when selecting the pipe diameters according to the recommendations described above.

A higher water velocity generates higher pressure losses. The coefficient of unit pressure losses depends on the value of the absolute roughness coefficient k. It is different for different materials and the smaller its value (i.e. the surface is smoother), the lower the water flow pressure losses and the lower the possibility of sludge formation and material destruction.

Over time, the pipes are inlayed by the deposition of chemical compounds that precipitate from the water or are the result of corrosion of the pipe material (chemical deposits and hydrobiological growth - biofilm). This increases the roughness of the inner pipe walls and at the same time reduces the cross-sectional area of the pipes. An increase in hydraulic head losses, i.e. a decrease in the flow rate (hydraulic efficiency) of water pipes, is highly undesirable [14, 15]. The roughness coefficient for copper pipes is k=0.0015 mm [15].

4 The importance of biofilm development

Water supply systems can be a breeding ground for pathogenic bacteria and parasites. Particular attention has been paid in recent years to the risks associated with the operation of hot water systems where conditions are favorable to the development of Legionella bacteria. The temperature range of 35 - 46^{0}C, long stagnation time of water and the presence of biofilm and sediments on surfaces in contact with water facilitate the multiplication of Legionella bacteria in installations [2, 9]. The type of installation material and associated characteristics such as corrosion resistance, surface roughness and chemical composition have a significant influence on the development of the biofilm. On the basis of numerous studies [2, 9], the inhibitory effect of copper ions on the multiplication of Legionella bacteria is found in copper installations.

The impact of speed is also very important. On the one hand, high velocity is desirable due to shear forces that reduce the upper layers of the biofilm, and on the other hand, it promotes the formation of biofilm, as it ensures the supply of nutrients [9]. Too low a water velocity promotes the development of a biofilm resistant to adverse environmental conditions, in which microorganisms multiply. Hot water and circulation water pipes in which there is a constant flow of water are not exposed to stagnation of water, as are cold water pipes. From a legionnaire protection point of view, it is recommended to ensure a maximum speed of 1-2 m/s in copper cold and hot water pipes without circulation. It should be added that thermal disinfection, as recommended by the regulations [6], is carried out without any negative effects on the durability of the copper installation [3].

5 Analysis of water speed measurements in halls

Measurements of the hot water flow velocity were carried out in two production halls with different water load pattern. Water flow velocity was measured with a portable ultrasonic flow meter (PORTAFLOW C) on the pipe delivering water to hot water tank. Data was archived every 10 seconds. Halls are localized in Bolesławiec and in Wrocław. These are objects with different degrees of soiling work and other sanitary equipment. In this type of buildings, water is distributed horizontally. Table 2 summarises the characteristics of the production halls where the measurements were made.

Table 2. Characteristics of the production halls

Location	Bolesławiec	Wrocław
Number of work shift	3	3
Number of people on shift Totally	I – 145 II – 65 III – 25 Σ= 235	I – 300 II – 250 III – 200 Σ= 750
Number of working days per week	5	6
Degrees of soiling work	high soiling	low soiling
Sanitary equipment (delivered hot water): Shower	Yes	No
Sink	Yes	Yes
Average daily hot water consumption [m^3]	16	20
Average daily hot water consumption per person [l/person]	68	26
Average weekly hot water consumption [m^3]	104	133

The results of the research of the water flow velocity during one week in a production halls are presented on Fig. 1 (hall in Bolesławiec) and Fig. 2 (hall in Wrocław).

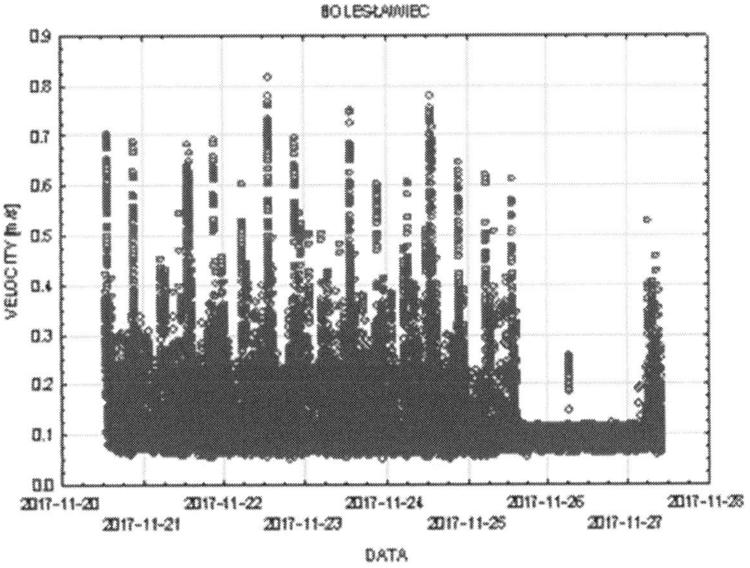

Fig. 1. Distribution of water velocity in production hall with high soiling in Bolesławiec

Table 3 presents the average, minimum and maximum values of the measured value of water flow velocity and standard deviation.

The both diagrams show that water velocities are below value 1 m/s recommended for copper pipes. Considerably larger water velocity is observed after the shifts on weekdays, than during weekend. In the hall in Bolesławiec, a marked increase in speed was observed after each finished shift. In Bolesławiec (fig. 1) water velocity is not higher than 0.82 m/s on weekdays while during the weekend average velocity 0.27 m/s. Between shifts water velocity is rather low, about 0.3 m/s. In Wrocław (fig. 2) water velocity is not higher than 0.55 m/s on weekdays and 0.15 m/s during a weekend when fewer people are at work. These velocities show that the system is oversized in this hall.

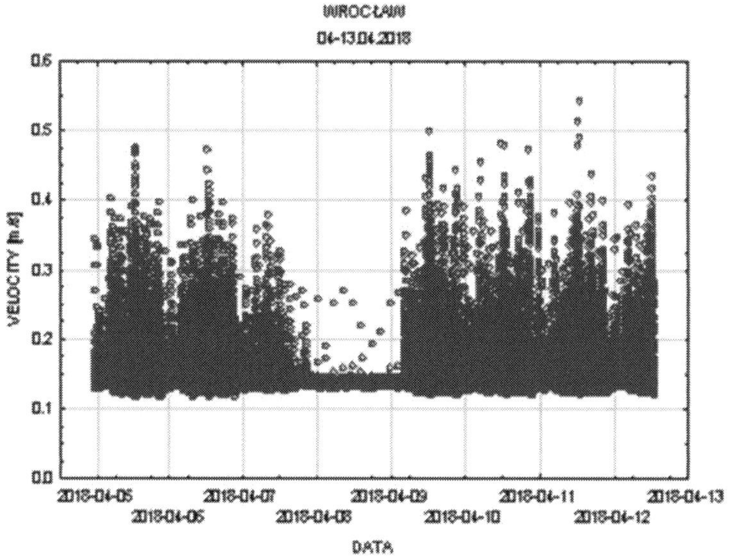

Fig. 2. Distribution of water velocity in production hall with low soiling in Wrocław

Table 3. Summary of flow velocity

Location		Average value m/s	Min. value m/s	Max. value m/s	Standard deviation
Bolesławiec	working days	0.128	0.000	0.816	0.083
	weekend	0.095	0.055	0.618	0.038
Wrocław	working days	0.155	0.117	0.542	0.042
	weekend	0.139	0.125	0.378	0.014

6 Conclusion

The water flow rates in the system are very important for many factors. The water velocity criteria are given mainly depending on the installation material. In addition, most design guidelines and standard [10] specify the criteria for water flow velocity, broken down into distribution pipes, risers and service lines to draw-off points. The approval of appropriate water velocity criteria is the first step to ensure the correct operation of the system. However, the proper construction, commissioning and maintenance of the water supply system will have an impact on water quality and human health.

Taking into account all water flow velocity criteria and their influence on the operation of the installation, speed criteria (1 m/s in main distribution lines and 2 m/s in connections from riser to draw-off points) should be used for selecting the diameter of copper pipes.

The results of measurements in production halls show that the water flow velocities are below than recommended value 1 m/s and are at weekends are much lower than assumed.

The average water flow velocities are in the range not recommended for water supply systems. These outcomes of velocities show that the water system is oversized in Wrocław. This is due to the method of determining the water flow rate. There is no clear method to calculate the water flow in the production halls.

Low water velocities in the period between work shifts and periods of water stagnation in the halls make it possible to conclude that, from the point of view of microbiological contamination, copper pipes are an appropriate solution in this type facilities, as it has an inhibitory effect on the proliferation of Legionella bacteria.

The participation in the conference was co-financed by the Faculty of Environmental Engineering, Wroclaw University of Science and Technology, Poland (award by Ministry of Science and Higher Education in years 2017–2018 No. 0401/0007/17).

References

1. Dasdemir A., Ural T., Ertürk M., Keçebas A.: Optimal economic thickness of pipe insulation considering different pipe materials for HVAC pipe applications, Applied Thermal Engineering, **121**, pp 242–254, (2017)
2. Toczyłowska B.: The impact of copper water pipelines on water quality, Polish Copper Promotion Center, Wrocław, Poland, (2004)
3. Technical requirements COBRTI INSTAL, Notebook 10. Guidelines for the design and use of copper pipe installations. Information Center "Installation technique in construction, (2003)
4. Falewicz P., Drela I., Kuczkowska S.: Aggressiveness of corrosive waters, Protection against corrosion, Poland, **4-5**, pp 187-191, (2008)
5. Dudkiewicz E., Żabnieńska-Góra A.: The importance of flow-rate to working of hot and cold water copper installation, Rynek Instalacyjny, **11,** pp 41-43, (2016)
6. Regulation of the Minister of Infrastructure of 12 April 2002 on the technical conditions that buildings and their location, 75 pos. 690 (with z later changes)
7. Regulation of the Minister of Health of 13 November 2015 on the quality of water intended for human consumption (dz. U, poz. 1989)
8. PN-EN 12502:2006 Protection of metallic materials against corrosion - Guidelines for assessing the risk of corrosion in water distribution and storage systems – Part 1: General provisions; Part 2: Factors affecting copper and copper alloys.
9. Toczyłowska B.: The role of biofilm in preventing and fighting with legionella bacteria in water supply system, Water supply and sewage installations - design, construction, operation (Seidel-Przywecki Warszawa 2015)
10. PN-92/B-01706 Water supply installations - Requirements in the design
11. PN-EN 806-3:2006 Specifications for installations inside buildings conveying water for human consumption – Part 3: Pipe sizing – Simplified method
12. DIN 1988-300:2012-05 Technische Regel für Trinkwasser-Installationen (TRWI) Ermittlung der Rohrdurchmesser Technische Regel des DVGW.
13. Górecki A.: Corrosiveness of tap water and installation material, Rynek Instalacyjny, **9**, pp 62-67, (2011)
14. Kotowski, A.: Hydraulic Analysis of Phenomena Reducing Pipeline Flowability. Ochrona Środowiska, **32**, 1, pp. 27–32, (2010)
15. Stefański W.: The influence of increasing the roughness and reducing the pipe diameter on hydraulic resistance, Instal, **9,** pp 66-74, (2006)

Application of selected methods of digital simulation to analyse dynamic impact of driven pneumatic wheel on soil in agrotechnical treatments and forestry to limit negative impact on the environment

Włodzimierz Malesa[1,*], and *Marek* Kopacz[1]

[1]Warsaw University of Technology, ul. Łukasiewicza 17, 09-400 Płock, Poland

Abstract. Loads transmitted by mechanical vehicles to soil result in its deformation and compaction. Increasingly bigger and more efficient machines affect the base they move on to a significantly greater extent. The purpose of this study with the use of digital simulation is to reduce expenses on costly experimental studies with the use of off-road vehicles or specially developed laboratory machines. The simulations were conducted using the Mechanical Event Simulation module, which is a part of the Autodesk Simulation Mechanical software. For each step, the simulation program provides the value of stress in all nodes of the base grid. The stress values obtained via dynamic simulations are consistent with the values registered under laboratory conditions. When classical experiments are replaced with numerical methods, not only the cost is reduced, but also the impact of weather conditions on measurement results is avoided and the time required for the study is shortened. The use of the described methods allows for design and selection parameters of exploitation pneumatics drive mechanism, which will significantly reduce the negative impact on the natural environment by machines in agrotechnical treatments and forestry. This is related to the environmental protection issues.

1 Introduction

Loads transmitted by mechanical vehicles to soil result in its deformation and compaction. Increasingly bigger and more efficient machines affect the base they move on to a significantly greater extent. Studies confirm that multiple passages of machines result in consolidation of soil, which results in significantly decreased yield of plant cultivation [5, 8].

The machines' ability to function effectively in the field depends on the physical properties of the base. Thus, mechanical properties of soil, as well as the size and type of impact of the wheel driving mechanism on the base must be considered when off-road vehicles are designed.

* Corresponding author: wlodzimierz.malesa@pw.edu.pl

It is assumed that consolidation of soil results in its regular deformation. In turn, the presence of tangential forces results in horizontal displacement of soil. The averaged pressure value relative to the surface of contact between the wheel driving mechanism and the soil, evenly distributed on the entire surface of contact between the wheel and the base, is taken for the analysis of stresses and deformations.

Horizontal soil deformation has the greatest impact on the driving force value, while vertical soil deformation of soil - on the rolling resistance value.

The traction properties of the wheel driving mechanism cooperating with a given soil are assessed mainly based on traction forces. The traction forces include: the rolling resistance (P_f), the driving force (P_j) and the drawbar pull (P_u). The relation between these forces is defined by the equation [13]:

$$P_u = P_j - P_f \tag{1}$$

The basic evaluation criterion of traction properties of the wheel driving mechanism is value of pulling force (P_u). Engine power determines value of driving force of vehicle. Shear strength of the soil allows to transmission of driving force on circumference of a wheel on a soil acts as driving force (P_j).

Physical of soil properties have influence on traction properties such as: cohesion, angle of internal friction and coefficients characterizing soil deformation under influence of stress propagation.

2 Application of the finite element method and digital simulation for analysis of the vehicle-land system

To analyse the impact of the wheel driving mechanism on the soil, Simulation Mechanical system by AutoDesk was used. The system contains the "Mechanical Event Simulation" module, which allows the user to model the properties of plastic-elastic materials and to adjust the nature of cooperation between the wheel driving mechanism and the base.

This module allows elements with non-linear strength properties. The geometric shape of models [11] may be prepared in the software environment, or a compatible file created using different software, e.g. the Autodesk Inventor, may be imported. Based on the geometric model, a grid of finite elements was created and the boundary conditions were defined. The system allows reading the resulting values (components of the stress status) in each point of a three-dimensional grid of finite elements; it also generates a map of results in a graphic form. In the case of dynamic systems, the results were saved as animation showing the movement of parts, as well as deformation and stress of the material. The values of stress and deformation components may be analysed by adding relevant viewing planes.

The designing process and simulation of the vehicle-land system was performed in a stepwise manner, as follows [3, 6, 8]:

- creating a digital model of the wheel driving mechanism and the base,
- defining mathematical models describing the behaviour of materials,
- selecting parameters defining the properties of materials,
- configuring loads affecting the system,
- defining zones and type of contact between the driving mechanism and the base,
- defining boundary conditions,
- selecting parameters and generating the FEM grid,
- simulation,
- analysis of the obtained results.

The vehicle-land system in this work is composed of two basic elements: the drive wheel and the strip of soil it moves on. A tire for agricultural machines (with dimensional signatures 14.9-28) was adopted for modelling. Such tires manufactured by Stomil and Good-Year were used in experimental studies conducted at the Faculty of Civil Engineering, Mechanics and Petrochemistry at the Warsaw University of Technology in a soil channel [4]. Use of the same tire geometry allowed comparing the digital simulation results with empirical results.

The digital model was exported to the Simulation Mechanical software. "joint" elements were generated in the system. These are special elements of the FEM grid used to connect elements of digital models to transfer the load [2, 9]

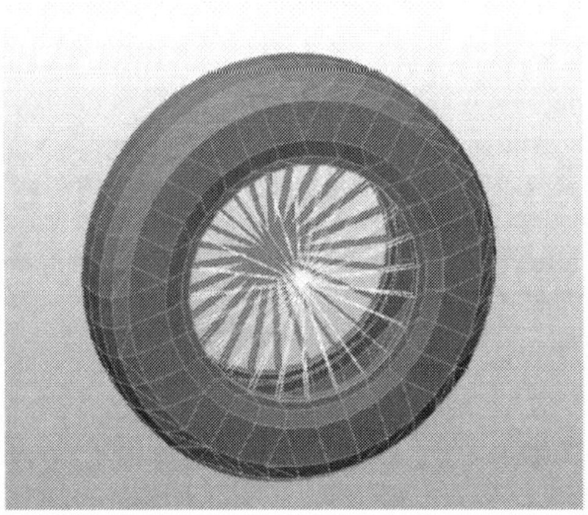

Fig. 1. View of nodes transferring loads from the wheel axis to the tire

The soil is represented by a cuboid element, which reflects the actual laboratory case in the soil channel. The base model required custom preparation of the grid in order to minimise the calculation time and to register the results in strictly specified points without the loss of the assumed accuracy. The grid was generated using the projection mechanism. The geometry of the grid, denser in the measurement points, was drawn on one of the side walls.

The flat grid was multiplicated into a three-dimensional form, as shown in the figure below.

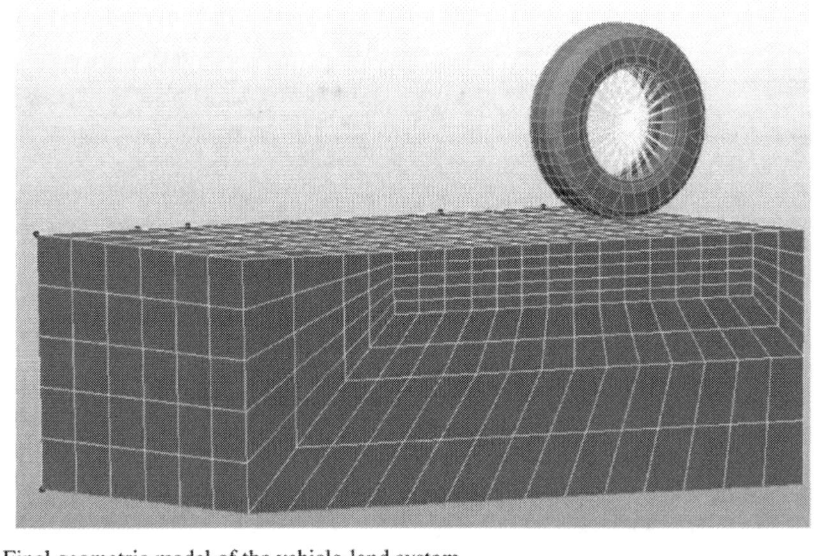

Fig. 2. Final geometric model of the vehicle-land system

The load and kinematic parameters were modelled in the examined system with the following specification:

- pressure: 7 kN and 10 kN,
- rotational speed of the wheel: 0.4 rps, i.e. the linear speed was set to approx. 7 km/h (the basic value, without slippage).

The internal surface of the tire was subjected to pressure of 0.15 MPa [4] which corresponded to the value of air pressure in a pneumatic tire. Parameters characterising the relevant material in the plastic-elastic model, e.g. density, Young's modulus and Poisson's ratio, were assigned to the individual elements of the studied system [15].

The data corresponded to a sandy-clay soil, identical to that used under laboratory conditions. The wheel slippage on the base was an additional parameter taken into account in the digital studies. Simulations were performed for the values of 10%, 15%, 20%, 25%, and 30%. The studies determined the type of contact between touching surfaces and the stiffness of the contact, i.e. software parameters ensuring the required cooperation between elements in contact (tire and soil). According to the software manufacturer's recommendations, the contact stiffness is selected based on observation of the model's behaviour upon modification of the order of magnitude [1]. Each simulation presents 2 seconds of the wheel's movement. During the first second, the wheel is initially at rest and then it starts moving, in the next second - when the driving conditions are stabilised - the calculated values are verified. 20 calculation steps were assumed for one second, and thus each of the simulations included a total of 40 steps.

3 Numerical calculations of the vehicle-land model

The simulations were conducted using the Mechanical Event Simulation module, which is a part of the Autodesk Simulation Mechanical software. This module allows analysis of dynamic mechanical systems made of mutually cooperating elements. The cases of dynamic simulations are much more complicated than static simulations. Apart from selecting the relevant parameters, they require appropriate preparation of finite element grids, to make the cooperation between the mobile parts and the value of mutual interactions as close to reality as possible.

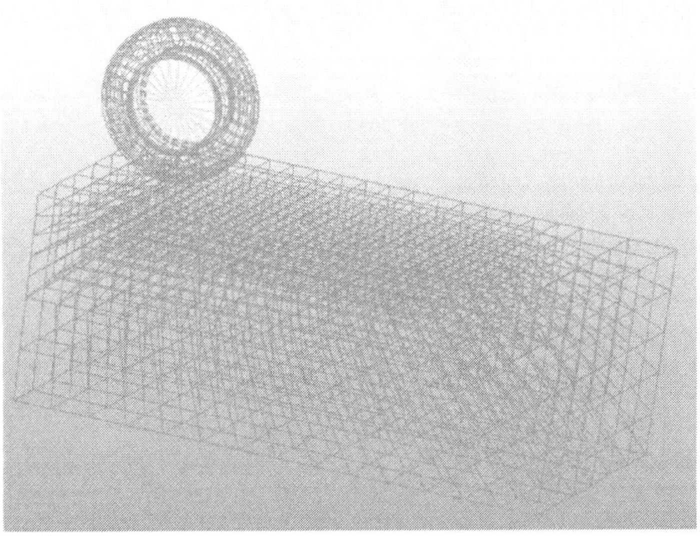

Fig. 3. View of the model grid

The most important element of the system was the soil (plastic-elastic material), where the stress distribution was studied. The wheel grid was generated using a tool that automatically divided the part into finite elements based on a set maximum size of the element. 3 760 calculation nodes, forming 9 203 elements, were generated in the entire model.

After the simulations, the results were presented as color maps of the reduced stress distribution. Despite the fact that the software averages the results for the graphic presentation, the accurate values may be read only in the grid's nodes. Therefore, prior preparation of the grid in such a manner to make the selected nodes overlap the planned measurement points was essential.

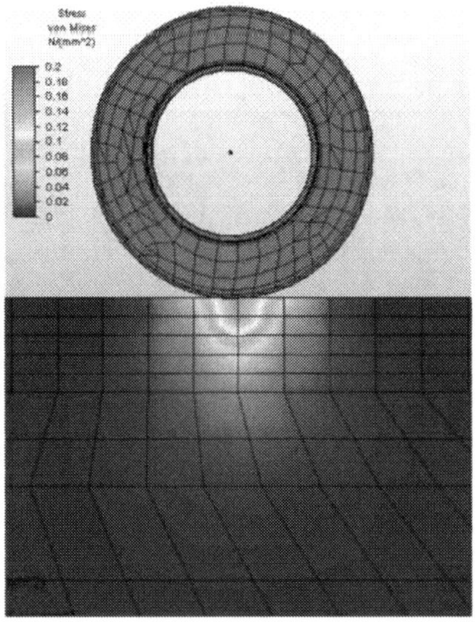

Fig. 4. Sample section along the wheel movement track, presenting the distribution of reduced stress in the soil

4 Analysis of the obtained results

The unknown slippage of the wheel rolling on the soil may be determined based on values calculated by the simulation software. The selection of the computing grid point makes it possible to depict the location of this point in every step of the simulation. For the first second, the wheel is initially at rest and then it starts moving; its movement is variable then. Hence, in order to determine the actual linear speed, the next second of each simulation, when the movement of the wheel is uniform and the wheel axis is moving at V_r speed, is taken into account. Accounting for the dimension of the rolling diameter of the tire (D = 1358 mm) [4] as well as the preset angular velocity (n = 0.4 rps), the theoretical speed of wheel axis V_t was determined. The digital model uses unloaded tire diameter, which was changed in the MES system under the influence of applied external load [15]. As a result, the value of slippage s was determined as defined by the following equation:

$$V_t = \omega R = 2\pi n R \left[\frac{mm}{s}\right] \quad (2)$$

$$s = \frac{V_t - V_r}{V_t} \quad (3)$$

where :

V_t – the theoretical speed of the wheel axis (a case where the temporary centre of the wheel rotation is located in the point of contact between the wheel and the base),

V_r – the actual speed of the wheel axis (the case where the temporary centre of rotation is located above the point of contact between the tire and the base),

R – the rolling radius of the drive wheel,

n, ω – angular velocity,

s – slippage.

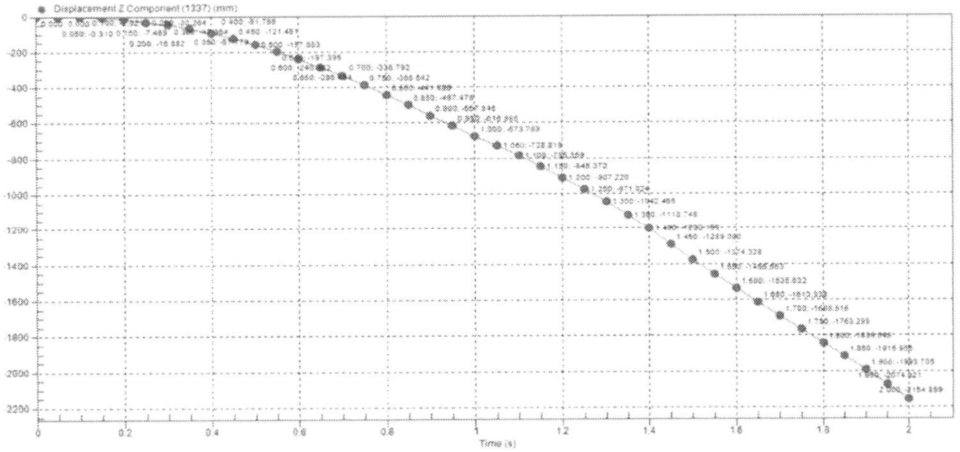

Fig. 5. Sample Sample diagram of movement of the wheel centre along the Z axis

The stress values generated by the system were compared to values obtained in the laboratory tests. During experimental studies in the soil canal [5], sensors measuring pressure under the wheel were located at the depth of 100, 200, 300, 400, and 500 mm below the soil surface. The soil model grid was adjusted to allow reading simulation results at the same depths. Sample results for sandy-clay soil and clay with pressure of 7 kN, with internal pressure in the tire is equal 0,15 MPa (the same as the laboratory tests) [4] are presented in table 1 and table 2.

Table 1. Stresses in sandy-clay soil, pressure force 7 kN

Depth [mm]	100	200	300	400	500
slippage [%]	reduced stress values [Kpa]				
10	51.6	36.4	24.1	16.4	10.7
15	82.6	57.9	37.9	25.3	16.2
20	94.3	66.2	43.5	28.9	18.4
25	96.3	67.4	44.2	29.4	18.9
30	98.0	69.2	45.5	30.4	19.5

The stress values obtained via dynamic simulations are consistent with the values registered under laboratory conditions. A tendency to increase the stress upon the increase in slippage was observed (Table 1, Fig. 6), which is in agreement with the theory and empirical studies [14] of the vehicle-land system. This is connected with the phenomenon of reduction of the drive wheel linear speed upon the increase in slippage.

The consistency of calculation results obtained using the finite element method with results obtained under laboratory conditions was confirmed for the dynamic effect the wheel driving mechanism exerted on the soil. The base-recorded stress values decrease along with the increase in measurement depth, and increase along with the increase in slippage.

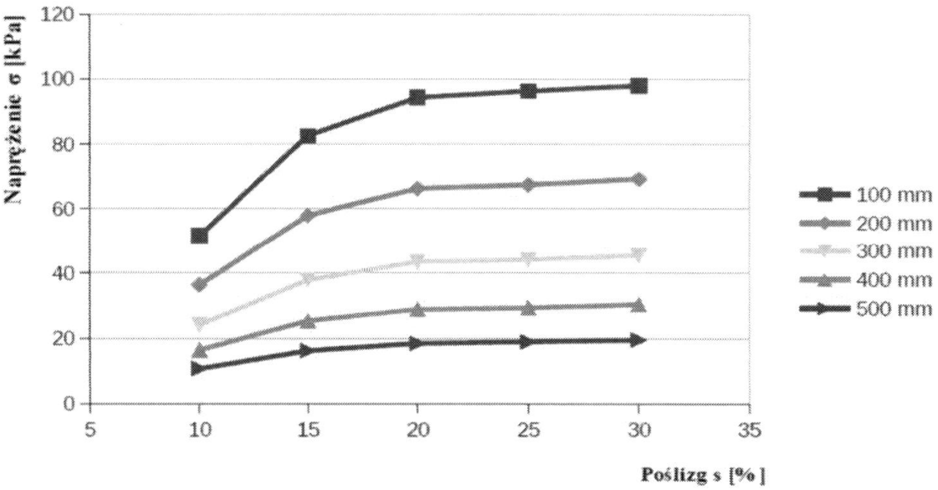

Fig. 6. Relation between stress and wheel slippage. Sandy-clay soil. Measurement depths 100-500 mm. Pressure force 7 kN.

Hence, to reduce pressure and deformations in the base, the angular velocity of the vehicles' drive wheels should be adjusted in such a way to minimize the slippage. To obtain this, monitoring and automation systems may be used. In addition, use of FEM allows designing such a shape of the tire tread so that the slippage is as low as possible under particular dynamic conditions.

The performed approximation of calculated stress generated in the soil under the impact of the wheel driving mechanism as a function of the measurement depth makes it possible to confirm a good consistency between the obtained results and Goriaczkin's theoretical mathematical model [4].

$$p = k \cdot z^n \tag{4}$$

where :

- p – stress [kPa],
- k – calculation factor of wheel model,
- z – depth of measure [mm],
- n – power factor.

For all simulated cases, the correlation coefficient has the value of at least 0.927. The results of approximation allow stating that the relation between stress in the soil under the impact of the dynamic load of the wheel driving mechanism and the wheel slippage is consistent with the logarithmic function [4, 11]. The obtained correlation coefficients for this approximation case have the average value of 0.933. However, one must pay attention to a similar accuracy of approximation with the power function; in this case the average value of the correlation coefficient was 0.921. Assuming that the results obtained via numerical simulation are consistent with the actual values, the power nature of the relation between stress generated in the soil and the wheel slippage cannot be excluded.

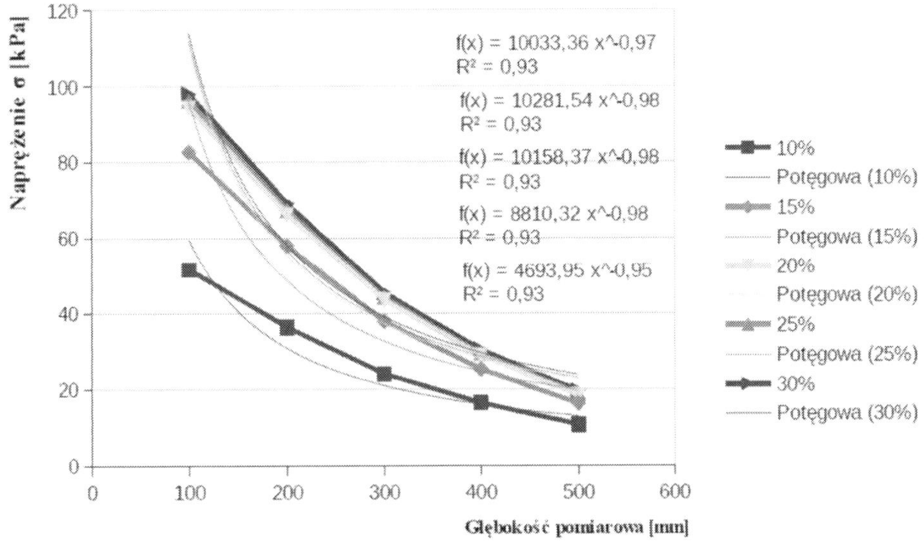

Fig. 7. Relation between stress and measurement depth, along with regression curves (sandy-clay soil, pressure force 7 kN, slippage from 10% to 30%)

Digital studies prove that the impact of slippage on stresses in the soil is considerable only at small depths. Stresses have a similar value below 0.4 m.

5 Conclusion

The purpose of this study with the use of digital simulation is to reduce expenses on costly experimental studies with the use of off-road vehicles or specially developed laboratory machines. When such experiments are replaced with numerical methods, not only the cost is reduced, but also the impact of weather conditions on measurement results is avoided and the time required for the study is shortened.

However, digital simulation in the case of a vehicle-land system dynamic load is a relatively complicated task. The correctness of obtained calculation results is affected by numerous factors, including the type and density of the FEM grid, the correctness of adopted material features of the base and elements being parts of the drive wheel, etc. The preparation of high-accuracy models is time-consuming and requires good knowledge of the software. Furthermore, complicated dynamic cases require high computing power, and efficient computer systems are necessary to perform fast simulations.

The developed digital model allowed calculation of stresses, which may be deemed credible based on the previous works, as well as data from experimental studies on similar cases. During the evaluation of stresses, subjective factor was also observed i.e. the estimated selection of calculation steps, in which the wheel axis has a relevant position as compared to the soil's grid of finite elements. A very high density grid would probably eliminate this problem. However, the computing capacities of the available workstation did not allow us to develop a more accurate model.

The use of the methods allows for design and selection parameters of exploitation pneumatics drive mechanism. It leads to significantly reduce the negative impact on the natural environment by machines in agrotechnical treatments and forestry. The methods allow to improvement of the natural environment.

References

1. Bekker M.G., Introduction to terrain-vehicle system, The University of Michigan Press, Ann Arbor, (1969)
2. Błaszkiewicz Z., A method for the determination of the contact area beteween tire and the ground. Journal Of Terramechanics,. **27**, 4, (1990)
3. Król K., Metoda elementów skończonych w obliczeniach konstrukcji, Wydawnictwo Politechniki Radomskiej, Radom, (2006)
4. Jakliński L., Mechanika układu pojazd-teren w teorii i badaniach, Oficyna Wydawnicza Politechniki Warszawskiej, (2006)
5. Lee J. H., Finite element modeling of interfacial forces and contact stresses of pneumatic tire on fresk snow for combined longitudinal and lateral slips, Jouranl of Terramechanics, **48**, 3, June, (2011)
6. Malesa W., Korzybski W., Zastosowanie MES w obliczeniach parametrów współpracy opony z glebą z uwzględnieniem wybranych aspektów dynamiki układu, Mechanik **1**, pp. 73, 9 , (2010).
7. Muro T., O'Brein J., Terramechanics. Land Locomotion Mechanics, A. A. Belkema Publishers, (2004)
8. Olszewski Z., Waluś K., Berdychowski M., Modelowanie cech geometrycznych opony pneumatycznej z zastosowaniem metody elementów skończonych. „Czasopismo Logistyka", (2014)
9. Ozaki S., Hinata K., Senatore C., Iagnemma K., Finite element analysis of periodic ripple formation under rigid wheel, Jouranl of Terramechanics, **61**, 3, October (2016)
10. Pytka J., Dynamics of wheel-soil systems, CRC Press, (2013)
11. Stasiak W.: Analiza modeli opisujących rozkład naprężeń w glebie, Journal of Research and Applications in Agricultural Engineering, Poznań, **53**, 2, str. 39-43, (2008)
12. Stasiak W., Modele propagacji nacisków w glebie generowanych przez oponę, praca doktorska, Politechnika Warszawska, (2003)
13. Stołyński A., Mechanika układu pojazd-teren, Wydawnictwo Ministerstwa Obrony Narodowej, 1966.
14. Wong J. Y., Terramechanics and off-road vehicle engineering, Elsevier Ltd., (2010)
15. Zabielski M., Model symulacji propagacji naprężeń w glebie generowanych przez opony maszyn rolniczych, praca doktorska, Politechnika Warszawska, (2015)

E3S Web of Conferences **46**, 00009 (2018) https://doi.org/10.1051/e3sconf/20184600009
3rd International Conference on Energy and Environmental Protection

Bottom ash obtained from biomass burning in fluidised-bed boilers as a mortar component

Waldemar Kępys[1,*]

[1]AGH University of Science and Technology, Faculty of Mining and Geoengineering, Mickiewicza 30 Av. 30-065 Crakow, Poland

Abstract. The production of energy from biomass causes generation of solid waste, in the forms of fly ash and bottom ash. Owing to both economic considerations and environmental protection, it is required to recover those types of waste. The physical and economic properties of bottom ash indicate that waste ash can constitute a substitute of sand in the production of mortars. Consequently, tests were performed on the influence of bottom ash, used as sand substitute, on the mechanical properties of mortar. The test results indicated a possibility of using bottom ash as a mortar component

1 Introduction

Electrical and heat energy generation in Poland is based on fossil fuel burning, similarly to many other countries. The intention to reduce CO_2 emission caused the necessity to replace coal by biomass in the co-burning process, or otherwise coal was completely withdrawn for the sake of biomass to be burned in power boilers. Similarly to the case of coal, biomass burning causes the generation of side products. Those products contain bottom ash, constituting part of the batch fed to the fluidised-bed boiler and/or non-combustible substances collected from the boiler, as well as flue gases consisting of fine particles collected from exhaust gas by dust removing equipment, together with the gas purification products. Regardless of the type of waste, it should be, first of all, recovered, in accordance with the hierarchy of waste control operations.

The process of ash generation during coal or biomass burning has been presented in detail in literature [1, 2]. The ash properties depend on a number of factors, including the following: type of fuel, method of fuel preparation, type of the boiler, burning process parameters, method of ash recovery and storage, as well as the method of exhaust-gas purification to remove solid and gas pollution [3].

One can distinguish several groups of biomasses that are most often used in power generation: wood and woody biomasses, herbaceous and agricultural biomasses, aquatic biomasses, or animal biomasses [4, 5]. The very type of biomass, its origin, and the method of plant growing influence the physical and chemical properties of combustion products. Consequently, those properties are crucial for subsequent waste handling. Nevertheless, when comparing the properties of the ashes generated from coal and biomass burning in

* Corresponding author: kepys@agh.edu.pl

power plants, we should conclude that the basic element content is the same in terms of quality. However, differences occur in the proportions of particular elements and chemical compounds [6, 7].

The applications of coal-burning products have been known around the world for a number of years. Such waste is primarily used in construction industry as partial cement replacement, mineral additive to cement production, adsorbent, light-weight aggregate, mine backfill, road sub-base, and there are many other industrial uses of ashes [6, 8]. According to [9, 10], the application of bottom ashes generated from coal burning, as partial replacement of sand in concrete, influences the workability, setting times, bleeding rate, and plastic shrinkage of fresh concrete. In the case of hardened concrete, bottom ash affects its density, strength, porosity, and durability. The identified reduction of concrete resistance was mainly caused by higher porosity and higher water demand when bottom ash was used. Besides, [11] stated that the use of bottom ash as fine aggregate in concrete, in the proportion of up to 50% of the replacement level, did not negatively affect the concrete strength and durability.

The burning of biomass alone generated waste whose physical and chemical properties differed from those of waste generated by coal burning. Such findings require implementation of a series of tests to establish the usability of various types of waste in the applications under discussion. The published test results refer to the use of waste from biomass burning in the building industry and road construction, as fertiliser or land reclamation additive or in mining [12, 13]. The limitations of the application of ashes generated from biomass burning result from the following: efficiency of biomass combustion, difficulty in delivering and preparing sufficient amounts of biomasses, cost, and technological requirements [14]. Consequently, ashes are often dumped and that is not beneficial for the environment, owing to the occupation of land and negative influence on soil, water, and air.

For that reason, research was implemented to identify the uses of bottom ash obtained from biomass burning in the building industry, as substitute of sand in the production of mortars.

2 Materials and experimental methods

In tests, used biomass bottom ash (BBA), generated as a result of biomass burning in a fluidised-bed boiler (CFB). Biomass had the form of wood chips and pellets made of sunflower shells. Those materials were mixed before delivery to the boiler. The proportion of wood chips was 70-80% wt. and the pellets constituted the rest. Waste was subjected to physical, chemical, and mineralogical testing. The grain content was determined by sieve analysis in accordance with the EN 933-1 Standard, while real density was determined by the pycnometric helium technique, using a Micrometrics Multivolume Pycnometer. The chemical composition was determined by using a mass spectrometer, with inductively coupled plasma mass spectrometry (ICP-MS) and inductively coupled plasma atomic emission spectroscopy (ICP-AES). Crystalline phases were detected by X-ray diffraction (XRD). To determine the influence of bottom ash on the soil and water environment, a leaching test was conducted, in accordance with *the* EN 12457-2 *Standard.* Distilled water was mixed with waste in the proportion of 1:10. The pH index was established for the aqueous extract and the content of pollution was identified as well, with the use of ICP-AES and ICP-MS. Chloride was analyzed by using the Volhard titration method. The loss of ignition was determined in accordance with the EN 1744-1 Standard, in the temperature of 975°C, during one hour.

Analysts conducted a series of tests of the usability of BBA as artificial aggregate by replacing sand with ash in the production of building materials. Usability was determined on the basis of the examination of mechanical properties of mortars. First, control mortar was produced, with plastic consistency, containing 1 part of cement mass and 3 parts of sand mass, with the water-cement proportion of 0.5. Next test mortars were produced. The sand portions were replaced with BBA, in the proportions of 25, 50, and 75% wt., respectively. Besides, mortar with a 100% BBA content was mixed. Portland cement CEM I 42.5R and standardised sand were applied. The water-to-cement proportion was the same in each mortar type and it amounted to 0.5. The mortar contents, with marking, are specified in Table 3. The mortars were mixed in accordance with the EN 196-1 Standard and placed in the box moulds, with the dimensions of 40x40x160 mm. The concrete bars were kept in moulds during 24 hours and, after recovery from moulds, in water until strength testing. The compression and flexural strength tests were conducted after 2, 7, and 28 days from sample production, in accordance with the EN 196-1 Standard.

3 Results and discussion

3.1 The properties of bottom ashes generated from biomass burning

Grain examinations were conducted owing to the influence of the aggregate grain structure on workability and on mechanical strength through compactness of the bodies [19]. The BBA grain content is presented in Fig. 1. The grains of the sizes below 2 mm constituted 95% of the quantity. According to the EN 13139 "Aggregate for Mortars" Standard, BBA belongs to fine aggregates (0/2 mm). The loss of ignition amounted to 0.3% for BBA, which indicated an insignificant content of unburned biomass in ash. That was much less in comparison to blast-furnace slag, being a popular artificial aggregate. As to slag, the allowed loss of ignition amounts to 3% (under the EN 13139 Standard), while the EN 450-1 Standard allows for loss of ignition at 7% in slag. The BBA grain density amounted to 2.58 g/cm^3 and that of sand to 2.63 g/cm^3.

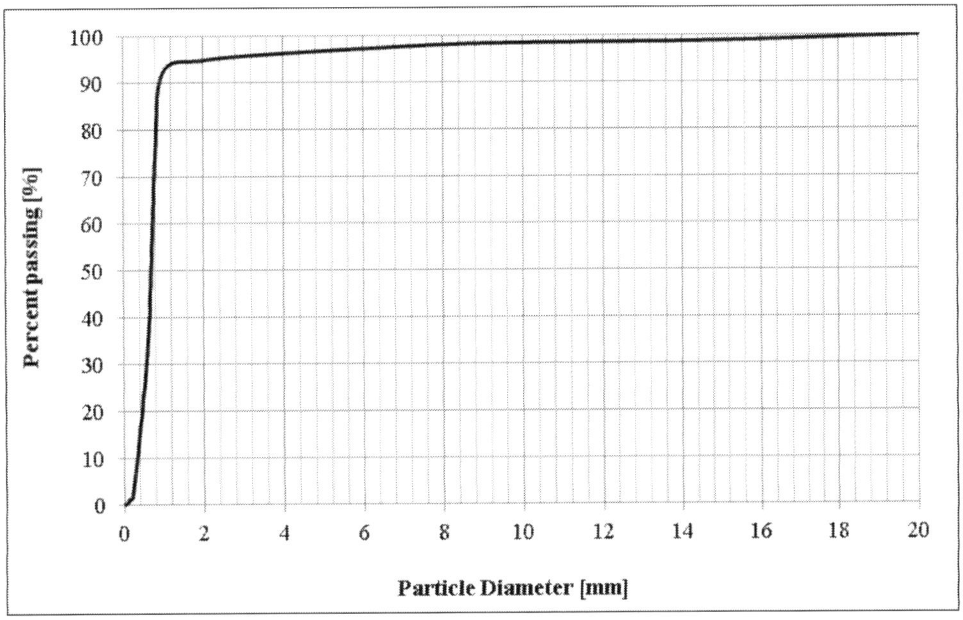

Fig. 1. Particle size distribution of BBA

SiO_2 was the main BBA component. Its share amounted to 84% wt. (tab. 1) The contents of the remaining components, e.g. those of CaO, Al_2O_3, and P_2O_5, were at the level of several percent each. Only the SO_3 content displayed 0.71% wt. (with the allowed 3% for ashes used in concrete mixing, according to the EN 450-1 Standard). That figure is essential because sulphates can cause an excessive increase of mortar swelling. The SiO_2 content originates directly from the use of sand as the batch material in the boiler and its origin is associated with the transfer of sand and dirt to biomass during harvesting. That was further confirmed by phase content analysis (Fig. 2) where SiO_2 quartz was a dominating component, accompanied by insignificant quantities of other forms of silica (crystobalite and keatite), generated during quartz sand baking. BBA also contained the residues of silt minerals: montmorillonite $Al_2O_3 \cdot SiO_2 \cdot nH_2O$ and trace quantities of anhydrite $CaSO_4$ and hematite Fe_2O_3.

Table 1. Results of simulation tests for selected construction variants

Parameters (%wt)									
SiO_2	CaO	Al_2O_3	P_2O_5	Fe_2O_3	MgO	K_2O	SO_3	Mn_2O_5	Na_2O
84.11	5.47	3.85	2.00	1.21	1.52	0.71	0.71	0.18	0.09

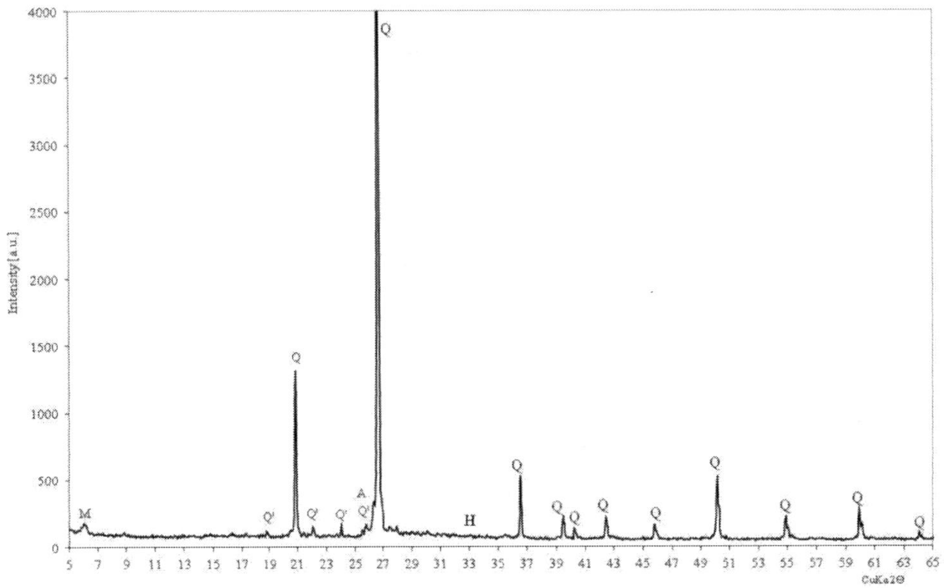

Fig. 2. X-ray diffraction of BBA (Q – quartz, Q' – other forms of silica, M – clay minerals, H – hematite, A – anhydrite)

The results of testing substance leaching from BBA are presented in Table 2. The data indicate a very small content of soluble components: chlorides, sulphates, or heavy metals found in BBA. When bottom ash is used to replace sand in mortar, pollution leaching is essential from the viewpoint of mortar's influence on the soil and water environment. Besides, the presence of chlorides negatively influences mortar durability, as well as binding time [15]. Consequently, the allowed water-soluble chloride content in aggregates used for the production of mortar may not exceed 0.15 % wt.

Table 2. Leachnig test

Concentration in leachates [mg/dm³]			
pH	9.64	Ni^{+2}	0.00052
Na^{+}	1.38	Co^{+2}	0.00006
K^{+}	60.45	Pb^{+2}	0.0003
Ca^{+2}	30.34	Hg^{+2}	0.0014
Mg^{+2}	1.6	Cd^{2+}	0.00026
Ba^{+2}	0.01	Cr^{3+}	0.021
Sr^{+2}	0.14	As^{3+}	0.0047
Zn^{+2}	<0.002	Cl^{-}	2.5
Cu^{+2}	0.003	SO_4^{-2}	89.00

3.2 Mortar properties

Table 3 presents the results of mortar compression tests, depending on mortar composition. Generally, sand replacement by BBA causes a reduction of strength, in comparison to that of control mortar (0BBA). That trend is maintained with the increase of the BBA content in mortar. After two days of seasoning, the strength of mortar containing BBA was from 27 to 22.5 MPa, with the value of control mortar strength of 29.1 MPa. The former results were by 7 to 22% lower from those of the control sample (Fig. 3). After 7 days, a considerable increase of strength occurred in all mortars, from 68 to 62%, in comparison to the results obtained after 2 days. The control mortar strength amounted to 48.9 MPa, while that of BBA mortars from 45.3 MPa (25BBA) to 36.4 MPa (100BBA), which corresponded to the reduced strength from 7 to 25%, in respect of the control mortar. The strength tests after 28 days indicated a further increase of compression strength in all mortars, although the increase of strength, in comparison to the results after 7 days, reached 17% (25BBA), 11% (50BBA), 10% (75BBA), and 8% (100BBA), with the 20% increase of strength in the control mortar. After 28 days, the mortar compression strength was reduced, depending on the BBA content: from 53 MPa (25% of BBA) to 41.8 MPa (75% of BBA), while that of the mortar with BBA alone amounted to 39.4 MPa, with 58 MPa for the control mortar.

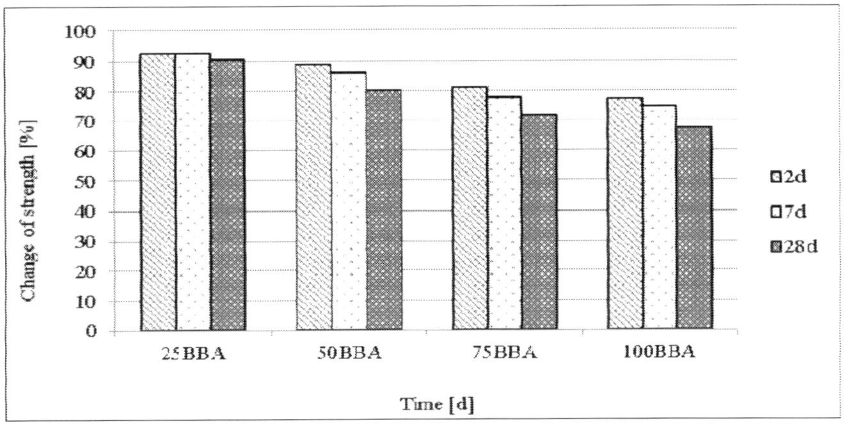

Fig. 3. Changes in the mortar compression strengths in comparison to that of the control mortar

Table 3. Mortar content and the flexural and compression strengths

Sample	Components				BBA/ Sand [%]	Flexural strength [MPa]	Compressive strength [MPa]		
	Cement [g]	Sand [g]	BBA [g]	Water [cm^3]		28d	2d	7d	28d
0BBA	450	1350	0	225	0/100	8.98	29.1	48.9	58.4
25BBA	450	1012.5	337.5	225	25/75	8.57	27.0	45.3	53.0
50BBA	450	675	675	225	50/50	8.25	25.8	42.0	46.8
75BBA	450	337.5	1012.5	225	75/25	7.36	23.6	38.0	41.8
100BBA	450	0	1350	225	100/0	6.79	22.5	36.4	39.4

4 Conclusion

Mineral waste is generally applied as artificial aggregate in the production of various types of aggregate mixes for road construction or mortar and concrete mixing. That allows for reducing the consumption of natural aggregates, with the additional disposal of waste. Both effects are beneficial for the natural environment.

This paper presents research intended to determine the possibilities of using bottom ash originating from biomass burning in a fluidised-bed boiler in mortar production. The grain content tests indicated that ash was classified as fine aggregate, with a 95% content of grain of up to 2 mm in size. SiO_2 was the main ahs component, with the 84% wt. content and insignificant shares of sulphates and chlorides. Bottom ash from biomass constituted a substitute of sand in cement mortar, with the ash content from 25% to 100%. The mortar compression tests indicated that the compression strength was reducing with the increase of the bottom ash proportion in mortar, in comparison to the control mortar (free of ash). Nevertheless, it is possible to use bottom ash as sand replacement in mortars, with a proper selection of mortar-mix proportions, depending on future application.

This study was conducted under statutory research of the Department of Environmental Engineering and Mineral Raw Material Processing (Project No. 11.11.100.482), Faculty of Mining and Geoengineering, AGH University of Science and Technology in Kraków, Poland.

References

1. Vassilev S.V., Baxter D., Andersen L. K., Vassileva Ch. G.: An overview of the behaviour of biomass during combustion: Part I. Phase-mineral transformations of organic and inorganic matter, Fuel, **112**, pp. 391–449, (2013)
2. Vassilev S.V., Baxter D., Andersen L. K., Vassileva Ch. G.: An overview of the behaviour of biomass during combustion: Part II. Ash fusion and ash formation mechanisms of biomass types, Fuel, **117**, pp.152–183, (2014)
3. Grammelis P., Skodras G., Kakaras E.: Effects of biomass co-firing with coal on ash properties. Part I: Characterisation and PSD, Fuel, **85**, pp. 2310–2315, (2006)
4. Vassilev S.V., Baxter D., Andersen L. K., Vassileva Ch. G.: An overview of the chemical composition of biomass, Fuel, **89**, 5, pp. 913-933, (2010)
5. Tortosa Masiá A.A., Buhre B.J.P., Gupta R.P., Wall T.F.: Characterising ash of biomass and waste, Fuel Processing Technology, **88**, pp. 1071–1081, (2007)
6. Ahmaruzzaman M.: A review on the utilization of fly ash. Progress in Energy and Combustion Science, **36**, pp. 327–363, (2010)

7. Williams A., Jones J.M., Ma L., Pourkashanian M.: Pollutants from the combustion of solid biomass fuels, Progress in Energy and Combustion Science, **38**, pp. 113–137, (2012)
8. Kurama H., Kaya M.: Usage of coal combustion bottom ash in concrete mixture, Construction and Building Materials, **22**, pp. 1922–1928, (2008)
9. Singh M., Siddique R.: Effect of coal bottom ash as partial replacement of sand on properties of concrete. Resources, Conservation and Recycling, **72**, pp. 20– 32, (2013)
10. Andrade L.B., Rocha J.C., Cheriaf M.: Influence of coal bottom ash as fine aggregate on fresh properties of concrete, Construction and Building Materials, **23**, pp. 609–614, (2009)
11. Singh M., Siddique R.: Properties of concrete containing high volumes of coal bottom ash as fine aggregate. Journal of Cleaner Production, **91**, pp. 269-278, (2015)
12. Wang S., Miller A., Llamazos E., Fonseca F., Baxter L.: Biomass fly ash in concrete: Mixture proportioning and mechanical properties. Fuel, **87**, pp. 365–371, (2008)
13. Kępys W., Pomykała R.: Research into the Usefulness of Ash from the Co-Combustion of Lignite and Biomass in Mining Technologies. Polish Journal of Environmental Studies, **23**, pp. 1381-1384, (2014)
14. Cabrera M., Galvin A.P., Agrela F., Carvajal M.D., Ayuso J.: Characterisation and technical feasibility of using biomass bottom ash for civil infrastructures. Construction and Building Materials, **58**, pp. 234–244, (2014)
15. Modolo R.C.E., Ferreira V.M., Tarelho L.A., Labrincha J.A., Senff L., Silva L. Mortar formulations with bottom ash from biomass combustion. Construction and Building Materials, **45**, pp. 275–281, (2013)

Coal combustion modelling in a frontal pulverized coal-fired boiler

Paweł Madejski[1,*]

[1]AGH University of Science and Technology, Faculty of Mechanical Engineering, Department of Power Systems and Environmental Protection Facilities, Mickiewicza 30 Av. 30-059 Kraków, Poland

Abstract. The paper presents results of numerical modelling of pulverized coal combustion process in the coal-fired boiler. In the numerical model, coal combustion process includes particle heating, devolatilization, char combustion, as well as turbulent flow and radiative heat transfer was modelled. Presented modelling results were carried out using the Open Source CFD code – Code_Saturne created and developed by EDF R&D and were used to study the combustion of coal in power plant boiler with the objective of simulating the operational conditions and identifying factors of inefficiency. The behaviour of the flow of air and pulverized coal through the burners was modelled, and the three-dimensional flue gas flow through the combustion chamber and heat exchangers was reproduced in the simulation.

1 Introduction

Currently, the coal fired power plants are facing new big challenge resulting from environmental concerns. The units must meet strict limits for emissions of gaseous pollutants keeping efficiency on the high level. Applying methods for reducing pollutant emission in the wrong way can greatly affect the process quality and reduce production efficiency.

The development of computer modelling techniques, especially the CFD methods (Computational Fluid Dynamics) allows to conduct the accurate analysis of coal combustion process and are widely used in modelling of front-fired boilers [1-2] as well as tangentially-fired boilers [3-5]. A complete information about the combustion process resulting from CFD model is helpful in the verification of combustion process quality. The analysis results of combustion process using CFD modelling can be used in optimization processes eg. to proper prediction of NOx reduction using primary or secondary methods [6, 7], to verify actual operating conditions and them impact on the thermo-flow processes inside the coal boilers [8-10], as well as to analyse different types of fuels as biomass, waste etc. [11-13].

* Corresponding author: madejski@agh.edu.pl

2 Characteristic of front-fired pulverized coal boiler

The numerical simulation was performed for the front-fired pulverized coal boiler, with a natural circulation. The nominal temperature of live and reheated steam produced in the boiler is equal to 540°C at the pressure equal to 13.5 MPa and 2.5 MPa of live and reheated steam respectively. The fuel is a hard coal with an average calorific value around of 20.500 MJ/kg.

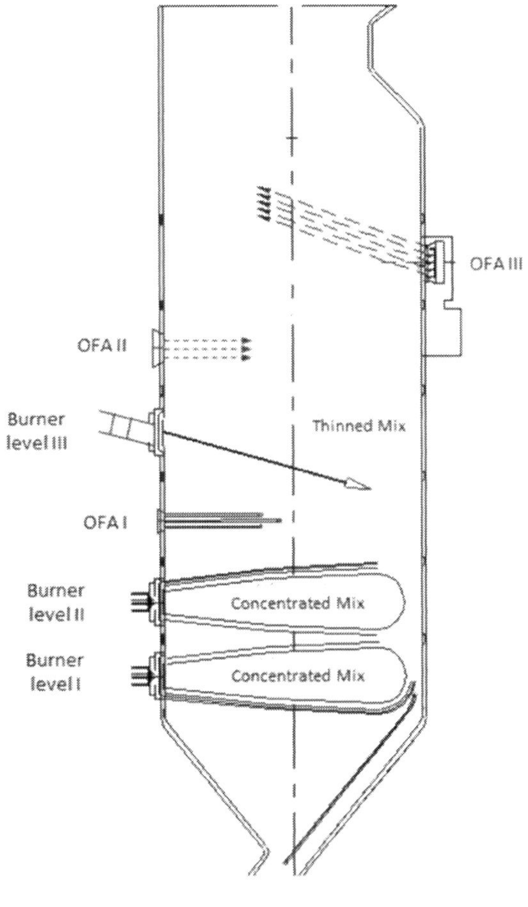

Fig. 1. Firing system of a frontal pulverized coal-fired boiler

The general scheme of analysed boiler with low-NOx furnace system is depicted in Figure 1. The low-NOx combustion system is characterized by air and fuel staging. The pulverized coal supplies boiler by 12 swirl low-NOx burners (burner level I and II) divided into two rows and by 12 burners (burner level III) located in the upper part of windbox (Figure 1). This system consists also of two levels of Over Fired Air nozzles (OFA) located at the front wall (6 OFA ports) and rear boiler wall (10 OFA ports). Boiler has an additional air inlets in order to avoid low-oxygen corrosion processes (protective air) and to decrease UBC content (bottom air). The share of protective and bottom air in total amount of air supplied to the boiler can be up to 20% and has to be taken into account in simulations. The number and location of all boiler inlets are presented in Table 1.

Table 1. Total number of all inlets in the boiler

Inlet	Side	Inlet numbers
Low-NOx burner	Front	2 x 6
Burner	Front	12
OFA	Front	6
OFA	Rear	10
Bottom air	Bottom	20
Protective air	Rear	15
Protective air	Left	5
Protective air	Right	5

To perform simulation, the properties of coal were defined based on the results from proximate and elemental analysis. The characterization of bituminous coal burned in Power Plant is presented in Table 2.

Table 2. Results of proximate and elemental analtysis used in coal combustion modeling (AR – As Received)

Proximate analysis	Value, wt% (AR)
Ash	22.34
Volatile Matter	25.77
Moisture	12.75
Fixed Carbon	39.14
Elemental analysis	**Value, wt% (AR)**
C	84.7
H	5.39
O	1.55
N	1.55
S	1.23

3 Operating conditions

The simulation was carried out for steady-state boiler operating and for the boiler load close to the nominal value. Operating conditions were determined based on the measured values registered by on-line measurement system (Table 3).

Table 3. Furnace operating conditions used in simulation

Name	Mass flow rate, kg/s	Temperature, K
Coal	24.18	386
Primary air	75.73	386
Secondary air	89.9	573
Protective air + bottom air	40.41	573

The share of coal and air for the individual inlets was determined based on the knowledge and experience from previous numerical studies and on-site tests. The particle size distribution was calculated based on the sieve analysis results. The percent of particles passing through two different sieves were measured for each coal mill. The average values for all 6 pulverized coal mills (R_{88}=11.5%, R_{200}=0.7%) and Rosin-Rammler law were used to calculate coal particle distribution:

$$P(d_i) = 1 - \exp(-d_i/D_m)^n \tag{1}$$

mwhere :

n – factor of spread in particle sizes,
d_i – coal particle diameter of class i, m,
D_m – t mean coal particle diameter, m.

Five classes of particle diameter (d_1=4.9 μm, d_2=16.9 μm, d_3=43.7 μm, d_4=90 μm, d_5=150 μm) were used for burners on Level I and II (concentrated coal-air mixture) and for burners on Level III (thinned coal-air mixture).

4 Modelling of pulverized coal-fired boiler operation

The simulation of pulverized coal combustion in a power plant boiler was performed in the Open Source CFD code – Code_Saturne, created and developed by EDF R&D [14,15]. The calculations were conducted for steady-state conditions and for boundary conditions prepared based on the available technical documentation of a boiler and measured values of boiler operating conditions.

To conduct calculations of the pulverized-coal fired boiler, the following processes as coal combustion including a devolatilization process, char and gas phase combustion as well as turbulent flow and radiative heat transfer need to be modelled. In the developed model presented in the paper, the mixture fraction approach together with using of Probability Density Function (PDF function) was adopted. Turbulent flow was modelled using standard k-ε model and to calculate radiative heat transfer the Discrete Ordinate model with 32 directions of heat transfer has been applied.

4.1 Geometrical model and numerical mesh

This three dimensional geometrical model of front-fired pulverized coal boiler analysed in the paper was created based on the technical documentations. Figure 2 shows the model of two passes boiler including low-emission firing system with main swirl burners and burners in form of drop tubes, water-walls, protective air nozzles, bottom air and Over-Fire Air nozzles located on the front wall. Additionally at the outlet of a first boiler pass superheater consisting of plates distributed over the width of the whole boiler was created.

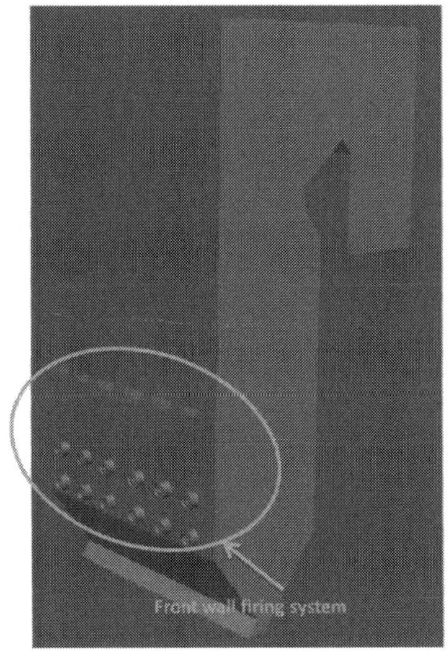

Fig. 2. Three dimensional geometric model of boiler

Numerical mesh consisted of about 3 million control volumes, mainly made of hexahedral elements type, was used to conduct calculations. In order to ensure a proper distribution of pulverized coal and air, a dedicated numerical mesh of burners located on the front boiler wall was created.

4.2 Coal combustion modelling

To conduct calculations of pulverized coal combustion and furnace operation, the following processes need to be modeled: turbulent flow, radiative heat transport, coal combustion and gas phase combustion. The pulverized coal combustion process can be divided into devolatilization and char combustion and gas-phase combustion process. The devolatilization process is described by two-reaction rates model (Kobayashi model) with calculated yield values of light and heavy volatiles Y_1 and Y_2 [15]. The gas phase combustion process was simulated using mixture fraction/PDF model [15,16]. To simulate reactions of gas phase combustion, the transport equation for mixture fraction variable f is solved. Based on the calculated local value of f and using PDF Beta function [16,17], the mass fraction of fuel, product and oxidizer can be calculated.

4.3 Boundary conditions

The boundary conditions used in the developed boiler model have been defined on the basis of the measured values when the boiler was operating at constant load. The simulation also includes non-uniform fuel and air boiler distribution define as the different values of the mass flow rate at individual boiler inlets. Mass fractions of pulverized coal with different particle diameter was calculated using results from sieve analysis and Rosin-Rammler distribution represented by Eq. (1). The boundary conditions at boiler water-walls and superheater walls were determined using balance calculations of boiler operation, and the temperature at the walls together with emissivity factors are presented in Table 4.

Table 4. Boundary conditions at the boiler walls

Boundary surface	Temperature, °C	Emissivity
Evaporator wall	390	0.7
Burners wall	730	1.0
Superheaters wall	577	0.4

5 Results of boiler operation modelling

The simulation of boiler operation was performed for steady state conditions and for input data defined and presented in previous sections. Carrying out the calculations of the boiler using a CFD modeling allows obtaining results in the form of distributions of key parameters such as velocity and gas temperature inside the boiler (Figure 4).

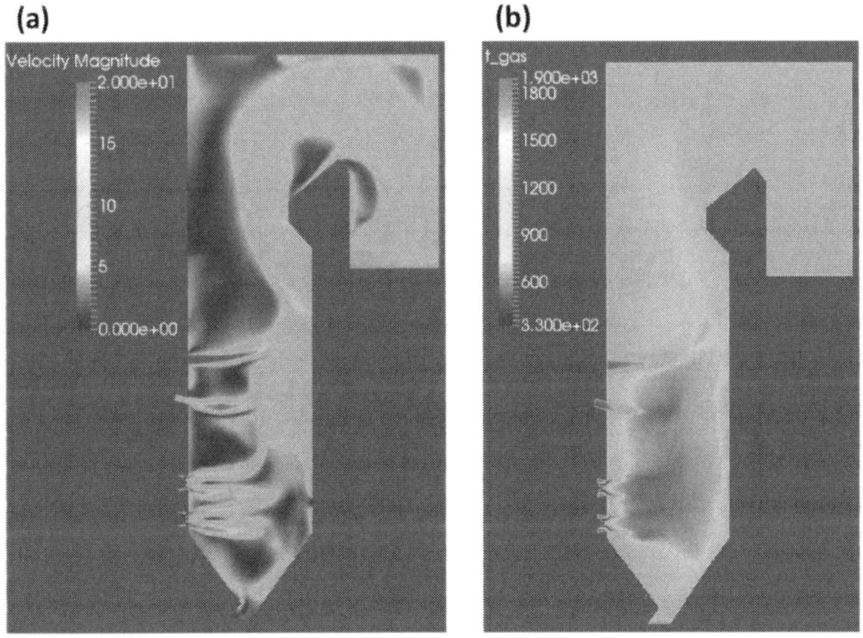

Fig. 3. Velocity (a) and gas temperature (b) distribution in a vertical plane of the boiler

Velocity distribution presented in Figure 4a allows identifying the exhaust gas flow path in the boiler and zones where the velocity reaches the highest values. The very high velocity of flue gases may be a danger of erosion hazard and can be the reason for too intensive convective heat transfer in local boiler zones. The temperature distribution inside the boiler is presented in Figure 4b, and allows identifying the zones where the greatest temperature exists. The highest temperatures occur in combustion zone, and also in the zone where the pulverized coal is supplied using drop tube burners.

6 Conclusions

The paper presents results of numerical simulation of front-fired pulverized coal boiler operation carried out using Computational Fluid Dynamics modeling. The calculations were performed using Code Saturne code, which is the Open Source CFD software. In the developed model, the devolatilization process was described by two-reaction rates model (Kobayashi model) with calculated yield values of light and heavy volatiles. The gas phase combustion process was simulated using mixture fraction/PDF model. For turbulent flow modeling the standard k-ε model was used and for radiative heat transfer, the Discrete Ordinate model with 32 directions of heat transfer has been applied. The results in form of velocity and temperature distribution can be used in order to answer some operational questions regarding optimization and combustion process efficiency inside large steam boilers. Thanks to the simulation results are possible to increase the understanding and detailed investigating of all processes inside the boiler and to determine and verify safe operating conditions.

This research was partially funded from GEKON program by the National Center of Research and Development and the National Fund for Environmental Protection and Water Management under research and development project No. GEKON1/O2/213655/9/2014. Author thanks for the support and the ability to carry out this work in collaboration with EDF R&D, Fluid Dynamics Power Generation and Environment in Chatou France.

References

1. Minghou X., Azevedo J.L.T., Carvalho M.G., Modeling of a front wall fired utility boiler for different operating conditions, Comput Method Appl M, **190**, **pp.** 3581-3590, (2001)
2. Sanghyun P., Jungeun A. K., Changkook R., Taeyoung Ch., Won Y., Young-Ju K., Ho-Young P., Hee-Chun L., Combustion and heat transfer characteristics of oxy-coal combustion in a 100 MWe front-wall-fired furnace, Fuel, **106**, pp. 718-729, (2013)
3. Belosevic S., Sijercic M., Oka S., Tucakovic D., Three-dimensional modeling of utility boiler pulverized coal tangentially fired furnace, Int J Heat Mass Tran, **49**, pp. 3371-3380, (2006)
4. Asotani T., Yamashita T., Tominga H., Uesugi Y., Itaya Y., Mori S., Prediction if ignition behavior in a tangentially fired pulverized coal boiler using CFD, Fuel, **87**, pp. 482-490, (2008)
5. Fan J., Qian L., Ma Y., Sun P., Cen K., Computational modeling of pulverized coal combustion processes in tangentially fired furnaces, Chem Eng J. **81**, pp. 261-269, (2001)
6. Dal Secco S., Juan O., Louis-Louisy M., Lucas Y-V., Plion P., Porcheron L., Using a genetic algorithm and CFD to identify low NOx configurations in an industrial boiler, Fuel, **158**, pp. 672–683, (2015)
7. Modliński N., Numerical simulation of SNCR (selective non-catalytic reduction) process in coal fired grate boiler, Energy, **92**, pp. 67–76, (2015)
8. Madejski P., Taler D., Analysis of temperature and stress distribution of superheater tubes after attemperation or sootblower activation. Energ Convers Manage, **71**, pp. 131-137, (2013)

9. Madejski P., Janda T., Taler J., Nabagło D., Węzik R., Mazur M., Analysis of fouling degree of individual heating surfaces in a pulverized coal fired boiler, J Energ Resour-ASME, **140**, pp. 1-8, (2018)
10. Madejski P., Taler D., Taler J., Numerical model of a steam Superheater with a complex shape of the tube cross section using control volume based finite element method, Energ Convers Manage, **118**, pp. 179-192, (2016)
11. Pająk T., Jurczyk M., Initial operating experience with the new Polish waste-to-energy plants, Waste Management, **6** Waste-to-energy, Eds: Karl J. Thomé-Kozmiensky, Stephanie Thiel (Neuruppin, TK Verlag Karl Thomé-Kozmiensky 2016)
12. Turlej T., Banaś M., Management of communal sewage waste in Poland, International Multidisciplinary Scientific GeoConference SGEM, **978**, pp. 627–634, (2017)
13. Bhuiyan A.A., Naser J., CFD modelling of co-firing of biomass with coal under oxy-fuel combustion in a large scale power plant, Fuel, **159**, pp. 150-168, (2015)
14. Archambeau F., Mechitoua N., Sakiz M., Code_Saturne: A finite volume code for the computation of turbulent incompressible flows – industrial applications, IJFV International Journal On Finite Volumes, **1, pp.** 1-64, (2004)
15. Code Saturne 3.3.0 Theory Guide, Fluid Dynamics, Power Generation and Environment Department, Single Phase Thermal-Hydraulics Group, EDF Lab Chatou, France May, (2014)
16. Pope S. B., The probability approach to the modelling of turbulent reacting flows, Combust Flame, **27**, pp. 299–312, (1976)
17. Poinsot T., Veynante D., *Theoretical and numerical combustion*, second ed., (R.T. Edwards Inc., Philadelphia 2005)

Construction of a numerical model of suspension in a numerical simulation of sedimentation

Krzysztof Kołodziejczyk[1,*]

[1]AGH University of Science and Technology, Mickiewicza 30 Av. 30-059 Krakow, Poland

Abstract. Modeling of multiphase systems, which includes suspensions, is an issue that is continually developed. There are no procedures at the moment that would clearly determine the way in which suspension is defined in numerical simulations. The article presents an analysis of the selection of a numerical model and the definition of the suspension with a polydisperse particle composition.

1 Introduction

The concept of a multiphase system is very extensive, and shall include both suspensions, i.e., systems comprising a dispersed suspension phase and systems including a liquid in various states of aggregation (e.g. water/steam systems). A system of two not mixed liquids should be regarded as a multiphase (two-phase) too, where a clear boundary allows one to determine the boundary (e.g. air/water systems) [1].

A range of studies on modeling of multiphase systems [2-4] have appeared over the last decade, including studies on the use of numerical methods in the design of slurry settling devices [5].

The largest group consists of models created based on the finite volume method [6] or the method of finite elements [7]. The models that can be divided into two classes: Euler-Lagrang and Euler-Euler. Models belonging to the Euler-Lagrangian Particle Tracking (Discrete Phase Model) class allow the modeling of multi-phase systems with a dispersion phase in a granular form, assuming a low concentration (free sedimentation).

These models are characterized in that in the continuous phase of the liquid, analyzed by the Euler method, the movement of each grain of the disperse phase is calculated using the Lagrangian method. As a result, each grain is individually modeled – this allows one to conduct numerical simulations of suspension where the dispersion phase can have any composition of the particle sizes. This way of defining a multiphase system has a significant limitation resulting from the number of grains of the dispersion phase that we can model. In practice, the dispersion is limited to a few or tens of thousands of grains. Therefore, models of this class will be dedicated to calculations such as the spread of pollutants or tracking particle traces and their propagation rather than modeling of systems in which the concentration of the dispersion phase plays an important role.

* Corresponding author: krkolodz@agh.edu.pl

Among the models belonging to the second group (Euler-Euler models), the VOF, Mixture and Eulerian models should be distinguished. In these models, each phase (phase dispersed and continuous) treated as a continuous means described using Euler method. In the calculation area, each of the phases is represented by its volume fraction. Depending on the model, the interactions between the phases are considered differently. Therefore, each model is dedicated to a different group of multi-phase systems or allows to analyze the system in other respects.

In high concentrations, where in one cubic centimeter there can be up to several millions of grains, Euler-Euler class models are used, where the continuous and dispersion phases can be treated as continuous phase with varying degrees of participation by volume.

2 The choice of the computational model

By setting out to model a multiphase system, one must assume a fundamental element to determine which class numerical model will be the most suitable one to carry out simulations of interest. As has been mentioned, the currently available models belong to two classes: Euler-Lagrange and Euler-Euler. In the case of numeric methods for modeling the work of a settling device, the basic criterion of choice of the model class will be the share of the dispersion phase.

The analysis of the selection of the model class in a multiphase system was performed for a sample industrial suspension. In the analysis, a suspension of carbon from a coal enrichment plant after stripping coal processing was used. The share of the disperse phase for its initial concentration $s_0 = 45$ kg/m^3 and the density of solid part $\rho_d = 2299$ kg/m^3 was, according to equation (1), $\varphi_d = 0.0195$. As the volume of the dispersion fraction is almost 2%, one can classify the sedimentation process which will occur in the suspension as a process of hindered sedimentation [8].

$$\varphi_d = \frac{s_0}{\rho_d} \tag{1}$$

An additional criterion that can be used here to calculate the average distance between particles of the disperse phase, which one can determine from the load of suspended particles and density ratio in the phases. The grain load of the suspension is determined from the relationship (2) and is $\beta = 0.0459$.

$$\beta = \frac{\varphi_d \cdot \rho_d}{\varphi_c \cdot \rho_c} \tag{2}$$

where :

d – index of the disperse phase,

c – index of the continuous phase.

The ratio of the density of the disperse phase and the continuous phase shall be $\gamma = 2.299$(3).

$$\gamma = \frac{\rho_d}{\rho_c} \tag{3}$$

Taking into account the above ratios (2) and (3) one can determine the average distance between particles in suspension from the equation:

$$\frac{L}{d_d} = \left(\frac{\pi}{6}\frac{1+\frac{\beta}{\gamma}}{\frac{\beta}{\gamma}}\right)^{\frac{1}{3}} \tag{4}$$

In the case considered, from the equation (4), we obtain: $L/d_d = 2.99$.

If the average distance between the particles is greater than 8 then there is confidence that each grain falls freely, independently of the other grains, and the Lagrangian DPM (Discrete Phase Model) model can be deemed appropriate to describe the motion of the grains. However, in the system where the average distance between particles is less than 8, one should consider one of the VOF (Volume of Fluid), Mixture or Eulerian models. Since the VOF model allows modeling multiphase systems where both the continuous phase and the dispersion phases are liquid, it cannot be used for modeling the sedimentation of coal slurry.

The Stokes number is a criterion for the use of a Mixture or Eulerian model. It is defined as the product of the reaction time for the particle and the system, where the response time of the particles is calculated from the relationship:

$$\tau_d = \frac{\rho_d \cdot d_d^2}{18 \cdot \mu_c} \tag{5}$$

From the equation (5) for grain $d_d = 7.89$ µm we obtain: $\tau_d = 8 * 10^{-6}$ s, where d_d is the diameter of the boundary grain.

The response time of the ts system is calculated based on the dominant flow rate and the dimension of the characteristic L_s system (usually the way from input to output) according to equation:

$$t_s = \frac{L_s}{V_s} \tag{6}$$

Ultimately, the Stokes number will be calculated according to (7).

$$S_t = \frac{\tau_d}{t_s} \tag{7}$$

The limit value of the Stokes number is 1. For S<<t we can use any model of a multiphase system, and for the St above 1, we shall use the DPM or Eulerian model.

In the case of modeling static sedimentation, a system without flow is analyzed, therefore, the dominant component will be the movement of particles of the disperse phase, thus making the Eulerian a more appropriate model to describe the process of static sedimentation. Due to the volume fraction of the disperse phase, the DPM model cannot be used.

In the case of modeling the operation of a settler, where flow occurs, the Stokes number will depend on the zone of sedimentation for which it will be calculated. In the refining zone the flow rate is normally greater than the rate of sedimentation, thus the Stokes number will have a value less than 1. In turn, in the layer of sludge, flow rate will tend to zero, which in turn will result in a Stokes number greater than 1. As a result, one can draw a conclusion that the most appropriate model for use in numerical simulations of a settler operation will be an Eulerian model.

3 Numerical model of a multiphase system

The Eulerian model, used to model multi-phase systems, allow the implementation of the simulation in which there is one main phase (solid) and at least one disperse phase. In this model, there is no limit to the number of dispersion phases beyond the limits of hardware consisting in the availability of memory and computing power. The Eulerian model, among all models of multiphase systems, is the one with the greatest demand for computing power and RAM. However, it is a model that offers the widest possibilities when calculating the interaction between the phases.

A characteristic feature of this model is that for each phase, a continuity equation and a momentum equation are solved. In this model, it is assumed that the pressure at any point is the same for all phases. Other sizes are calculated individually for each phase. In this model, it is possible to take account of various types of inter-phase interactions as well as calculation of flow by one of the turbulence models for the various phases.

Phases of the Eulerian model are treated as interpenetrating continua, representing a multiphase system, in which each phase has its share of the volume described as φ_q, where the total volume of shares of the various phases in the whole system is 1.

As a result, in Euler-Euler class models, a system of equations is solved to describe the flow. In numerical calculations, the same material with different physical properties, e.g. the diameter of grains [9] should also be considered a separate phase. Due to the fact that the disperse fraction in this model may be composed of grains of a specified diameter, it is possible to obtain multiple system by building several dispersion phases with varying grain composition. However, this results in multiplication of the number of equations that must be solved for such a system. The main advantage of this model is that it can be used in the full range of shares by volume of each phase [3].

In Eulerian class models, the balance of the q phase for a multiphase mixture [3] is described by the equation (8).

$$\frac{\partial}{\partial t}(\varphi_q\rho_q) + \nabla(\varphi_q\rho_q\vec{v}_q) = \sum_{p=1}^{n}(\dot{m}_{pq} - \dot{m}_{qp}) + S_q \tag{8}$$

where :

- φ_q – volume share of phase q,
- $\vec{v}_q$ – velocity of phase q,
- ρ_q – density of phase q,
- $\dot{m}_{pq}$ – mass flow between p and q phases = 1, 2, 3, ..., n,
- S_q – additional (external) source of mass for phase q, variable or constant.

When the sum of the volume of the individual phases of 1 (9)is maintained.

$$\sum_{q=1}^{n}\varphi_q = 1 \tag{9}$$

The equation of the velocity balance for the q phase mixture in an Eulerian model is described by the vector equation:

$$\begin{aligned}&\frac{\partial}{\partial t}(\varphi_q\rho_q\vec{v}_q) + \nabla(\varphi_q\rho_q\vec{v}_q\vec{v}_q)\\ &= -\varphi_q\nabla p + \nabla\cdot\vec{\vec{\tau}}_q + r_q\rho_q\vec{g} + \sum_{p=1}^{n}(\vec{R}_{pq} + \dot{m}_{pq}\vec{v}_{pq} - \dot{m}_{qp}\vec{v}_{qp})\\ &+ (\vec{F}_q + \vec{F}_{lift,q} + \vec{F}_{vm,q})\end{aligned} \tag{10}$$

where :

- p – pressure,
- $\vec{\vec{\tau}}_q$ – phase q stress tensor,
- $\vec{g}$ – acceleration of gravity,
- $\vec{R}_{pq}$ – inter-phase interactions resulting from the presence of other phases, dependent on the friction, pressure, cohesive interactions etc.,
- $\vec{v}_{qp}$ – inter-phase velocity,
- $\vec{F}_q$ – external mass forces acting on the phase q,
- $\vec{F}_{lift,q}$ – lift acting on the phase q,
- $\vec{F}_{vm,q}$ – force associated with the apparent weight of phase q,
- $\dot{m}_{pq}$ – flow of mass from phase p to phase q,
- $\dot{m}_{qp}$ – flow of the phase q to phase p,
- n – total number of phases of the system.

In the calculation of a multiphase system for each phase, a set of equations set forth above is solved. In the case of computation for the range of a turbulence model, the equations are supplemented with the equations describing the turbulence. In effect, the modeling of a multiphase system using several dispersion phases results in a very high demand for computing power and a very long calculation times.

The continuity equation for one fraction, assuming no mass transfer between the phases and a lack of sources of weight will be described by the equation (11).

$$\frac{\partial}{\partial t}(\varphi_q \rho_q) + \nabla(\varphi_q \rho_q \vec{v}_q) = 0 \tag{11}$$

The sum of the volume shares of each volume fraction φq must equal 1 according to equation (10).

In addition, for incompressible fluids, the continuity equation (11) will have the form:

$$\sum_q \nabla(\varphi_q \vec{v}_q) = 0 \tag{12}$$

The equation for fluid is described by the dependency (13).

$$\frac{\partial}{\partial t}(\varphi_q \rho_q \vec{v}_q) + \nabla(\varphi_q \rho_q \vec{v}_q \vec{v}_q) = -\varphi_q \nabla p_q + \nabla \cdot \vec{\vec{\tau}}_q + \varphi_q \rho_q \vec{g} + \sum_{p=1}^{n} (\vec{R}_{pq}) + (\vec{F}_q + \vec{F}_{lift,q} + \vec{F}_{vm,q}) \tag{13}$$

The forces acting on the phase of the q, as a result of the presence of other phases, are described by the equation (14).

$$\vec{R}_{pq} = c_{pq}(\vec{v}_p - \vec{v}_q) \tag{14}$$

The rate c_{pq} is described by the formula (15).

$$c_{pq} = \frac{C_D}{8} A_{pq} \rho_q |\vec{v}_p - \vec{v}_q| \tag{15}$$

where the resistance coefficient CD will depend on the Reynolds number calculated from one of the equations (16) or (17):

$$C_D = \frac{24}{Re} \text{ for } Re \ll 1 \tag{16}$$

$$C_D = 0{,}44 \text{ for } 1000 \le Re \le 1 \div 2 \text{ x } 105 \tag{17}$$

For Reynolds numbers in the range 0.1 < Re < 1000, it can be designated using one of the models:

The Schiller Naumann Drag Model (18) [10] – dedicated for calculations of multiphase systems with a low share dispersion phase.

$$C_D = \frac{24}{Re}(1 + 0{,}15\, Re^{0{,}687}) \tag{18}$$

Wen Yu Drag Model (19) [11].

$$C_D = \varphi_c^{-1{,}65} \max\left(\frac{24}{Re'}(1 + 0{,}15 Re^{0{,}687});\ 0{,}44\right) \qquad Re' = \varphi_c Re \tag{19}$$

where :

φ_c – volume of the continuous phase.

Gidaspow Drag Model (20)[12].

$$C_D = C_D(\text{Wen Yu}), \varphi_c > 0{,}8$$

$$c_{pq} = 150\frac{(1-\varphi_c)^2\mu_c}{\varphi_c d_p^2} + \frac{7}{4}\frac{(1-\varphi_c)\rho_c|\vec{v}_c - \vec{v}_d|}{d_p}; \ \varphi_c < 0{,}8 \quad (20)$$

4 Definition of the numerical model of a grain composition of the polydisperse suspension

In real life systems, except in individual cases, suspensions with polydisperse grain composition occur [13]. An example of a grain composition described by a particle size distribution curve is presented in Figure 1. The use of particle size distribution described using a graph in numerical calculations would be very cumbersome, since in practice the grain composition is described using a random variable. The distribution of particle sizes most commonly used here is the log normal parameter distribution with two parameters: m and sigma. In numerical simulations of the process of sedimentation, a fully polydisperse size distribution can be defined only in Euler-Lagrangian class models, i.e., in the case of free sedimentation with low concentration slurries.

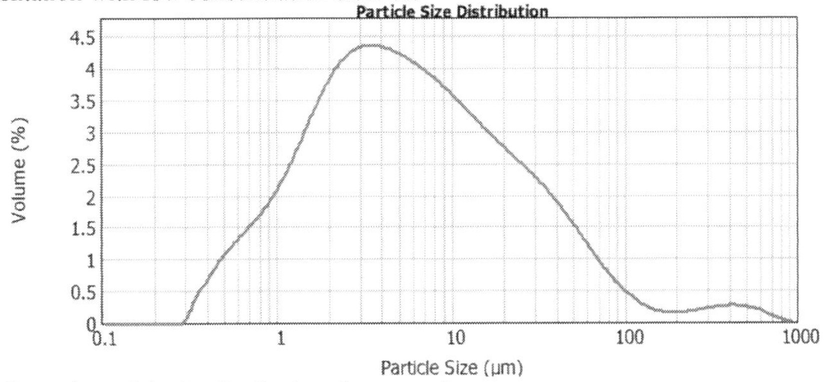

Fig. 1. Example particle size distribution of a suspension

In sedimentary systems, where Euler-Euler class models are used, the dispersion fraction is defined as a separate phase or a set of phases, each with a defined volume share and a defined particle size. Tests modeling the sedimentation using a monodisperse particle size distribution fraction have not provided satisfactory results. The studies and analyses carried out so far [14] indicate that defining a dispersion fraction based on a particle size distribution described by a histogram comprising several grain classes is the solution that gives promising results. An example of a particle size distribution in a slurry described by grain sizes is presented in table 1.

Table 1. An example of grain composition described using grain classes

Grain class µm	**Representative grain diameter** µm	**Share** %
0÷2	1.0	17.9
2÷5	3.5	20.1
5÷12	8.5	23.0
12÷30	21.0	20.4
>30	60.0	18.6

Based on the analysis carried out [14], it should be noted that in numerical simulations of the polydispersity dispersion, the use of the monodisperse numerical suspension model with substitution parameters does not produce positive results, therefore it is necessary to use multiphase systems in numerical simulations. The presented method of defining a multiphase system based on the grain composition of the suspension allows obtaining near-laboratory results of sedimentation in numerical simulations, with a change in the concentration of the suspension at the overflow and outflow, including the change in grain composition at the overflow and outflow.

5 Conclusion

The analysis shows that taking into consideration the purpose of conducting numerical simulations of sedimentation, i.e., to obtain a concentration distribution in a settler, numerical simulations of settling device operation should use Euler-Euler class models. These models allow you to obtain information on the volume share of the individual fractions (the main and dispersion fraction), which is, in effect, the distribution of the concentrations of the suspension in the device. As a result of the analysis, it should be noted that the most appropriate model for numerical simulation of sedimentation will be the Eulerian model. At the same time, it should be noted that there is a necessity of further work to determine the number of grain fractions sufficient to conduct reliable simulations, burdened an error of no more than a few percent.

References

1. W. P. Kowalski, Osadniki wielostrumieniowe. Kraków: AGH (2004)
2. A. Prosperetti i G. Tryggvason, Computational Methods for Multiphase Flow, 1st edition. Cambridge ; New York: Cambridge University Press (2009)
3. B. Gicala, Modelowanie numeryczne przepływów wielofazowych z fazą dyspersyjną. Podstawy teoretyczne i zastosowanie. Gliwice: Instytut Techniki Górniczej KOMAG (2011)
4. G. Tryggvason, R. Scardovelli, i S. Zaleski, Direct Numerical Simulations of Gas-Liquid Multiphase Flows. Cambridge, UK ; New York, NY: Cambridge University Press (2011)
5. S. Berres, R. Bürger, i E. M. Tory, „Applications of polydisperse sedimentation models", Chem. Eng. J., **111**, nr 2–3, pp. 105–117, (2005)
6. R. Eymard, T. Gallouët, i R. Herbin, „Finite volume methods", **7**, B.-H. of N. Analysis, Red. Elsevier, pp. 713–1018, (2000)
7. O. C. Zienkiewicz, R. L. Taylor, i R. L. Taylor, The finite element method, t. 3. McGraw-hill London (1977)
8. J. Bandrowski, H. Merta, i J. Zioło, Sedymentacja zawiesin: zasady i projektowanie. Gliwice: Politechnika Śląska (2001)
9. M. Banas, „The dependence of sedimentation efficiency on susupension concentration", Inzynieria Chem. Proces., **25**, 3, p. 659–664, (2004)
10. L. Schiller i A. Naumann, VDI Zeits, t. 77. (1933)
11. C. Y. Wen i Y. H. Yu, „Mechanics of Fluidization", Chem Eng Prog Symp, **62**, pp. 100–111, (1966)
12. D. GIDASPOW, Multiphase Flow and Fluidization. San Diego: Academic Press (1994)
13. M. Banas, P. Warzecha, i B. Hilger, „DISTRIBUTION OF TURBIDITY IN THE WATER TREATMENT PLANT SETTLER", *14th SGEM GeoConference on Ecology, Economics, Education And Legislation*, **2**, pp. 193-200 (2014)

14. K. Kołodziejczyk, „Projektowanie osadnika wielostrumieniowego z zastosowaniem numerycznej symulacji przepływu", Przem. Chem., **96**, 8, (2017)

E3S Web of Conferences **46**, 00012 (2018) https://doi.org/10.1051/e3sconf/20184600012
3rd International Conference on Energy and Environmental Protection

Design and experimental research of a quadrocopter flying robot

Grzegorz Suchanek[1,*] and *Wojciech* Ciesielka[1]

[1]AGH University of Science and Technology, Mickiewicza Av. 30 30-059 Cracow, Poland

Abstract. The drone is an unmanned aerial vehicle. Currently, many commercially available remote controlled flying toys are used to be called drones. This is an erroneous nomenclature, because the drone must have an autonomus flight function implemented. Due to it's simple mechanical construction, the most popular drones are in the form of a multirotor, in which arrangement the engines are placed in one plane. One of the most important advantages of this type of robots is the ability to maintain a certain position in space. Today, this allowed for e.g. taking photos from the air or inspecting hard-to-reach places. For use for environmental protection purposes, drone equipped with appropriate sensors and instrumentation may be used to monitor air pollution. The mechanical part of a quadrocopter flying robot was based on a TAROT frame with a 450mm engines spacing. The frame has been expanded with a dedicated set of legs to raise the clearance up to 150mm. Four dedicated EMAX MT2213 electric motors were installed on the frame, which are the main drive. They are characterized by the propeller hub-free-mounting, which minimizes possible imbalances. A single engine cooperating with a dedicated 10-inch propeller and a 4.5-inch pitch generates a maximum thrust of a 0.85kg. In the case of this system, it sums to a total of 3.4 kg. The weight of the ready to flight robot is 1.35 kg. To power the robot, a lithium-polymer battery with a capacity of 2.2 Ah is used, providing flight time of about 8 minutes. The basic work mode of the robot is a manual one, which means a self leveling mode with manual control. In addition to this mode, an autonomous navigation mode using GPS coordinates has been implemented. This navigation mode was also been tested during field tests. The operation of this navigation mode is very similar to the position maintaining mode, but operates on a larger scale. The robot in this mode is vectorically controlled, performing forwards/backwards and sideways movements to the set location.

1 Introduction

Currently, new articles on topics related to drones are constantly appearing in domestic and foreign literature. These studies cover various aspects. The descriptions of drone control units are a constantly emerging topic. The proposed controllers differ both in the complexity of the electronic part and the algorithms used. The simplest hobbyists

* Corresponding author: suchanek@agh.edu.pl

controllers use the popular Arduino platform [1]. Furthermore, even research is conducted on the minimum requirements for the processing power of the control unit microcontroller [2]. More complex control units utilizes e.g. ARM microcontrollers [3] or the Raspberry Pi platform [4].

In the case of a multirotor flying robot, the mechanical construction of the frame has no less importance, as the optimal durability and appropriate materials translate into the overall performance of the drone. Optimization works in this direction are also carried out [5]. Other drone components, such as motors [6] are also optimized.

In addition to the publications related to the mechanical and electronic parts, they are also publications covering the mathematical description of the multi-rotor [7]. These issues relate to, for example, the mathematical description of the dynamics of unmanned aircraft [8] or the use of mathematical models for control purposes [9].

In addition to primary military usage, drones have found wide application in many industries. Using drones, it is possible to inspect elements of a building that are hard to reach, such as a roof without having to mount scaffoldings. Drones are also great for inspecting dangerous areas, for example buildings in danger of collapsing or electrical power lines [10]. After equipping the drone with appropriate sensors and instrumentation, it is also possible to conduct environmental pollution tests [11].

The drone can also be used for transport purposes. The drone can transport parcels, medical supplies or samples of biological materials to the laboratory [12]. In this last case, the authors present how to mix the content of samples during transport, using appropriate maneuvers.

With the damming the risks associated with the use of drones by persons without specialized knowledge related to the airspace, the need arose to develop effective methods for detecting and protecting against drones. The analysis of detection methods of unmanned aerial vehicles is presented in [13]. Such systems operate on the principle of image and sound detection generated by the drone during operation. Other solutions are based on the principle of detecting WiFi network or control equipment signals. In [14], a fast video system utilizing a rotating camera for tracking a drone was presented.

The purpose of this work is to design, build and carry out field verification tests of a quadrocopter flying robot. Information recorded during flights can be used in environmental management systems [15,16] and contribute to its sustainable developmentDepletion of energy from conventional sources and a continuous increase in energy consumption necessitates the search for new energy sources, preferably from renewable ones [1,5]. One of such sources is biomass [3].

2 Design of the mechanical part of the system

A popular TAROT FY-450 frame with engine spacing of 450mm was used to build the robot's mechanical structure. The frame has been expanded by a dedicated landing gear with a height of 150mm, increasing the original ground clearance.

For the main motors of the drone, a dedicated for the multirotors EMAX MT2213 motors were selected. They are characterized by precise machining and hub-free-mounting of propellers, minimizing any unbalances of the propeller mounting. Such engine can deliver a maximum of 850g thrust (for this design it sums to a total of 3400g) with 10-inch propellers with a 4.5-inch pitch and 3S (11.1V) power supply. The weight of the motor including the nut and propeller is 70 grams. The motors are controlled by the classic FastPWM 30A ESC's with SimonK multirotor firmware, accepting a control signal with a frequency up to 400 Hz. The weight of one regulator is about 42g.

The last element of the mechanical part is a LiPo 3S1P battery with a capacity of 2200mAh with a weight of 160g, providing flight time of 8 minutes. The total take-off weight (including the weight of flight controller) is approximately 1350g.

3 Design of the electronic part of the system

The main part of the control unit, which is the most important part of the project, is the ARM LPC4088 microcontroller from NXP with the Cortex-M4 core clocked at 120MHz. Due to the operation characteristics of the control equipment, the control unit was equipped with an additional auxiliary microcontroller, also NXP ARM LPC1114 chip with a Cortex-M0 core clocked at 50MHz.

The other important parts of the control unit are, inertial sensors (IMU), on the basis of which the stabilization of the platform takes place. For this purpose, two advanced MPU9250 IMU's from InvenSense were selected, each including: a 3-axis gyroscope, a 3-axis accelerometer and a 3-axis magnetometer. In addition to that, the inertial unit has been expanded with a precision pressure sensor MS5611 from TE Connectivity, a precision temperature sensor ATD7320 from Analog Devices and an additional LSM303C magnetometer from ST Microelectronics. All Sensors are mounted on a separate PCB connected to the main PCB using a flexible FFC conductor. This type of connection reduces the transmission of vibrations to the sensor board. The installation of sensors on a separate module ensures also an easy replacement of the sensors set without the need to replace the control board.

In addition to that, the control unit was equipped with an μUSB connector enabling to change it's configuration from the computer. The user interface was built using a color OLED display, a joystick, and two additional buttons. The possibility of recording measurement data, necessary to verify the correct operation of sensors, prepared filters and regulators, was ensured by equipping the system with a NandFlash high-speed memory module with a capacity of 256MB. With a sampling rate of 500Hz, this allows to flight data be recorded for about 7 minutes. The non-volatile memory used here contrary to the RAM type of memory, creates a black box of the system in a natural way, allowing to determine the cause of the system failure. The μSD memory card connector enables easy transfering of collected data to a computer, via a card, where it can be easily converted into a CSV file. The measuring unit has also been equipped with additional connectors, allowing to extend its capabilities in the future, including o an optical sensor that makes it easier to hold the position, and an external CAN bus. The use of the well-known automotive bus enables easy connection of additional external environmental sensors, without grounding loop problems known to the market control units, introducing additional interfernces. Such communication are conducted digitally with the differential bus, which additionally eliminates the problems with introducing more interference to the measurement sensors.

4 Design of the informatic part of the system

The software part of the system consists of several prepared programs. The most important of them is the software of the flight control unit, which was entirely prepared in C language. This software is divided into two microcontrollers: the main system with the Cortex-M4 core and the supporting Cortex-M0 core. As was mentioned earlier, the use of an additional core was necessary due to the specificity of the operation and the variety of manufacturers of available control equipment.

The main microcontroller takes control of all measuring sensors and flight stabilization in all modes of the system operation. After turning on the power supply of the unit, all internal microcontroller subsystems are initialized. This process in particular, includes the preparation of communication interfaces used. After sending the configuration parameters to each of the sensors, the configuration is read again to check if the sensors have correctly accepted the changes. If any of the sensors has not been initialized correctly, this indicates that it has been damaged. A critical error message will appear on the built-in display and further operation will not be possible. In the next step, the calibration data is read. This process covers both the configuration of the control unit itself from the internal memory of the main microcontroller as well as the sensor calibration data from the external memory on the interchangeable sensor PCB. If the data could not be read, the default values are used, and the possibility of arming the drone is blocked until the repair/re-calibration is done.

After initialization, the main control interrupt is started at the frequency of 500Hz, in which the measurement data from the sensors are collected and processed cyclically. In this interrupt, appropriate control signals are also generated, filtration of sampled signals is carried out, algorithms for stabilizing the robot are performed in each of the operating modes, and all data are written in the black box. Finally, the main program goes to an infinite loop. If the robot remains un-armed, the user interface is operated on the built-in display and communication with the PC is processed.

After arming the robot, the autopilot program is executed at a lower frequency of 10Hz. This frequency was selected due to the operating frequency of the GPS receiver used. The purpose of the autopilot is to supervise the control interrupt in the experimental navigation mode. This navigation mode use the roll and pitch axes. This mode of operation is possible to implement in a multirotor robot, because it has the ability to move sideways, not only forward and backward. In this case, the set direction (yaw) remains constant.

5 Verification tests

In order to accurately verify the correct operation of all algorithms, the verification tests were carried out in several successive stages, parallel to the work progress of the software part.

The first stage was the verification of the operation of used filters in *Matlab*, using the registered measurement data. This made it possible to verify the correct operation of all used measuring sensors, control equipment and the selection of optimal filter's parameters. The orientation of the measuring axes of all the sensors used was also set and checked. After this stage, all the measuring sensors and engine speed controllers used were calibrated.

The next stage was to analyze the black box on the drone, after implementing the filters and regulators. In this way, the operation of the implemented automatic control systems was verified, comparing the obtained results to a model created in parallel in the *Simulink* engineering environment.

The final stage was field tests. Based on them, the correct operation of regulators and filters during real flight tests was checked, which allowed to assess what further modifications will be needed to increase the stability of the platform, as well as enable more precise selection of controller parameters. During field tests, the data processing efficiency on the control unit was also verified.

The first of the tested modes was the altitude hold mode using atmospheric pressure measurement. Figure 1 shows the path traveled by the robot in this mode.

The robot in the altitude hold mode slowly moved chaotically without stabilizing the position while maintaining the set altitude. The starting position (0, 0) on the path map is the location where the robot was at the start of flight data recording.

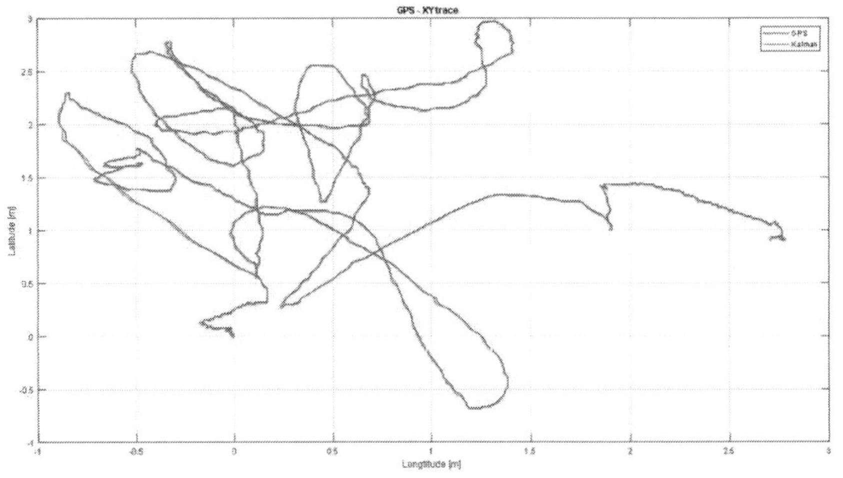

Fig. 1. A traveled path by the robot in altitude hold mode

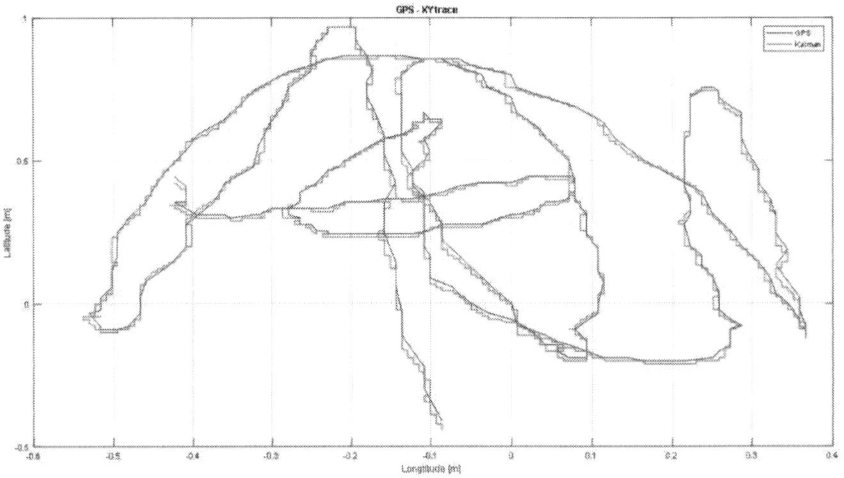

Fig. 2. A traveled path by the robot in position hold mode

The operation of the GPS position maintenance mode is very similar to the altitude maintenance mode, with the difference that in this case the position with the use of the GPS receiver is additionally stabilized. In this mode (figure 2), the robot did not move in straight line further than 0.9m from the set position. The moving area can be enclosed in a rectangle measuring 0.9m by 1.4m.

The implemented navigation mode combines the operation of the autopilot with previously described modes: position hold and altitude hold. The way in which navigation works is analogous to the position maintaining mode, but operates on a larger scale. The robot in this mode is vectorically controlled to the set point, performing forward/backward as well as sideways moves (Figure 3).

Figure 4 shows a picture of the robot during operation in navigation mode – a trip from point to point.

A robot programmed in this way can be used to carry out stationary environmental measurements (after landing) in hard to reach areas, including roofs, recesses, towers or chimneys.

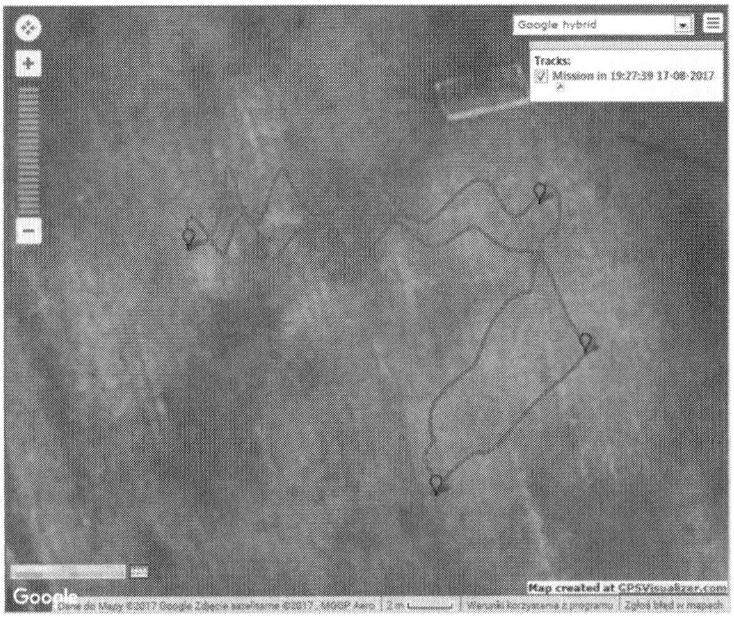

Fig. 3. A travelled path by the robot in navigation mode - map

Fig. 4. Robot during operation in navigation mode - a trip from point to point

Currently, quadcopter is equipped with a MS5611 temperature/pressure sensor and a SVAN945A sound level meter. The effectiveness of the proposed solutions was confirmed during field tests. For example, temperature values of 28.74°C, atmospheric pressure 985.85hPa and equivalent sound level L_{Aeq} 38.7dB were recorded in the environment test point, respectively. Due to the strictly limited size of this article, detailed research results will be presented during the conference.

6 Conclusion

The new research station is the most technically advanced of all stations for testing cutting process built at the Institute of Mechanical Engineering at Warsaw University of Technology in Płock. It gives the possibility to conduct a wide range of tests for variable construction and working parameters of the cutting unit as well as for variable parameters of the plant itself, e.g.: stem's diameters or moisture. The obtained results allow for a better understanding of the cutting process of energy plants carried out with a two-disc cutting unit. The research results will also enable the construction of more efficient and energy-saving cutting units for agricultural machines.

References

1. Bonisławski A., Juchniewicz M., Piotrowski R.: Projekt techniczny i budowa platformy latającej typu quadrocopter, Pomiary Automatyka Robotyka, **1**, pp. 91-97, (2014)
2. Szczęsny S., Golijanek-Jędrzejczyk A., Rafiński L., Stranc P.: Robot typu quadrocopter sterowany mikrokontrolerami o niewielkiej mocy obliczeniowej, Zeszyty Naukowe Wydziału Elektrotechniki i Automatyki Politechniki Gdańskiej, **46**, pp. 93-96, (2015)
3. Gheorghiţă D., Vîntu I., Mirea L., Brăescu C.: Quadcopter Control System. Modelling and Implementation, IEEE, *19th International Conference on System Theory, Control and Computing (ICSTCC), October 14-16*, pp. 421-426, (2015)
4. Strzelczyk P., Macek-Kamińska K.: Kontroler lotu dla bezzałogowych obiektów latających, Pomiary Automatyka Robotyka, **4**, pp. 69-73, (2015)
5. Velan Y. M.: Cost Effective Design and Development of Manned Drone, *IEEE, First International Conference on Recent Advances in Aerospace Engineering (ICRAAE)*, (2016)
6. Islam S., Husain I., Mikail R.: Slotless Ligthweight Motor for Drone Applications, IEEE, pp. 5041-5048, (2017)
7. Bieniak R., Gudzenko O.: Control Methods Design for a Model of Asymmetrical Quadrocopter, Journal of Automation, Mobile Robotics & Intelligent Systems,**10**, 2(2016)
8. Bieńczak R., Komorek A., Kowalik R.: Matematyczny opis dynamiki ruchu bezzałogowego statku powietrznego, Technika Transportu Szynowego, **12**, pp. 157-162, (2015)
9. Kusalewicz A., Armiński K., Zubowicz T.: Użytkowy model matematyczny quadrocoptera do celów sterowania, Zeszyty Naukowe Wydziału Elektrotechniki i Automatyki Politechniki Gdańskiej, **51**, pp. 103-106, (2016)
10. Kinjo H., Morita M., Sato S., et al.: Infrastructure (Transmission line) Check Autonomous Flight Drone (1), IEEE, 978-1-5090-6664-3/17/, ICIIMBS 2017, Track2 Artifical Inteliigence, Robotics and Human-Computer Interaction, pp. 206-209, (2017)
11. Berner B., Chojnacki J.: Monitorowanie zanieczyszczeń środowiska za pomocą dronów, Autobusy, **7-8**, pp. 57-60, (2017)
12. Katariya M., Chung D. C. K., Minife T., et al.: Drone inflight mixing of biochemical samples, Analytical Biochemistry, **545**, pp. 1-3, (2018)
13. Fellner A., Mańka B., Mańka A.: Analiza aktualnych metod detekcji bezzałogowych statków powietrznych (dronów), Technika Transportu Szynowego, **12**, pp. 484-488, (2015)

14. Park J., Kim D. H., Shin Y. S., Lee S.: A Comparison of Convolutional Object Detectors for Real-time Drone Tracking Using a PTZ Camera, *17th International Conference on Control, Automation and Systems (ICCAS 2017)*, pp 696-699, (2017)
15. Ciesielka W.: Management of environmental noise – Cracow example, Polish Journal of Environmental Studies, **16**, 3B, pp. 53–58, (2007)
16. Ciesielka W.: Management of environmental noise – the Cracow example, Archives of Acoustics, **32**, 4, pp. 983–994, (2007)

Dew point temperature of wet air stream at variable pressure

Bogusław Ptaszyński[1,*], *Zbigniew* Kuczera[1], *Rafał* Łuczak[1] and *Piotr* Życzkowski[1]

[1]AGH University of Science and Technology, Faculty of Mining and Geoengineering, Mickiewicza 30 Av. 30-059 Cracow, Poland

Abstract. One of the examples of how atmospheric air could be utilized includes its use as a carrier of mechanical energy transmitted by various devices, e.g. air compressors, constituting parts of various technical systems. In many of them, atmospheric air is subjected to significant changes in static pressure. This research paper specifies the effect of the variable static pressure on the dew point temperature of wet air. This issue was presented both analytically and graphically. The knowledge will allow to effectively identify these areas in which the water vapour contained in the air will be subject to phase changes, unfavourable for humans.

1 Introduction and objective of study

Streams of gas mixtures are utilized in numerous fields of technology. They are subjected to specific transformations for various purposes and in various devices. If one of the gaseous components that make up the mixture which is undergoing a given process is subject to a phase change, in thermodynamics such a mixture is called the wet gas [5]. The gaseous components that do not change their state of matter are called the dry gas, while the mixture component that can change its physical state is called moisture. In the vast majority of cases, moisture is present in the wet gas in the form of superheated steam. Knowledge of the temperature at which moisture changes its physical state plays a key role in many issues, for example, a stream of exhaust gas of combustion engines, which contains water vapour that can condense in the exhaust system. It is essential to make sure that this phenomenon does not occur in the exhaust manifold. The ventilation uses wet air that, thermodynamically, is the wet gas, in which water is the moisture. In many technical processes, it is required that the water present in the air stream in the form of superheated steam does not change its state of matter to liquid. In industrial ventilation, it is essential that during the flow of wet air through the ducts, condensation does not occur, as it is a source of corrosion or bacteria proliferation in the ducts. In such cases, the knowledge of the dew point temperature of the wet air is necessary in order to be able to thermally insulate the air transporting ducts effectively. The literature provides for various Formulas for the so-called dew point temperature of wet air [1,7]. Such dependencies, however, do not take into account the change in static pressure in the flow pathway of the wet air stream.

* Corresponding author: bptaszynski@gmail.com

The purpose of this research paper is to determine the effect of the variability of static pressure on the dew point temperature of the wet air.

2 Dew point temperature of wet air at constant pressure

The dew point of the wet air in isobaric cooling processes is presented in thermodynamics using the p = f(T) plots for the water [5] or for the wet air [6]. Figure 1 illustrates a plot of isobaric cooling of the wet air for constant air pressure, at a constant water content in the form of superheated vapour.

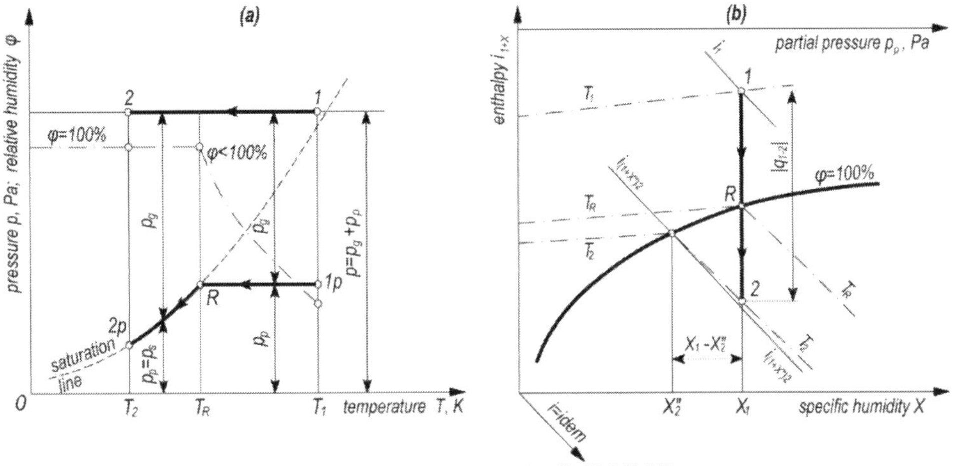

Fig. 1. Isobaric cooling of wet air presented in system a) p,T, b) i,X [6]

Point 1p in Figure 1a presents the thermodynamic state of steam, determined by the temperature of air and steam T, and the partial pressure of water vapour in the air p_p. The air pressure p is the sum of the partial pressures of dry air and water vapour. Point 1 represents the thermodynamic state of wet air. The conversion of volumetric cooling of such air at the constant pressure is represented by line 1-2. In this process, the initial state of water (steam) is represented by point 1p, and the final state by point 2p. The dew point is denoted by R and its coordinate on the abscissa, called the dew point temperature, as T_R. This quantity is more frequently defined in the Celsius scale and denoted as t_R. The water vapour at the initial point is superheated because the partial pressure of the vapour p_p is lower than its saturation pressure read for the point determined from the intersection of the axis perpendicular to the abscissa, passing through point 1p, and the evaporation curve. The dew point R is the point of intersection of the line illustrating the isobaric cooling of the water vapour from point 1p and the evaporation curve. If the temperature of the wet air drops, the saturation pressure of the water vapour decreases, while the relative humidity of the air increases. At the dew point T_R, the initial partial vapour pressure equates to its saturation pressure and the value of the relative humidity reaches 100%. With further cooling of the wet air, the partial pressure of water vapour decreases and is equal to the water vapour saturation pressure. The partial pressure of the dry air for the temperatures $T < T_R$ increases accordingly, so that the condition of the cooling process's isobaric was met.

The process of such air cooling is also illustrated in Fig. 1b in the coordinates of the specific enthalpy i_{1+X} and the degree of moisture X [6]. Figure 1b illustrates that the final state of such cooling of wet air is demonstrated in point 2 and it is located in the haze air region. The point of intersection of the T_2 isotherm with the 100% relative humidity curve allows to read the value of the degree of moisturizing the air with the dry saturated steam

X_2''. On this basis, a generally known definition of the dew point temperature can be explained. It says that it is the highest air temperature at which the vapour contained in the air during its cooling undergoes condensation. However, such an imprecise definition suggests that $t_R \geq 0{,}01°C$. Therefore, Figure 1 should be completed with the information derived from the so-called water phase diagram, i.e. from the p-t diagram for H_2O [5]. The completion should take into account both the evaporation (condensation) curve and the sublimation (deposition) curve – Figure 2. Based on it, it is possible to say that the dew point R is a point with the temperature t_R, at which the actual partial vapour pressure is equal to the saturated vapour pressure. The saturated vapour pressure for the temperatures above zero results from the evaporation (condensation) curve, and for the temperatures below zero from the sublimation (deposition) curve. This definition confirms that each wet air has a dew point R and the corresponding dew point temperature t_R, which can also take negative values. It should also be remembered that for subzero temperatures, in addition to ice, supercooled water may also occur.

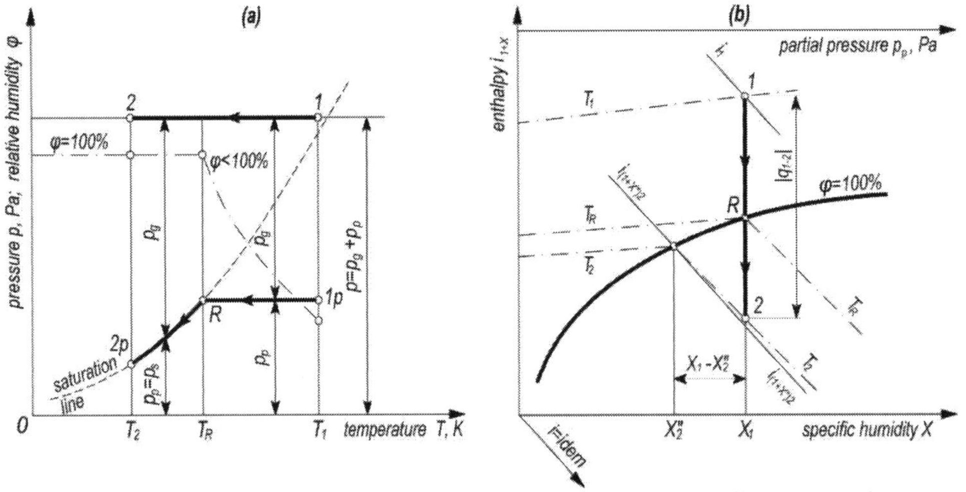

Fig. 2. Isobaric cooling of wet air with evaporation and sublimation curves illustrated in the system a) p,T, b) i,X

In the further analyzes, empirical Magnus Formula [4] was used for determining the saturated water vapour pressure E:

$$E = 610.6 \cdot 10^{\frac{b(T-273.15)}{T-c}}, [Pa] \tag{1}$$

where :

- b – constant dependent on air temperature and state of water [-],
- c – constant dependent on air temperature and state of water [K],
- T – air temperature [K].

For positive and negative values of temperatures of the air which is in contact with supercooled water, the constants in Formula (1) take values: b = 7.5; c = 35.86K. On the other hand, for negative values of temperatures of the air which is in contact with ice, the constants are equal to: b = 9.5; c = 7.66K. Using (1), in which $E = p_p$ for $T = T_R$, Formula for the dew point temperature can be developed as follows:

$$p_p = 610.6 \cdot 10^{\frac{b(T_R - 273.15)}{T_R - c}}, [Pa] \tag{2}$$

Having logarithmized and transformed (2), final Formula is obtained, which defines the temperature of the air dew point:

$$t_R = \frac{(273.15 - c)}{b \cdot \ln 10 - \ln \frac{p_p}{610.6}} \ln^{\frac{p_p}{610.6}}, [°C] \tag{3}$$

where the size c is dimensionless.

For positive and negative values of temperatures of the air which is in contact with supercooled water, the air dew point can be determined from Formula:

$$t_R = \frac{237.29}{17.26935 \cdot \ln 10 - \ln \frac{p_p}{610.6}} \ln^{\frac{p_p}{610.6}}, [°C] \tag{4}$$

For negative values of temperatures of the air which is in contact with ice, the air dew point temperature can be determined from Formula:

$$t_{R_L} = \frac{265.49}{21.87456 \cdot \ln 10 - \ln \frac{p_p}{610.6}} \ln^{\frac{p_p}{610.6}}, [°C] \tag{5}$$

For wet air with a low value of vapour pressure pp, the dew point values calculated according to (4) and (5) will differ. Examples of values for different vapour pressures are demonstrated in Table 1.

Table 1. Dew point temperature for different values of vapour pressure.

	Dew point temperature, °C				
	p_p=610.6 Pa	p_p=600 Pa	p_p=500 Pa	p_p=400 Pa	p_p=300 Pa
t_R acc. to Formula (4)	0.0	-0.240	-2.714	-5.673	-9.379
t_R acc. to Formula (5)	0.0	-0.212	-2.403	-5.036	-8.354

Relationships between the temperature values contained in Table 1 are confirmed by the positions of the terminal points 2* and 2** of the air cooling process illustrated in Figures 3 and 4. It should be noted that, most frequently, for such small pressure values of the water vapour in the air subjected to the process of volumetric cooling, we deal with the formation of rime and snow on solid surfaces or on particles present in the air, after reaching the temperature calculated according to (5). However, in some conditions of the wet air cooling process, excess moisture can be obtained in the form of supercooled water, but such physical state is a highly unstable.

3 Dew point temperature of wet air at variable pressure

In processes of cooling flowing wet air streams, static pressure may be subject to significant changes [3], which may lead to condensation of water vapour in the air stream. The presented Formulas for the dew point temperature (4) and (5) do not take into account the effect of changes in static air pressure. As it was demonstrated in [2], variable pressure induces a shift of curves and coordinate axes on the i-X plot (Mollier plot), describing the thermodynamic parameters of the wet air. The consequences of these changes are presented in Fig. 3.

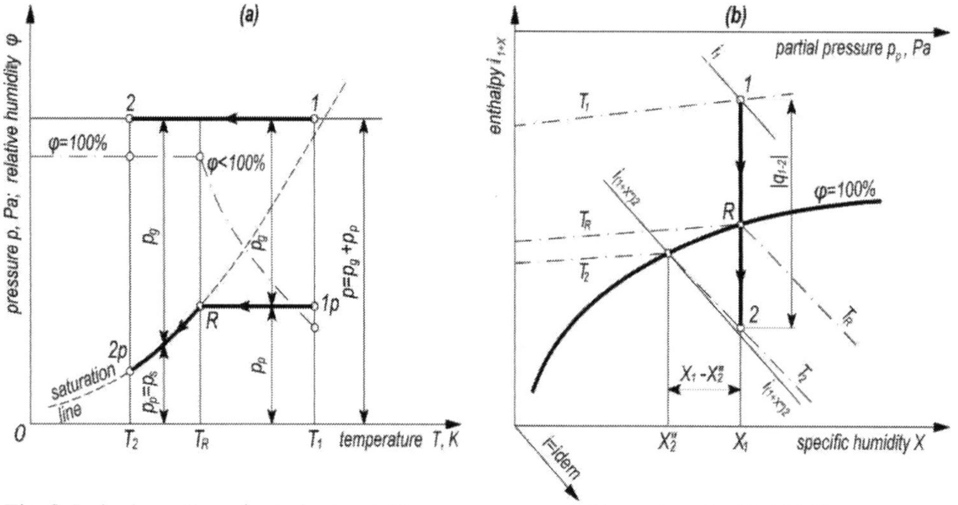

Fig. 3. Isobaric cooling of wet air at variable pressure presented in system a) p,T , b) i,X

It can be noticed that, with the change of pressure, the positions of the evaporation curve and the sublimation curve, connected at the node representing the triple point of water TR, change respectively. With a quick analysis of this plot, a conclusion can be drawn that there are several triple points of water. It is well-known that there is only one triple point of water, defined by its coordinates (T_{TR}, $p_{p(TR)}$). That is why all the TR points marked in Figure 3 have the same coordinates. The partial pressure of the triple point (611.2 Pa) can be read from the horizontal axis placed in the upper part of Figure 3b. Static pressure changes modify the auxiliary axis on the i-X plot, showing the values of the partial pressure of water vapour in the air. In Figure 3b, point 1 is marked, representing the thermodynamic state of wet air at the initial pressure and the corresponding dew points R for different end pressures. The temperature values of its dew points, read from the plot, are different and depend on the variability of pressure. This confirms the existence of the dependence of the dew point temperature on the static pressure of wet air. This dependence is also illustrated in the p-T diagram – Figure 3a. Here are also several TR triple points of water (for different pressures), but in this case, each of these points has the same coordinates. This means that each change in pressure modifies the scale at its ordinate axis, where the partial vapour pressure and the static pressure of the wet air are laid off. Point 1p, marked in Figure 3a, illustrating the thermodynamic state of wet air for the initial pressure and the conversion of its cooling, confirm the existence of the dependence between the dew point value of the wet air and its pressure.

Taking into account the dependence [2]:

$$\frac{p_p}{p_{p_1}} = \frac{p}{p_1} \tag{6}$$

where :

- P – static pressure of wet air [Pa],
- p_1 – static pressure of wet air in state 1 [Pa],
- p_p – partial pressure of water vapour in air for static pressure p [Pa],
- p_{p1} – partial pressure of water vapour in air for static pressure p_1 [Pa],

Formula for the dew point temperature was developed in the general form:

$$t_R = \frac{(237.15 - c)}{b \cdot \ln 10 - \left(\ln \frac{p_p}{610.6} + \ln \frac{p_1}{p} \right)} \left(\ln^{\frac{p_p}{610.6}} + \ln \frac{p_1}{p} \right), [°C] \tag{7}$$

where the size c is dimensionless.

For positive and negative values of temperatures of the air which is in contact with water (supercooled water in subzero temperatures), the value of the temperature of the air dew point which takes into account the dependence on pressure, takes the following form:

$$t_R = \frac{237.29}{17.26935 \cdot \ln 10 - \left(\ln \frac{p_p}{610.6} + \ln \frac{p_1}{p} \right)} \left(\ln^{\frac{p_p}{610.6}} + \ln \frac{p_1}{p} \right), [°C] \tag{8}$$

And for negative values of temperatures of the air which is in contact with ice, the value of the temperature of the air dew point which takes into account the dependence on pressure, takes the following form:

$$t_{R_L} = \frac{265.49}{21.87456 \cdot \ln 10 - \left(\ln \frac{p_p}{610.6} + \ln \frac{p_1}{p} \right)} \left(\ln^{\frac{p_p}{610.6}} + \ln \frac{p_1}{p} \right), [°C] \tag{9}$$

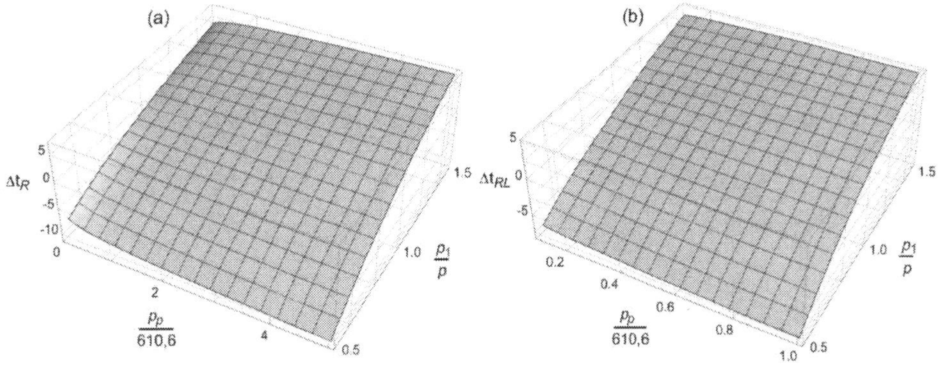

Fig. 4. Difference in temperature of air dew point a) Δt_R determined from Formulas (8) and (4), b) Δt_{RL} determined from Formulas (9) and (5)

Figure 4a presents the difference between the temperature of the air dew point Δt_R determined from (8) which takes into account the variable pressure, and the temperature of the dew point determined from (4) which does not take into account the variable pressure in the range of positive and negative values of air temperatures. Figure 4a illustrates that for the ratio $p_1/p = 1$, the value $\Delta t_R = 0$, regardless of the value of the partial pressure of water vapour in the air. If the air pressure p_1 is lower than the initial pressure p ($p_1/p < 1$), then $\Delta t_R < 0$, and when the pressure p_1 is greater than the initial pressure p, then $\Delta t_R > 0$. For the analyzed pressure range, these differences can even reach approximately -11°C. On the other hand, Figure 4b demonstrates the difference between the temperature of the air dew point Δt_{RL} determined from (9) which takes into account the variable pressure, and the temperature of the dew point determined from (5) which does not take into account the variable pressure in the range of negative values of air temperatures. It is also noticeable (as in Fig. 4a) that for the ratio $p_1/p = 1$, the value $\Delta t_{RL} = 0$, regardless of the value of the partial pressure of water vapour in the air. Also, if the air pressure p_1 is lower than the initial pressure p ($p_1/p < 1$), then $\Delta t_{RL} < 0$, and when the pressure p_1 is greater than the initial pressure p, then $\Delta t_{RL} > 0$. For the analyzed pressure range, these differences can even reach approximately - 7°C.

4 Conclusion

Based on the conducted research, the following conclusions can be formulated:

- The concept of the dew point temperature of the wet air can be explained based on Figure 2, which is an extension of Figure 1 contained in the Literature [6]. Thanks to the proposed changes, it is possible to correctly determine the concept of the dew point temperature for any wet air.
- Formulas for the values of temperature of the air dew point which are the most frequent in Literature, use the dependence between water vapour saturation pressure and water, which for the partial pressures of water vapour in the air which are lower than the pressure of the triple point of water, yields such values of this temperature which can only be obtained in special, very rare cases. However, these Formulas do not include the influence of variable air pressure on the determined value of the dew point temperature.
- Formulas for the dew point temperature of the wet air developed in this paper include the so-called sublimation curve and take into account the influence of variable pressure.
- Differences in the values of the dew point temperature, resulting from variability in air pressure, confirm the usefulness of the developed dependencies.

This research study was supported by the AGH University of Science and Technology statutory research No. 11.11.100.920.

References

1. Pełech A. *Wentylacja i Klimatyzacja. Podstawy*, Oficyna Wydawnicza Politechniki Wrocławskiej, Wrocław, (2008)
2. Ptaszyński B. Wykorzystanie wykresu Molliera w analizie przemian powietrza wilgotnego w wyrobiskach górniczych, Ciepłownictwo Ogrzewnictwo Wentylacja, **10,** 47, pp. 389-436, (2016)

3. Ptaszyński B., Łuczak R., Życzkowski P., Kuczera Z. Thermodynamic processes of the air flowing through a ventilation shaft in underground mines, Archives of Mining Sciences, **1**, pp. 149-163, (2018)
4. Roszczynialski W., Wacławik J. *Podstawy Aerologii Górniczej, Cz.I,* Skrypt AGH Nr 713, Kraków, (1979)
5. Styrylska T. *Termodynamika. Podręcznik dla studentów wyższych szkół technicznych,* Politechnika Krakowska, Kraków, (2004)
6. Szargut J. *Termodynamika Techniczna,* Wydawnictwo Politechniki Śląskiej, Gliwice, (2005)
7. Wacławik J. *Wentylacja kopalń,* Wydawnictwo AGH, Kraków, (2010)

Economical and technical aspects of using air heat pumps for hot water

Mariusz Szreder[1,*]

[1]Warsaw University of Technology, Faculty of Civil Engineering, Mechanics and Petrochemistry, Łukasiewicza 17, 09-400 Płock, Poland

Abstract. The paper presents selected technical and economic aspects related to the use of air heat pumps for the preparation of domestic hot water DHW. The research shows that heating the water in the 130 dm^3 storage tank with the A21/W30-50 profile lasted an average of 130 minutes, and the heat pump consumed about 2.2 kWh of energy per one heating cycle. Heating water in a 130 dm^3 storage tank from 25°C to 40°C lasted an average of 60 minutes, while heating water for another 5°C took 30 minutes. The operation of the heat pump in the higher temperature of the water in the storage tank is less effective, making the heating of water in the tank above 50°C is ineffective.

1 Introduction

The dominant trend in utility water heating installations is the continuous reduction of energy demand and increasing the use of free renewable energy. The whole fits in with the European Union strategy regarding the reduction of CO_2 emissions and the development of Renewable Energy Sources [1].

In recent years, heat pumps have become a relatively popular solution used to heat domestic hot water. This is mainly due to the low purchase price and the simplicity of assembly of this system. Compared to solar systems, their advantage is lower investment price and hot water production practically throughout the year [2].

Solar collectors and air heat pumps are especially recommended as an alternative to electric heating, fuel oil or liquid gas. The main argument is lower operating costs. Undoubtedly, such association is also a viable alternative in the absence of a natural gas connection. Supplementing the installation with a solid fuel boiler with an air heat pump and solar collectors is very beneficial due to the cumbersome and economically and ecologically unjustified exploitation of the boiler outside the heating season [5].

The integration of many systems is becoming more and more popular [3, 4]. The use of a heating system integrated with an air heat pump and solar collectors requires proper adaptation of the installation as well as an economic justification for the investment. The main justification for associating the solar collector system and the heat pump are the rising prices of traditional fuels or the lack of availability of cheaper heat energy carriers [6].

* Corresponding author: mariusz.szreder@pw.edu.pl

To sum up, the integration of solar collectors with a heat pump will be more expensive to invest at each of them separately, but in this way we get the cheapest and most stable heating system for hot water outside the heating season.

2 Construction of the air heat pump for domestic hot water

There are various technical solutions available on the market, intended exclusively for DHW heating, including an integral water tank in fig. 1 or intended for cooperation with an existing heating system in a building. A modular variant of the heat pump (without a tank) enables easy installation of the heat pump in a boiler room, in which there is e.g. a solid fuel boiler and a water tank, which does not need to be replaced with a new one.

In the first solution, it is possible to obtain higher coefficient of performance COP values due to the direct integration of the condenser with the hot water tank [8, 9].

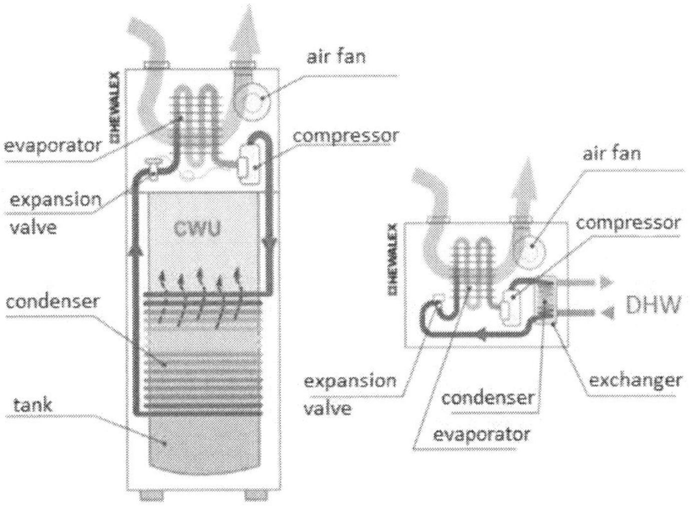

Fig. 1. Construction of air heat pump for DHW [9]

The heat pump fan sucks warm air from the room and pushes it through the evaporator, on which the air is cooled by approx. 5 ÷7K, and then it is removed outside the building or into the same room. Due to this small cooling of the air, the heat pump receives a large amount of heat from it, which is used to heat domestic hot water in the tank. In the summer, warm air is sucked in from outside the building, and the air cooled down in the evaporator can be directed to the inside of the building [11].

3 Research stand

The Hewalex DHW heat pump in the version with an external hot water tank was used for the preparation of research stand. This solution enables the use of storage tank with different water capacity, the method of installing a 2.5 kW DHW heat pump in the research stand is shown in Fig. 2.

Characteristic for this heat pump is to connect the upper source to the circulation circuit in the tank, not to the built-in coil. In the designed research stand, two tanks with a capacity of 130 dm^3 were used. This solution enables testing of a heat pump with variable heat load of the upper heat source.

The designed research stand has been equipped with the EKONTROL system offered by Hewalex. The measuring equipment enables the heat flow to be measured in the heat pump's upper source circuit. The research stand has been equipped with an additional G922-COP measuring module for testing the COP coefficient with access to the database at ekontrol.pl.

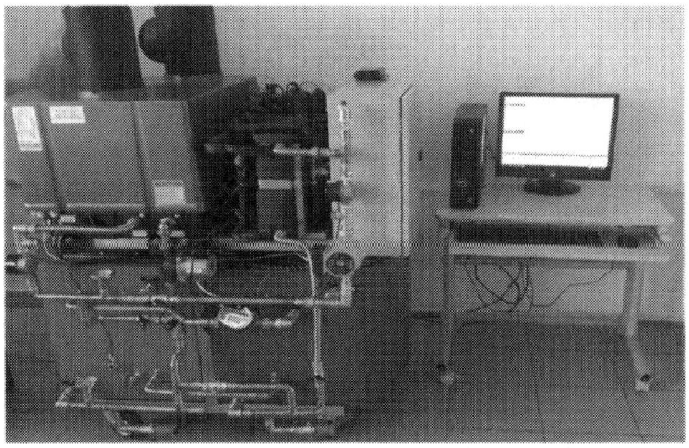

Fig. 2. Installation of a DHW heat pump in a research stand

4 Experimental results

The ventilation air of the laboratory room and outside air were used for the tests. The series of measurements was carried out for cases of heating water in storage tanks of 130 and 260 dm^3 from room temperature to a water temperature of 50°C. The measurements were carried out for 3 speeds of a standard 60 W circulation pump, yielding the average water flow rate, respectively 0.17 kg s^{-1} on 1st gear, 0.24 kg s^{-1} on 2nd gear and 0.27 kg s^{-1} on 3rd gear.

Fig. 3 shows changes in registered temperatures in the heat pump cycle for the use of outdoor air at 7°C for heating water in a 260 dm^3 storage tank to 50°C (A7/W30-50 profile) and setting the circulation pump to 3rd gear.

The increase in the temperature of the water in the storage tank gradually forced an increase in the temperature of water in the heating circuit, so that the temperature of condensation also gradually increased. As the temperature in the heating circuit increased, the thermal load of the heat pump has been gradually reduced, which is evidenced by the recorded increase in the evaporating temperature of the refrigerant at a constant air temperature at the inlet to the evaporator [10].

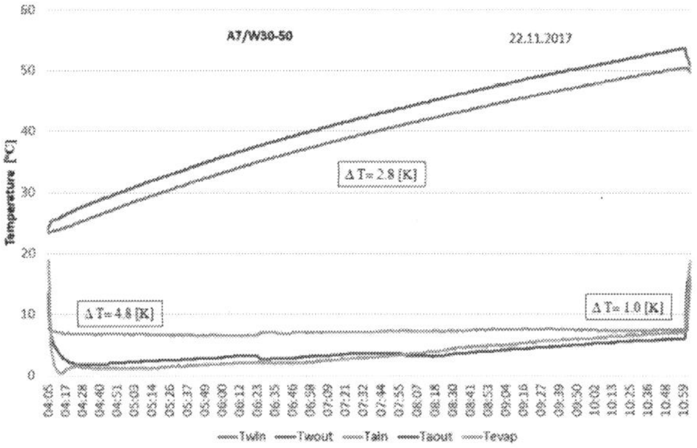

Fig. 3. Graph of temperature changes in the heat pump for outside air circulation of 7 °C and water mass flow 0,27 kg s^{-1} for a storage tank of 260 dm^3

On average, 20 minutes after the heat pump was switched on, the heat pump's operating conditions stabilized. In this case, the evaporating temperature of the refrigerant was 1.1°C at the beginning of the heating cycle and gradually increased to 7°C at the end of the heating cycle.

Temperature difference at the inlet and outlet of the water in the heating circuit for the water mass flow 0.27 kg s^{-1} amounted to ΔT_w=2.8K on average, while in the evaporator circuit it has changed from ΔT_a=4.8°K to ΔT_a=1K.

For the A7/W30-50 power supply profile, the heat pump produced 21.7 kWh of heat energy, consuming 5.8 kWh of electricity, and the average COP was 3.75. The registered measurement results show that the evaporating temperature of the refrigerant for stabilized operating conditions is close to the temperature of the air at the outlet from the evaporator.

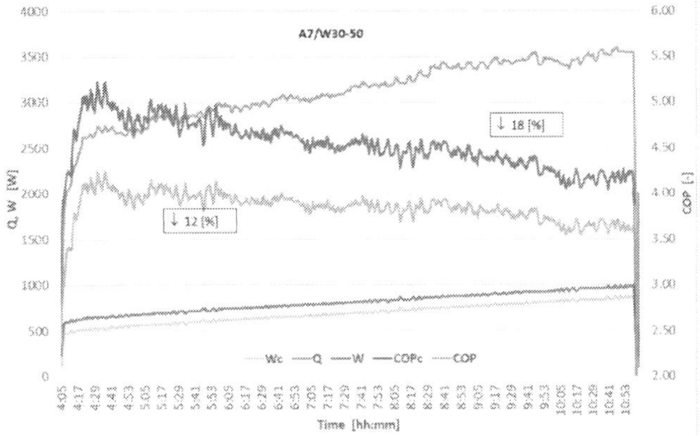

Fig. 4. Diagram of changes in the demand for electric power and registered heating power in the heat pump for the external air circulation of 7 °C and water flow rate of 0.24 kg s^{-1}

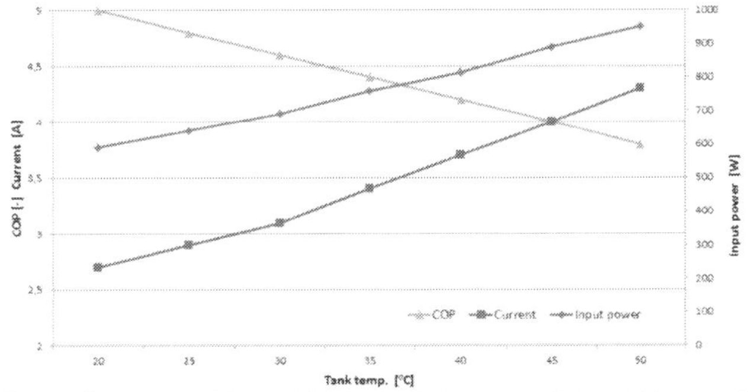

Fig. 5. List of recorded current values, electrical power consumption and COP for ventilation air 21°C

Momentary COP and heating power values were determined for each measuring cycle. Fig. 4 presents data registered in the G922-COP module. The demand of a rotary compressor for electric power varied in the range of 510 ÷ 860 W depending on the condensing temperature. In fixed conditions, the heat pump operated at a maximum heating power of 3520 W. As a result, COP values for the compressor were obtained in the range from 5.10 to 4.25 for stable conditions of the heat pump, calculated COP for the A7/W30-50 profile was 4.55. The demand of the circulating pump and heat pump control system for electric power was 120 W. Thus, the heat pump's COP for the A7/W30-50 profile dropped to 3.80 (a decrease by 16%) [7].

Fig. 5 shows how the demand for electric power and COP of the heat pump changes depending on the temperature of the water in the tank. The results of the measurements show that the air heat pump achieved the highest COP = 5.00 in the first phase of heating the water in the tank. The current consumption by the compressor was below 3 A, and the demand for electrical power was 600 W. In the final phase of heating, the efficiency of the heat pump fell to COP = 3.60, and the demand for electrical power increased to 950 W.

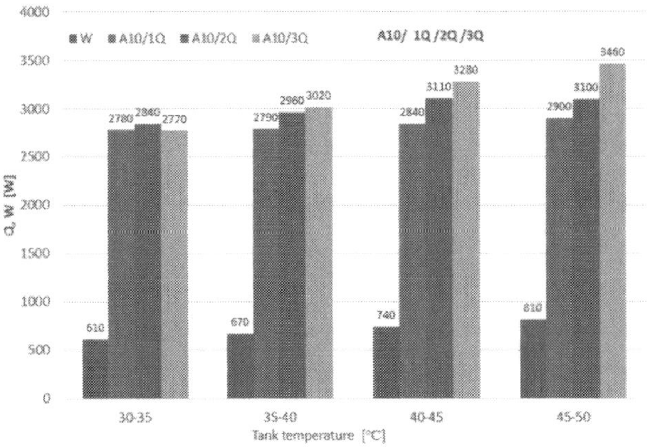

Fig. 6. Comparison of the average values of heating powers for selected water temperature ranges in the tank

At full heat load, the heat pump achieves maximum efficiency in the range of higher water temperatures in the heating circuit (increase by 25%). The air heat pump obtained for the A21/W30-50 profile the highest COP=4.55 in the first phase of water heating in the storage tank for the water mass flow 0.27 kg s^{-1}, and the demand for electrical power was 610 W. In the final heating phase, efficiency of the heat pump dropped to the value of COP=3.95, and the demand for electric power increased to 810 W. The values averaged for the whole measurement cycle were: heating power generated at 3440 W, with the demand for 840 W of electric power, one obtained COP=4.10.

5 Conclusion

During the laboratory tests, the measurement data were recorded on an ongoing basis in the EKONTROL system database. In the heat pump control system, the compressor start-up block is automatically activated if the temperature of the air supplying the lower source falls below 0 °C, due to the low efficiency of the device. This means that for negative ambient temperatures, the pump's operation with the use of outside air is ineffective. Pump operation in winter is possible with the use of ventilation air, but with the discharge of cooled air outside the building.

The research shows that heating the water in the 130 dm^3 storage tank with the A21/W30-50 profile lasted an average of 130 minutes, and the heat pump consumed about 2.2 kWh of energy per one heating cycle. Heating water in a 130 dm^3 storage tank from 25°C to 40°C lasted an average of 60 minutes, while heating water for another 5 °C took 30 minutes. The operation of the heat pump in the higher temperature of the water in the storage tank is less effective, making the heating of water in the tank above 50°C is uneconomical.

With 1 kWh of electric energy collected by the compressor, the presented heat pump is able to generate 3.6 kWh of heat energy in 75 minutes for the water mass flow of 0,17 kg s^{-1}. For the water mass flow of 0,27 kg s^{-1}, from 1 kWh one may generate 4,5 kWh of heat energy in 68 minutes.

Domestic hot water heat pumps may replace the use of solar collectors, however, it is important to remember both their advantages and limitations. The lowest operating costs are characteristic for a solar installation, which is also a less complex device, with proven reliability and durability. Heat pumps are also covered by shorter warranty periods, which, depending on the manufacturer, range from 2 to 5 years (for solar collectors it is usually 10 years).

An air heat pump for DHW is an ideal solution for heating systems equipped with solid fuel boilers. It is a maintenance-free and economical system for producing domestic hot water outside the heating season.

References

1. Anifantis A.S., Performance assessment of photovoltaic, ground source heat pump and hydrogen heat generator in a stand-alone system for greenhouse heating, Chemical Engineering Transactions, **58**, pp. 511-516, (2017)
2. Dong X, Tian Q, Li Z., Experimental investigation on heating performance of solar integrated air source heat pump. Applied Thermal Engineering, **123**, pp. 1013-1020, (2017)
3. Guo J., Wu J., Wang R., Li S., Experimental research and operation optimization of an air-source heat pump water heater, Applied Energy, **88**, pp. 4128–4138, (2011)
4. Jiang M., Wu J., Wang R., Research on the control laws of the electronic expansion valve for an air source heat pump water heater, Building and Environment, **46**, pp. 1954-1961, (2011)
5. Jung H., Kang H., Yoon W., Performance comparison between a single-stage and a cascade multi-functional heat pump for both air heating and hot water supply, Journal of Refrigeration, **36**, pp. 1431-1441, (2013)
6. Lachman P., Heat pumps in hybrid systems, InstalReporter, **4**, pp. 27-30, (2014)
7. Miara M., Efficiency of heat pumps in real conditions of use, InstalReporter, **2**, pp. 12-16, (2011)
8. Peng J., Li H., Zhang C., Performance comparison of air-source heat pump water heater with different expansion devices. Applied Thermal Engineering, **99**, pp. 1190-1200, (2016)
9. Piwowarczyk Sz., Air heat pump for hot water, InstalReporter, **9**, pp. 40-46, (2014)
10. Szreder M., Investigations into the influence of functional parameters of a heat pump on its thermal efficiency, TEKA Commission of Motorization and Energetics in Agricultures, **13**, pp. 191-196, (2013)
11. Szreder M., A field study of the performance of a heat pump installed in a low energy house, Applied Thermal Engineering, **71**, pp. 596-606, (2014)

Energy efficiency of ground-air heat exchanger in the ventilation and airconditioning systems

Rafał Łuczak[1,*], *Bogusław* Ptaszyński[1], *Zbigniew* Kuczera[1], and *Piotr* Życzkowski[1]

[1]AGH University of Science and Technology, Faculty of Mining and Geoengineering, Mickiewicza 30 av, 30-065 Cracow, Poland

Abstract. In the article, analysis of heat exchangers, working together with air-conditioning system, are presented. For an object with known requirement to hot and cold, air heat exchanger (ground type) is designed. For that defined system, the energy analysis of heat exchanger's energy work in yearly cycle, including a work of air treatment with full (cooling – desiccation and heating of air in the summer, heating and moisturizing in the winter) and not quite full (cooling of air in the summer, heating of air in the winter) air-conditioning are examined. Effects connected with a reduction of energy costs needed for heat treatment of air blown to the room are specified included the climatic conditions like air heating and cooling degree-hours.

1 Introduction

Nowadays, it can be seen a dynamic trend to minimized the building energy consumption. Passive and zero-energy buildings are more and more popular. It is connected with not only economy reasons, connected with building exploitation costs, but also paying an attention to environmental protection and reducing the emission of hazardous ashes and gaseous to the atmosphere. Limitation of energy consumption is possible, for example, by the increasing of heat insulation of building barrier, which reduced a heat loss by permeation. Buildings with energy-efficient central heating, ventilation and air-conditioning systems, as well as modernization of existing installations allow to reduce the costs of building exploitation. Greater awareness of users is connected with needed of the indoor air quality improved, which is associated with the comfort of staying indoors. Rooms for people, protect human against the weather conditions, but also create increased opportunities for adverse chemical and physical changes in the air. Air, which is supplied from the outside, does not meet the required parameters and it is necessary to pre-treat it before. The right comfort indoor is obtained by used air removing and providing fresh air with the right temperature and humidity. Building ventilation can cause the heat losses, and to reduce them air conditioning units with heat recovery were used. Such system can work with renewable energy sources or can be supported by air heat exchangers (ground type).

The aim of a article is to present the applicability of air heat exchanger (ground type) in the air conditioning system. The type of air-conditioning system has affect on the

[*] Corresponding author: rluczak@agh.edu.pl

arrangement and construction of the ground heat exchanger. In the article the air treatment processes in the centralized system CAV (Constant Air Volume) and in the two-stage system for air treatment Multisplit are described. The main goal of this article is a presentation of the necessary arguments when the decisions about usage of air heat exchanger (ground type), are taken into account. Examples (I-IV variants) allow to make a right decision at the phase of the project or modernization of the existing ventilation and air-conditioning, including the effectiveness of the applied solution.

2 Usage of air heat exchanger (ground type)

Air heat exchangers (ground type) are recommended in low-energy buildings or in buildings with air ventilation systems. The operation of the air heat exchanger (ground type) depends on heating and cooling of fresh air, which is supplied to the building. In the summer time (warm period), air in the ground exchanger is chilled out and after this process it is directed to the ventilation central. In the winter time, air in the ground exchanger is warmed and after this process it is directed to the heat exchanger or the mixing chamber in the ventilation central, where this air is additional heated from the air, which is removed from the room. An additional function of the ground heat exchanger cooperating with the ventilation or air-conditioning central is the supply air treatment and protection of a cross-flow or rotary exchanger from frosting in the winter time. Air flow through a recuperator in the central would be disadvantageous in the summer time, so this is the reason why those devices are equipped with bypassing the heat exchanger [2]. Air heat exchangers (ground type) as the open type exchangers, are used to support air treatment processes in mechanical ventilation systems. They work as the maintenance-free systems and require only seasonal cleaning. The air heat exchangers (ground type) are used in the ventilation central of single-family houses with relatively small supply air flow. Currently, usage of air heat exchangers (ground type) has been extended to the large-scale buildings and can be exploited in multi-family houses, office and commercial buildings (eg. supermarkets) and schools [8]. The heat flow, that can be exchange with the ground, depends on soil properties, its composition, humidity and size of the flowing air stream. According to [3], the source of heat, which is stored in the ground up to 10 m under its surface, is a solar radiation, so it is the reason why the ground must be cyclically regenerated and a heat from the depth of the earth does not have an influence on the operation of the ground exchanger. Because of increasing costs, the horizontal collector installations are made up to 2 meters deep. In Poland, at this depth a seasonal temperature changes are occurred with annual amplitude equal to several degrees Celsius. In the summer time, a ground temperature can achieve 17°C, and in the winter time – 5-6°C. In an example, the temperature of ground varies between 4°C and 12°C. The deeper the exchanger is placed, the higher and less variable the ground temperature is occured. In the polish conditions, a constant temperature equals 10°C is on the deep of 10 m [9].

The heat from ventilation air stream can be exchanged with a ground in the non-permeable and membrane way and with the participation of intermediate agent. In the non-permeable exchangers, air has a direct contact with a ground and we can distinguished gravel, comb, plate and modular exchangers. In the membrane exchangers, air flows, for example, in a plastic pipe or in a concrete channel in the ground do not contact with it directly, whereas in the exchangers with an intermediary factor, to transfer a heat stream between ground and air, glycol solution is commonly used [2, 4].

Depending on the used constructional solutions connected with laying the exchanger in ground, the following types of the air heat exchangers (ground type) can be distinguished [2, 5]:

- gravel exchanger – to the heat exchange, gravel bed with washed gravel and fine stone fraction is used,

- comb exchanger – it is a type of gravel exchanger,
- plate exchanger – air exchanges the heat stream with a ground through the exchanger, flowing between its plates,
- modular exchanger – it is built by modules made of wires with a semi-circular cross-section,
- tubular exchanger – it is built by a single pipe or a system of pipes, which are arranged in various configurations, e.g. in the form of a ring, meander or Tichelmann's system,
- glycol exchanger – air/glycol exchanger is used for heat exchange.

Researches and measurements of many ground-and-air exchangers, both non-permeable and membrane are carried out. Information about ground heat exchangers are available in the literature. The greatest efficiency of the air heat exchangers (ground type) is where the differences between external and internal temperatures are the highest – i.e. hot summer and cold winter. The need of heating the supply ventilation air also exists, for example, in the summer time, when the outside temperature unexpectedly and significantly decrease. The task of the ground heat exchanger is to reduce the variability of air parameters within a day [1]. Fig. 1 presents three most commonly used tubular exchangers: Tichelmann's system, meander and ring.

Fig. 1. Tubular Exchange, a – Tichelmann's system, b – meander system, c – ring system [6,8]

3 The air heat exchange (ground type) selection

Selection on several variants of ground exchangers are made for air-condition installation, working in the object which is designed for office purposes, in which 20 people may be presented at the same time. The usable area of the building is F=200 m^2, and its cubic capacity is V=600 m^3/s. The design heat load is Q_Z=12.6 kW and the calculated heat gains in the summer time is Q_L=10.16 kW. Assuming that the temperature of air in the object in the summer time is t_P=24°C, and the calculated temperature of outside air for the second climatic zone is t_Z=30°C, the supply air with temperature t_N=18°C, which will assimilate the heat gains in the building, is calculated. If the air-conditioning system works in Constant Air Volume (CAV), the supply air flows on the base of formula (1) will be $\dot{V}$=5040 m^3/h.

$$\dot{V} = \frac{Q_L}{\rho C_P (t_N - t_P)},\ \mathrm{m^3/s} \tag{1}$$

where :

C_P	–	specific heat of dry air [kJ/kgK], C_p=1.005 kJ/kgK,
Q_L	–	heat gains in the summer time [kW], Q_L=10.16 kW,
t_N	–	temperature of supply air [°C], t_N=18°C,
t_P	–	temperature of air in the room [°C], t_p=24 °C,
$\dot{V}$	–	supply air flow [m^3/s],
ρ	–	air density [kg/m^3], ρ=1.2 kg/m^3.

For design heat load and constancy of the supply air stream in the summer and winter time and temperature of air in the building t_p=20°C, after the transformation of formula (1), the temperature of supply air in the winter is calculated t_N=27.5°C (formula (2)).

$$t_N = \frac{Q_Z + \dot{V}\rho C_p t_p}{\dot{V}\rho C_p}, \text{°C} \tag{2}$$

where :

Q_Z – design heat load of the object [kW], Q_Z=12.6 kW,

$\dot{V}$ – 1.4 m³/s

The air-conditioning system is changed because of designated supply air stream (5040 m³/h), at which air exchange rate is 9 h⁻¹. The minimum supply air stream is determined because of hygienic reasons, the value is assumed as 30 m³/h for a person, i.e. total air hygienic stream is $\dot{V}_z$=600 m³/h. As a heat recovery process the air recirculation is used with a 1:1 mixing ratio. The recirculation usage except heat recovery from exhausted air, will allow to dry the air in the summer time, and moisturize it in the winter time. The air treatment processes in the summer and winter are presented in Figs. 2 and 3 respectively. The operation of air-conditioning system with full air treatment is presented in Fig. 2, however, for reduced supply air stream, the air cooling process is ended in the N1 point, in which air temperature is equal to the temperature in the object. Further reducing the air temperature, that lowering the heat stream in the object, takes place in the Multisplit air-conditioning system, which is installed in the building.

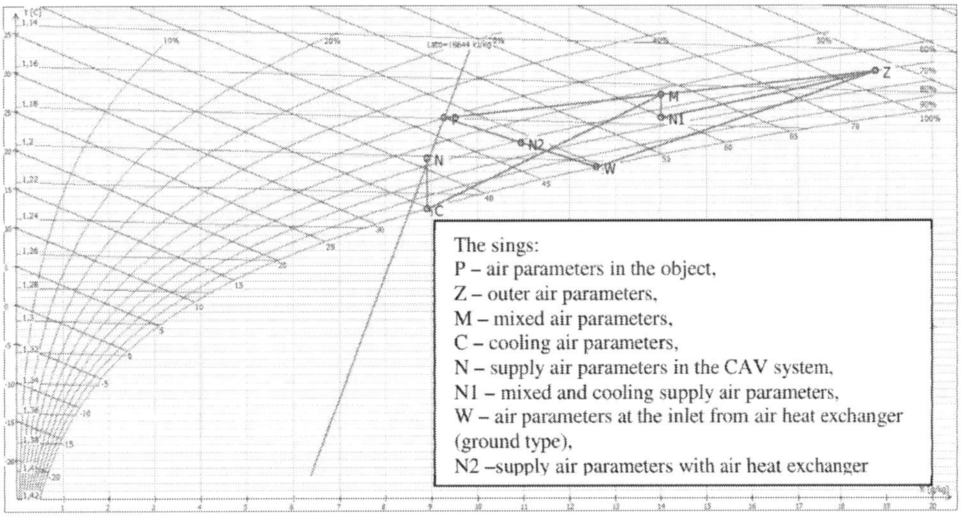

Fig. 2. Air changes in the air-conditioning system in the summer time

In the winter time, the outer air stream (600 m³/h) is mixed with the exhausted air stream, but the resulting M1 mixing point is located in the air zone saturated with moisture. This is a reason that outer air stream should be preheated by 5°C (Z1 point), and after this process it can be lead to the mixing chamber. As a result of preheating, point M2 is achieved, which is a mixture of air, where moisture condensation does not exist. Then, the air is heated (point O) and steam is humidified to the supply point (N) parameters. In order to reduce the costs of air treatment, the process of secondary heating in the air-conditioning central can be limited to 20°C (O1 point) and further heating – compensating for heat losses in the object, can be done in the Multisplit system, that is installed in the building.

For that formulated air-conditioning system in the office object, variants of the air heat exchangers (ground type) are specified. The fresh air stream is sucked in by the ground intake and is flown thorough the ground heat exchanger, where it is pre-heated (in the winter time) and pre-cooling (in the summer time).

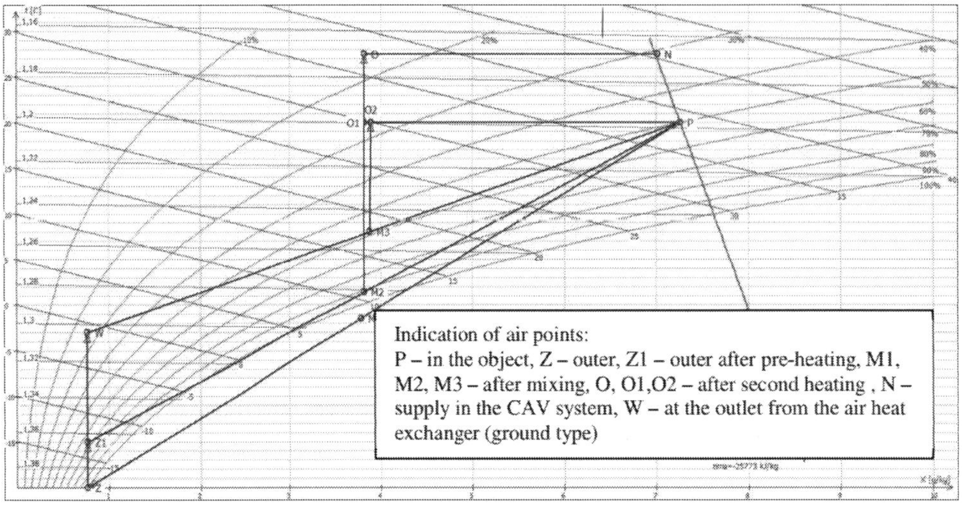

Fig. 3. Air transformation in the air-conditioning system in the winter time

On the basis of the dependence (3), the thermal power of the ground heat exchanger equal 3.82 kW is determined, assuming that air at the outlet from the exchanger has a temperature -3°C.

$$Q_{GPWC} = \dot{V}_Z \cdot \rho \cdot C_p \cdot (t_W - t_Z) \quad , \mathrm{W} \tag{3}$$

where :

Q_{GPWC} – thermal power of the ground heat exchanger [kW],

t_W – air temperature at the outlet from the ground heat exchanger [°C], t_W=-3°C,

t_Z – outer air temperature [°C], t_Z=-20°C

$\dot{V}_Z$ – outer air stream [m³/s], $\dot{V}_Z$ =0,17m³/s.

In the article the usage of the ground heat exchanger for the object, for which the outer air stream is 5040 m³/h, is considered (markets, schools and others). This example shows that the energy stream supplied to air from ground can be equal 34 MWh/a, but this idea entails large dimensions and costs of the exchanger itself. Such a large "thermal gains" from the ground, in the situation of modernization of existing ventilation and air-conditioning systems, reduce the efficiency of the heat exchangers in the central. Therefore, the work efficiency of such systems should be make at the stage of the system design, in which thermal power of air heater and coolers can be reduced.

4 Work analysis of the ground heat exchangers

The ground heat exchanger is an installation for the energy recovery in the ground for heating and cooling of ventilation air. According to [8], for a single-family house with a usable area of approx. 150 m², thermal power of the ground exchangers is 3-4 kW, which is connected with reducing a heater and cooler thermal power. The Rehau Company in its offer, apart from ground heater exchanger for ventilation in single-family houses, has large installations for large-volume buildings, i.e. schools, markets and offices. Table 1 contains technical date for ground heat exchangers placed at the depth of 1,8 m in the ground with general soil characteristic. For comparative purposes, four variants of the ground heat exchanger's work are selected. The first one includes the system, through which 5040 m^3/h of air flows and the exchanger cooperates with the CAV system. The three other systems are connected with working of ground heat exchanger and central air-conditioning system, in which hygienic air stream is treated, while the ground exchanger differs in the type of pipe laying in the ground. Variant I is an installation in the Tichelmann's form consisting 35 pipes with diameter equal 200 mm and a length of 35 m. Separator diameter is 700 mm, and a space between pipes is 0.5 m. In variants II-IV, an outer air stream 600 m^3/h is treated.

Table 1. Results of technical parameters of the ground heat exchangers and an amount of theirs thermal energy and heat stream collected/returned from the ground to the ventilation air

Parameter	Variant I Tichelmann's system	Variant II Tischelmann's system	Variant III meander system	Variant IV ring system
Volume stream of outlet air, m^3/h	5040	600	600	600
Length of exchanger, m	1330	182	120	120
Pipe diameter, m	Separator – 700 mm pipe of heat exchanger – 200mm	Separator – 500 mm pipe of heat exchanger – 200mm	Pipe of heat exchanger – 315mm (4x30m)	Pipe of heat exchanger – 315mm
Average speed of air flow, m/s	1.5	0.9	2.5	2.5
Loss of air pressure in the exchanger, Pa	47.1	4.0	40.7	41.5
Results of calculation for heating variant				
Minimum air temperature at the inlet of the ground heat exchanger, °C	-20.0	-20.0	-20.0	-20.0
Minimum air temperature at the outlet of the ground heat exchanger, °C	-3.0	-3.0	-2.9	-3.0
Maximum heating power, kW	28.7	3.78	3.80	3.78
Energy stream collected to the air, kWh/a	34003.0	4025.6	4060.5	4048.3
Energy stream recived from the air, kWh/a	2261.1	267.2	270.5	269.7
Results of calculation for cooling variant				
Maximum air temperature at the inlet of the ground heat exchanger, °C	31.0	31.0	31.0	31.0
Maximum air temperature at the outlet of the ground heat exchanger, °C	17.7	17.8	17.7	17.7
Maximum cooling power, kW	22.45	2.71	2.73	2.73
Energy stream recived from the air, kWh/a	23217.1	2429.6	2778.1	2769.5
Energy stream collected to the air, kWh/a	1414.8	59.2	169.0	168.5

The analyzed period of the exchanger work is divided into two half-periods, i.e. the heating period from 01.09 to 31.03 and the cooling period from 01.04 to 31.08. The calculation is made by using the Rehau program for selection of ground heat exchangers.

Table 1 summarizes the results of technical parameters of the ground heat exchangers and an amount of energy collected and returned to the ground by air during the whole year.

On the basis of calculation it can be said that, the most effective in terms of heat exchange and construction is the meander exchanger (variant III). The length of the pipe exchanger is 120 m, and a received heat stream in the winter time is 4000 kWh/year, and in the summer time – received cold is 2700 kWh/year.

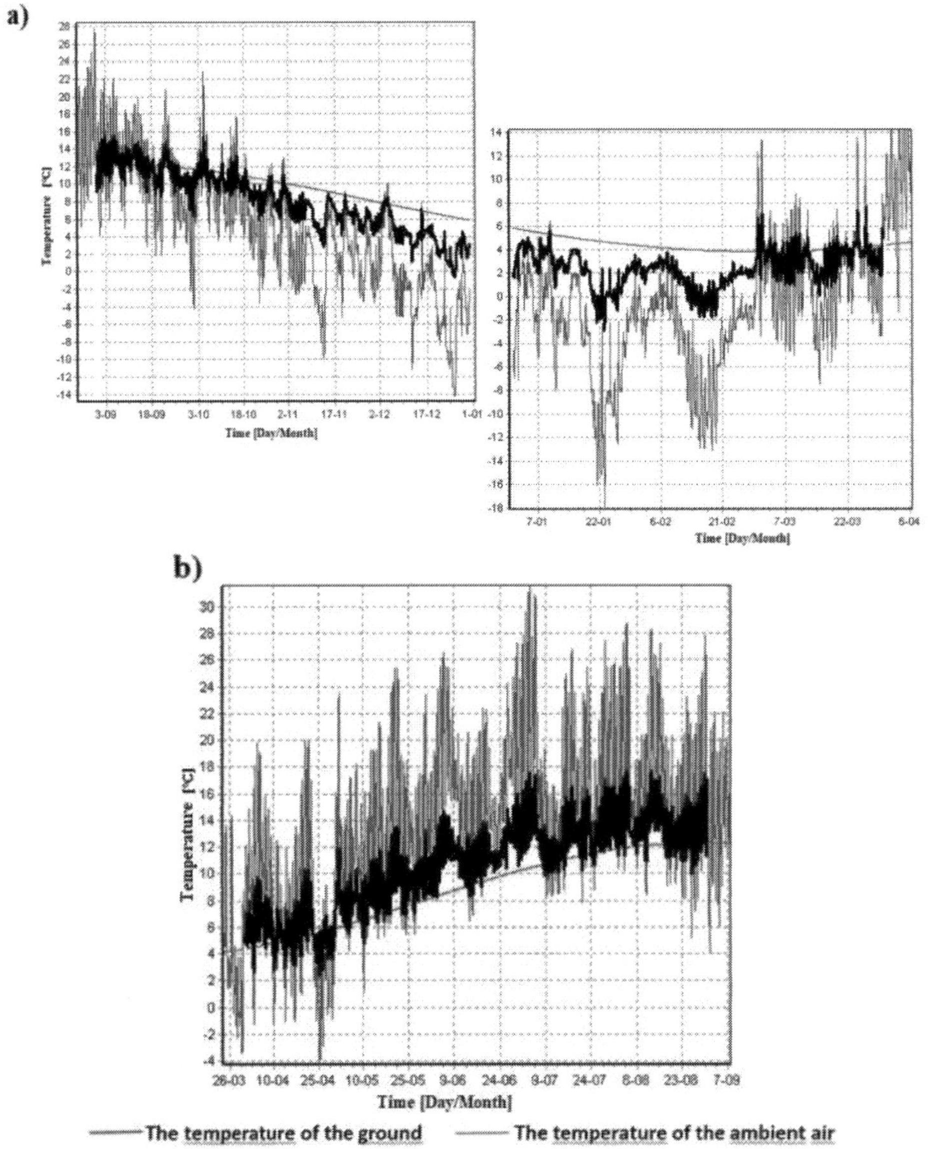

Fig. 4. Temperature distribution of ground, outer air and air from the ground exchanger a) in the winter time, b) in the summer time

In Fig. 4, the ground and air temperature distribution at its treatment in the ground exchanger in the meander system (the winter and summer time) are presented. It shows, that in the winter time, the short periods of time, in which thermal energy from the air stream is returned to the ground (March, September and October) are appeared. These quantities (summarized in table 1) do not affect the formulated application and additionally in this time, the exchanger does not work. In the summer time, it is allowed to switch off the ground exchanger in the situation, when air temperature at the outlet is lower than 5°C. The outside air, after passing through ground heat exchanger and mixing with the exhausted air from the object (mixing ratio 1:1), cannot have a temperature lower than 15°C and this situation determines the microclimate parameters. The ground exchanger usage (variant III) reduces the system demand for heating and cooling power. Taking into account the air parameters, which are determined by the points in Fig. 2, it follows that the thermal power of cooler (conversion Z-N1) is 3.92 kW and the cooling power of Multisplit system is 10.16 kW. In the situation, when additional air recirculation is used (processes P-Z-M-N1), the thermal power of cooler is reduced to 1.2 kW at the constant cooling power of the Multisplit system. Connection to the air heat exchanger (ground type) (process Z-W) makes that the outside air is cooling and drying and can reach the maximum temperature equal 17.6°C. The air temperature after mixing (process P-W-N2) is 20.8 °C, and this situation causes that the cooling power of the Multisplit system at the air cooler elimination is reduced (up to 8,89 kW).

In the winter time (Fig. 2), a work of the air-conditioning system with the air heat exchanger (ground type) (conversion Z-W-P-M3-O2) eliminates the usage of pre-heater, causes air humidification, its heating (point M3) and the heater power reduction (process M3-O2) to 4.75 kW at the constant heating power of the Multisplit system equals 12.6 kW. The cost of generating 4000 kWh of energy by using electricity in a 24-hour tariff is 2800 PLN. On the other hand, the cold production in the amount of 2700 kWh is 540 PLN for the Multisplit system and in the case of ice water is equal to 1200 PLN. It follows that the usage of exchanger brings annual savings on the level 3300-4000 PLN.

5 Conclusion

The air heat exchangers (ground type) are used in the ventilation and air-conditioning systems not only for small single-family houses but also are available for air-conditioning of large-area objects. However, the large air stream treatment requires a considerable area for the exchanger installation and long pipes, which increases the costs of investment. If the installation area is not a problem, it is appropriate to consider the consequences of exchanger cooperation with air-conditioning system, because of it is possible to avoid its overdimensioning.

The reduction of the outer air stream flowing through the ground exchanger allows for a significant reduction of the size of the exchanger, its surface, the length of pipes and the costs of construction. It can be used for smaller objects as it is discussed in the article. Usage of the exchanger contributes to reduce of energy consumption on heating and cooling of air stream and reduce of the thermal power of ventilation and air-conditioning systems. The ground heat exchangers are the most effective with the air recirculation process in the air-conditioning system and can reduce the annual air treatment costs about 3300-4000 PLN.

This research study was supported by the AGH University of Science and Technology statutory research No. 11.11.100.920

References

1. Besler M., Cepiński W., Fijewski M. Uzdatnianie powietrza w wymienniku gruntowym dla pomieszczeń o różnych wymaganiach. Rynek instalacyjny, pp. 38-40, styczeń/luty (2015)
2. Chmielewski K., Amanowicz Ł. Bezprzeponowe powietrzne gruntowe wymienniki ciepła w układach wentylacji mechanicznej. Rynek instalacyjny, pp. 76-80, maj (2017)
3. Sikora K. Pomiary temperatury w gruncie oraz w gruntowym wymienniku ciepła (GWC) w rocznym cyklu eksploatacyjnym. PAK, **56**, 3, pp. 265-267, (2010)
4. Szymański M., Wojtkowiak J. Analiza całorocznej pracy rurowego gruntowego wymiennika ciepła RGWC w układzie wentylacji mechanicznej budynku mieszkalnego. Ciepłownictwo, ogrzewnictwo, wentylacja, **11**, pp. 36-38, (2008)
5. Topolańska J., Krawczyk D. Gruntowe powietrzne wymienniki ciepła. Przegląd stosowanych rozwiązań. Rynek instalacyjny, pp. 78-81, kwiecień, (2014)
6. http://geoheat.eu
7. http://gruntowe-wymienniki.pl
8. https://www.rehau.com/pl-pl
9. http://termomodernizacja.pl/grunt-jako-dolne-zrodlo-energii-w-pompie-ciepla

Energy optimization of mechatronic systems with PMSG

Dariusz Tarnapowicz[1,*], and *Sergiej* German–Galkin[1]

[1]Maritime University of Szczecin, Wały Chrobrego 1-2, 70-500 Szczecin, Poland

Abstract. Currently, the improvement of efficiency in electricity systems is one of the key tasks facing the industry. Due to the energy criterion, the aim of the optimization of mechatronic systems is to reduce an adverse impact of the armature in the synchronous machine and limit losses in supply lines. In recent years, autonomous generating sets with Permanent Magnet Synchronous Generator (PMSG) and ecological LNG engines are more and more often installed in power systemsThe article presents the concept of using a voltage inverter operating in the power source mode in a parallel connection with PMSG for distribution and regulation of active and reactive power

1 Introduction

Energy issues are decisive when designing a power plant in autonomous facilities, especially uninhabited facilities. For the reasons of optimization, various energy indicators should be taken into account, e.g. increasing system efficiency, regulation of reactive power, ensuring electromagnetic compatibility with load and power grid, etc. These issues are solved in practice thanks to advances in electromechanics, and well as semiconductor and microprocessor technologies. Nevertheless, the analysis of the problem shows that in modern energy systems, there are still unused reserves for the increase of energy efficiency when converting electromechanical energy. The greatest possibilities to increase the efficiency of electromechanical energy conversion can be obtained thanks to the optimization of operating modes of the synchronous generator. This article is devoted to solving the problem of optimization, which realizes energy-saving modes of synchronous generator's operation, in which the adverse reaction of the generator armature is limited and the efficiency of the supply lines is increased.

One of the examples of the discussed mechatronic system may be the "Shore To Ship" (STS) system – power supply of ships from land during a berth at the port. The STS system enables to reduce exhaust emission in the port and limit the costs of electricity on the ship [1],[2]. The source of electric power can be an autonomous generator set with PMSG driven by LNG gas engine [3]. The generating set is usually located far from the quay with a variable load – ship. This is caused by a limited possibility of interference in the port infrastructure.

* Corresponding author: d.tarnapowicz@am.szczecin.pl

The functional scheme of the mechatronic system of a PMSG and an autonomous inverter (AI), describing in the article, is presented in Figure 1. In this case, the CSAI block is the autonomous inverter control system, and the ARL block represents the active and passive load with various parameters.

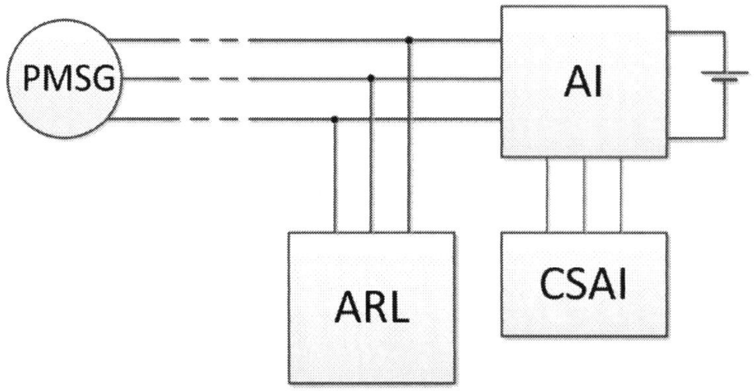

Fig. 1. Functional scheme of the PMSG and AI mechatronic system

The aspect presented in the paper is similar to the issue of using STATCOM (STATic COMpensator) and active filters. The difference is that compensation of reactive power is not the main objective, as is the case in STATCOM, but the compensation of an adverse impact of the PMSG armature.

To solve this problem it is necessary to make the analysis of the system consisting of three elements: PMSG, AI and ARL. This analysis is based on the classical theory [4],[6],[9] which arose before the development of technologies of power electronics.

The basis of analytical methods for the analysis of mechatronic systems is the theoretical study by A. Bulgakov [4], which is based on the fact that the electromagnetic and energetic properties of the inverter are determined when transferring the power from the inverter to the load. This is realized at the basic frequency, which is determined by the angular speed of the machine shaft, and the exchange of energy between passive elements takes place at the carrier frequency of transistors switched in the inverter. This exchange is accompanied by the pulsation of currents in the system, causing a slight distortion of current and additional (usually small) losses. On the basis of this analysis, typical working modes for the entire system can be described and examined.

The basis for the mathematical description of electromagnetic processes in the system (based on the basic frequency), is the spatial vector method [5], which enables the presentation of all state variables in spatial vectors in one coordinate system.

The mathematical conversion of real state variables for the system in the values of state variables in axes *d, q* is presented by equations (1) [6], [7] for electromotive force of the machine (similarly, conversion equations are realized for voltages and currents).

$$\begin{bmatrix} e_d \\ e_q \end{bmatrix} = \frac{2}{3} \begin{bmatrix} \cos\omega t & -\frac{1}{2}\sin\omega t & -\frac{\sqrt{3}}{2}\sin\omega t \\ -\sin\omega t & \frac{1}{2}\cos\omega t & \frac{\sqrt{3}}{2}\sin\omega t \end{bmatrix} \cdot \begin{bmatrix} e_A \\ e_B \\ e_C \end{bmatrix} \tag{1}$$

During the construction of the CSAI control block (designed to operate with a variable ARL load), the superior control of currents *Id, Iq* of the inverter is realized [8]. Electromagnetic processes in such a system can be calculated with the use of a scheme of the system shown in Figure 2.

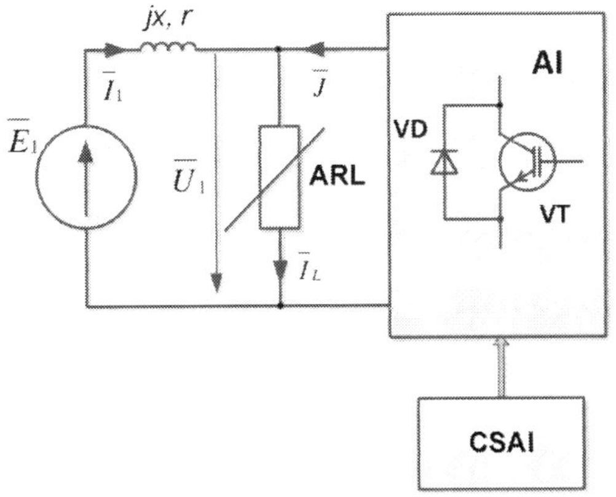

Fig. 2. Scheme for the GS and AI system with variable ARL load

PMSG was analyzed. The parameters of PMSG are presented in Table 1.

Table 1. Dew point temperature for different values of vapour pressure.

PMSG Parameter	$\omega_{m.Nom}$	$U^*_{1.Nom}$	r	L	• •	p	m
unit	*rad/s*	*V*	*Ohm*	*mH*	*Wb*		
volume	78.5	310	0.05	0.64	0.987	4	3

In the scheme presented in Figure 2, the autonomous AI inverter is a controlled current source, which is connected in parallel to the ARL load and the PMSG electromotive force . Inductive reactance x and resistance r in the scheme are internal parameters of the generator and transmission supply line.

The mathematical description of the scheme from Figure 2 can be presented as:

$$\bar{E}_1(t) = \bar{U}_1(t) + L\frac{d\bar{I}_1(t)}{dt} + r\bar{I}_1(t) + jx\bar{I}_1(t),$$
$$\bar{I}_L(t) = \bar{I}_1(t) + \bar{J}(t) \quad (2)$$

where :

- $\bar{E}_1$ – EMF vector of PMSG,
- $\bar{U}_1$ – voltage vector on ARL load,
- $\bar{I}_1$ – current vector in the PMSG armature winding,
- $\bar{J}$ – set current vector for the AI inverter,
- $\bar{I}_L$ – current vector in the ARL load,
- r, x – resistance, reactance of stator windings.

The EMF value at the PMSG output is determined by the rotational speed of the machine shaft and the magnetic flux:

$$\bar{E}_1(t) = p\omega_m(t)\cdot\bar{\Psi}_0 = \omega(t)\cdot\bar{\Psi}_0 \quad (3)$$

where :

p – number of pole pairs,

□$_m$ – generator shaft speed,

$\overline{\Psi}_0$ – vector of magnetic flux associated.

The mathematical description of the system (2) in the coordinates *d, q* can be presented as follows:

$$\begin{aligned} E_d &= U_{1d}(t) + L\frac{dI_{1d}(t)}{dt} + rI_{1d}(t) - xI_{1q}(t), \\ E_q &- U_{1q}(t) + L\frac{dI_{1q}(t)}{dt} + rI_{1q}(t) + xI_{1d}(t), \\ I_{Ld}(t) &= I_{1d}(t) + J_d(t),\ \ I_{Lq}(t) = I_{1q}(t) + J_q(t) \end{aligned} \tag{4}$$

Currents $I_{1d}(t)$, $I_{1q}(t)$ and voltage $U_{1q}(t)$ reflect the generator current $I_l(t)$ and $U_l(t)$ in axes *d, q*.

The quantities in equations (4) depend on time only in transient states. In the steady state, the system of equations (5) transforms into the form:

$$\begin{aligned} E_d &= U_{1d} + rI_{1d} - xI_{1q}, \\ E_q &= U_{1q} + rI_{1q} + xI_{1d}, \\ I_{Ld} &= I_{1d} + J_d,\ \ I_{Lq} = I_{1q} + J_q \end{aligned} \tag{5}$$

The presence of a controlled current source by the change in amplitude and phase makes it possible to compensate the PMSG reactive current and (if necessary) the active current (partially), and thus optimize the efficiency of the entire system.

The examination of electromagnetic processes determined in a certain system is realized on the basis of the vector analysis for two cases:

- In the first case, the inverter compensates inductance impact of the PMSG armature realized with the use of reactive power compensation. The vector diagram for this case is shown in Figure 3, and equations (5) with $x>>r$ are transformed into the form:

$$\begin{aligned} 0 &= U_{1d} - xI_{1q}, \\ E_1 &= U_{1q} + xI_{1d}, \\ I_{Ld} &= J_d, I_{Lq} = I_{1q} \end{aligned} \tag{6}$$

  In this case, only active power is taken from the machine. This minimizes the armature reaction, and reactive power is taken from the inverter. It should be noted that in this case, the inverter acts as an active filter, and a capacitor may be used instead of a DC source in the inverter's power supply circuit [9], [10], [11].
- In the second case (Figure 3, b), control currents are realized so that $I_{ld}=0$, a $I_{Lq} > I_l$. In the second case the inverter compensates inductance and resistance impact of the PMSG armature realized with the use of reactive power compensation and partially also active power compensation. Then the inverter additionally generates active power for the load and it should be powered from a DC power source. The mathematical description of the system in the case has the following form:

$$
\begin{aligned}
0 &= U_{1d} - xI_{1q}, \\
E_1 &= U_{1q} + xI_{1d}, \\
I_{Ld} &= J_d, I_{Lq} = I_{1q} + J_q
\end{aligned}
\tag{7}
$$

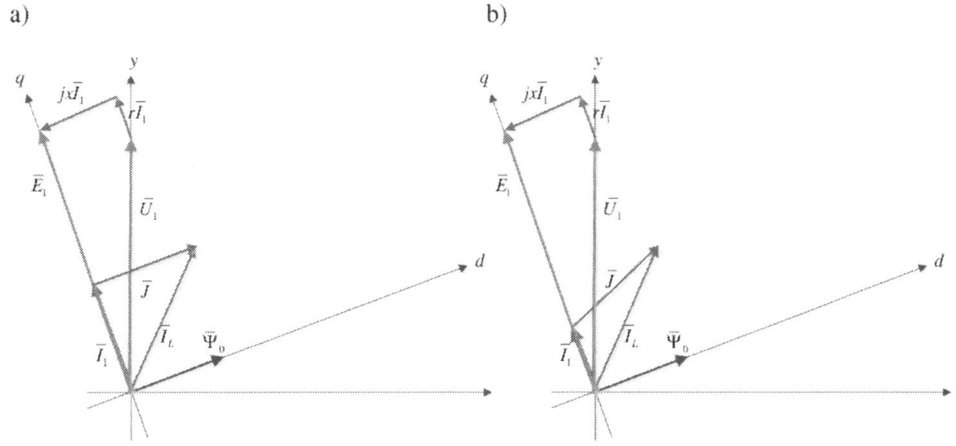

Fig. 3. Vector diagrams for PMSG and AI system with variable ARL load

Simulation researches for the mechatronic system with PMSG carried out in the Matlab-Simulink environment for the above-mentioned cases. Figure 4 presents a virtual simulation model of the system with variable ARL load.

The results of the system simulation for the considered cases are presented in Figure 5, Figure 6, Figure 7 and Figure 8.

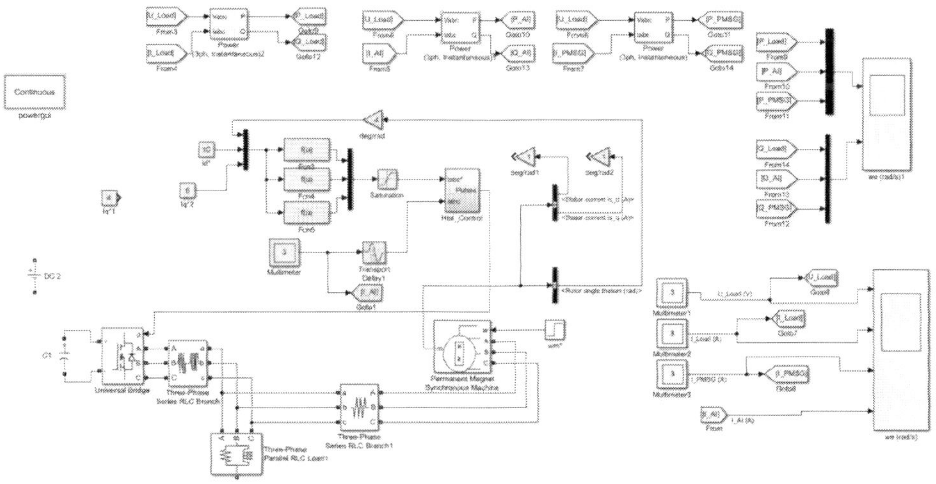

Fig. 4. Model of the mechatronic system with PMSG

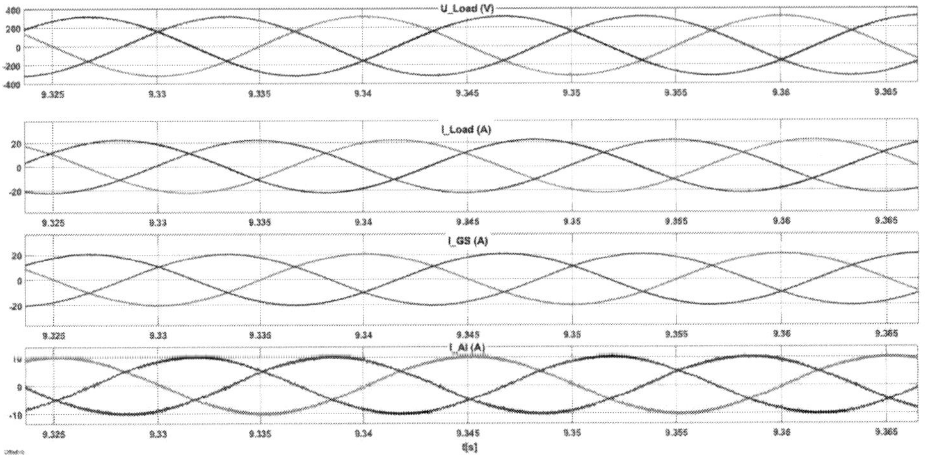

Fig. 5. Electromagnetic processes in a system with compensation of reactive power

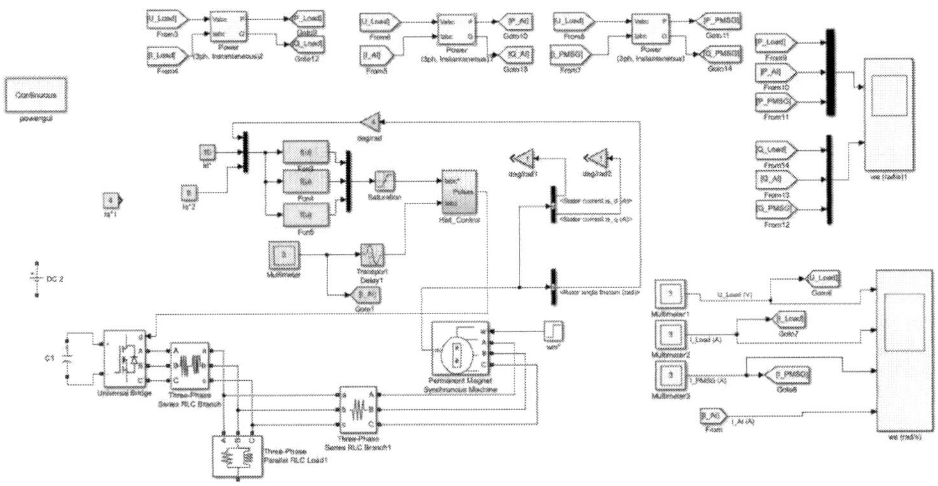

Fig. 6. Energy processes in a system with compensation of reactive power

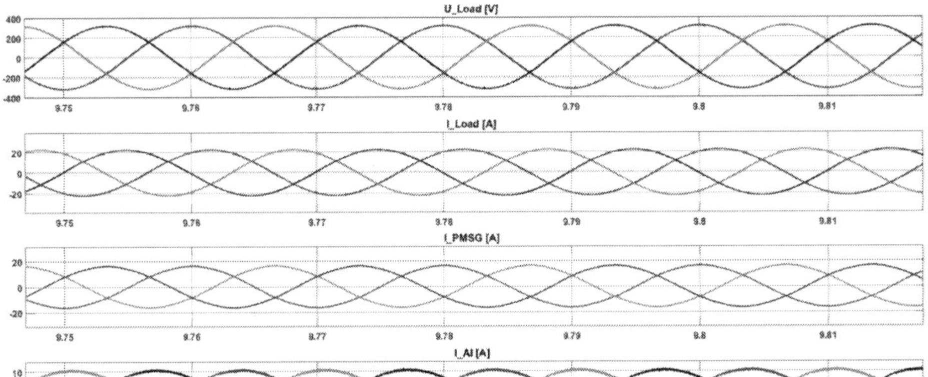

Fig. 7. Model of the mechatronic system with PMSG

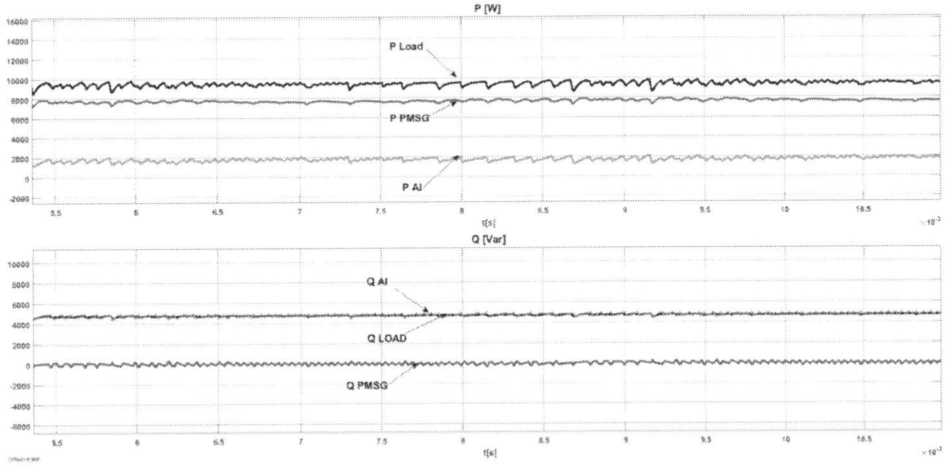

Fig. 8. Energy processes in the PMSG-AI system with compensation of reactive and active power

2 Conclusion

In recent years, autonomous generating sets with ecological LNG engines (example STS system) are more and more often installed in power systems. The use of a parallel active filter in the mechatronic system with PMSG (presented in the article) enabled the energy optimization of the synchronous machine and the power line. The results of simulation tests for the system confirmed the theoretical assumptions and the possibility of adjusting reactive and active power provided by PMSG.

This research outcome has been achieved under the research project: Nowoczesne technologie w systemach "Shore to Ship" No 2/S/IEiAO/16 financed from a subsidy of the Ministry of Science and Higher Education for statutory activities.

References

1. Borkowski, T., Nicewicz, G. and Tarnapowicz, D., The methodology used in defining air pollution from ships mooring in ports. Scientific Journals of the Maritime University of Szczecin. Poland, 2013, **36,** 2, pp. 17-22, (2013).
2. Nicewicz, G., Sosinski, M., Tarnapowicz, D. Identification of power factor in marine electrical grid. *International Multidisciplinary Scientific GeoConference-SGEM. Bulgaria*, pp. 391-398, (2014)
3. Tarnapowicz, D.,German-Galkin, S:. The use of generating sets with LNG gas engines in "Shore to Ship" systems. Management Systems in Production Engineering, Poland, **3**, 23, pp. 172-177, (2016)
4. Bulgakov A., A.: Novaya teoriya upravlyayemykh vypryamiteley. - M.: Nauka, Russia, pp 326, (1970)
5. Kovacs, K., Rac, I., Theil, H. Transiente Vorgänge in Wechselstrommaschinen, Verlag der Ungarischen Akademie der Wissenschaften, Budapest, Hungary, (1959)
6. Park, R.H.Tho-reaction theory of sinchronous machines. Transactions of the American Institute of Electrical Engineers, USA, **48**, pp. 716 – 727, (1929)
7. Gorev A.A. Transient processes of a synchronous machine. Leningrad Nauka, Russia, **502**, (1985)
8. German Galkin S. Virtual Laboratory of semiconductor systems in Matlab-Simulink wednesday. «Lan», St. Petersburg, Russia, (2013)

9. Akagi H. et al.: Generalized theory of instantaneus reactive power in three-phasecircuits. Conf. Rec. IPEC'83, Tokyo, (1993)
10. Kim, H, Blaabjerg, F., Back-Jensen, B. Instentaneous power compensation in three - phase systems using p-q-r theory. IEEE Transactions on Power Electronics, USA, **17**, 5, (2012)
11. Peng, F. Z., Lai, J.S. Generalized instantaneous reactive power theory for three-phase power systems," IEEE Transactions on Instrumentation and Measurement, USA, **45**, pp. 293 – 297, (1996)

Estimation of borehole heat exchanger thermal properties

Jerzy Wołoszyn[1,*]

[1]AGH University of Science and Technology, Department of Power Systems and Environmental Protection Facilities, Mickiewicza 30 Av. 30-065 Cracow, Poland

Abstract. An important issue in the design of installations with borehole heat exchangers (BHE) is knowledge of the thermo-physical properties of the grouting material and soil. The objectives of this work is to estimate the thermal conductivity and specific heat of the grouting material and the soil. The estimated values were determined on the basis of two numerical heat transport models in the borehole and using the results of the small sandbox experiment. The design of experiment technique (DOE) and the response surface methodology (RSM) were used to achieve the aim of this work. Numerical calculations were carried out with the use of the new finite element with multiple degrees of freedom (MDF) and a quasi-3D model in ANSYS package. The estimation requires minimization of four output parameters (quality criteria), and therefore it is a multiple-objectives optimization problem. Based on the DOE, RSM and multiple-objective optimization the material properties of BHE and soil was determined. For both models the discrepancies $Y1_{P1}$ are below 4 %, $Y2_{P1}$ are below 11.5%, $Y1_{P2}$ are below 6.5 % and $Y2_{P2}$ are below 11.5 %. The average discrepancies below 5 % based on the verification measurements with different operating parameters was obtained. It is noteworthy that for the two different models the same values of the estimated parameters were obtained.

1 Introduction

The ground source heat pump system (GSHP) is a commonly utilized renewable technology for heating and cooling of buildings [9]. The borehole thermal energy storage systems (BTES) also take on importance. To design such systems, the knowledge of the thermal properties of ground and BHE is crucial. Developments during the 1980s include design software which helped overcome under sizing problems. Current design software helps to obtain the needed inputs, especially the thermal conductivity of the ground. In paper by Mogensen [7] a method by which thermal conductivity of the ground and borehole resistance could be obtained in situ was presented. We now call it thermal response test (TRT). Further development of TRT was conduct in USA at the Oklahoma State University and in Sweden at Luleå University of Technology [9]. During TRT, one approach is to use active control of the heat input to maintain relatively uniform heat input. Alternatively, TRTs without uniform heat input may be analyzed with the parameter estimation approach [1]. In

* Corresponding author: jwoloszy@agh.edu.pl

work [9] authors mentioned that developments in test rig design have included: small suitcase-sized units; telemetry; temperature measurements using optical fiber; downhole measurements using wireless sensors and wired sensors; and a low-powered test rig using a heating cable and no pumping. The analysis procedures also were developed. In the analysis procedures the groundwater flow [4] and ambient losses are included [8]. Improved analysis of tests with variable heat flow rates with [4,12,13] or without parameter estimation are conducted, improved error analyses [4] and proposals to compute the mean temperature more accurately [3]. To estimate the thermal conductivities and diffusivities of soil and grouting material, both analytical and numerical models have been used. Most of papers report that, estimation procedures based on analytical models are simpler, faster, and more robust than those based on numerical methods [5]. On the other hand, procedures using numerical models, either one- or multi-dimensional, can estimate the heat capacity of grout; and the field testing time may be short. However, they require more input parameters and are more sensitive to the accuracy of the input parameters than algorithms based on analytical models [2,6].

The objective of this work, is to conduct estimation of the thermal conductivity and specific heat of the grouting material and the soil. This research was conducted using the finite element method (FEM), DOE technique and RSM. The ANSYS software with an implemented new MDF [11] and q3D numerical model of heat and mass transport in a BHE was used. This models has been verified via experimental basis and compared to other literature BHE models in work [10]. The estimation procedure was conducted according presented in figure 1 algorithm.

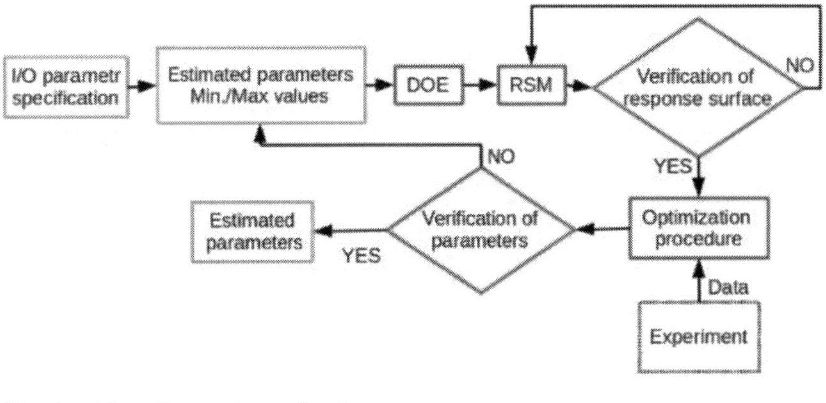

Fig. 1. The algorithm of parameters estimation

2 Description of the sandbox experiment and computational domain

The parameters estimation were done using measurements from a small laboratory sandbox experiment. The small laboratory sandbox with dimensions of 0.46 m x 0.46 m x 2.92 m has been constructed with a borehole heat exchanger of length 2.5 m centered horizontally along the length of the box. The vertical cross section of the box is a square with sides of 0.46 m. In figure 2 the developed sandbox tank was presented. An aluminum pipe with an outer diameter of 0.080 m and 0.002 m wall thickness serves as the borehole wall. An copper pipe with an outer diameter of 0.018 m and 0.0011 m wall thickness serves as the U-tube. The distance between pipes axes is equal 0.046 m.

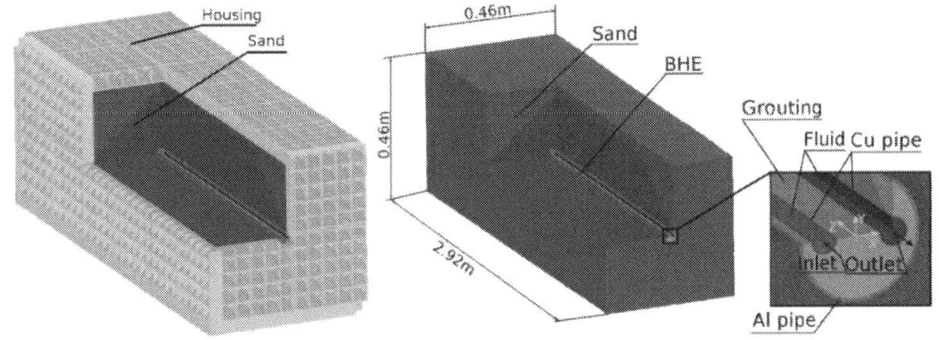

Fig. 2. Small laboratory sandbox and computational domain

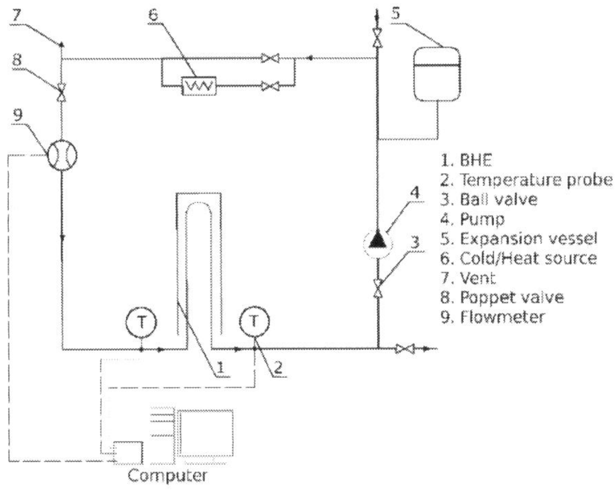

Fig. 3. Diagram of the testing unit for measurements

A testing unit to conduct the measurements (Fig. 3) was connected to the U-tube in the sandbox. Electric heating element supply heat to the circulating water. A pump circulates the water, and a flow meter was used to measure the volume flow rate of circulating water.

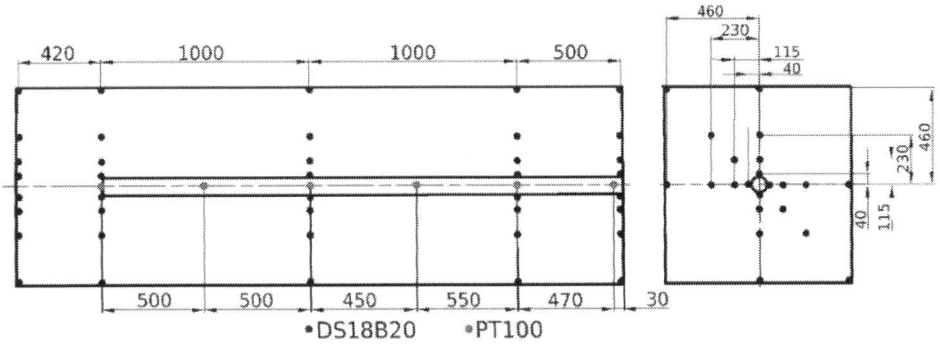

Fig. 4. Location of the temperature sensors, mm

The uncertainty of the measured flow rate is ±2 % of full range. The inlet and outlet fluid temperature measurements were conducted using PT100 temperature resistance sensors in 1/10B accuracy class. The BHE axis temperature in several location were measured using RTDs in 1/3B accuracy class. Locations of these RTDs are illustrated in figure 4. All of these RTDs were connected to a high-accuracy data acquisition system (NI PCI-6289 and NI

SCXI-1503). The temperature of sand was measured using 110 of DS18B20 temperature sensors with uncertainty of ±0.5 K. Locations of these temperature sensors are illustrated in Figure 4.

To simplify the computational domain, the polypropylene housing of the sandbox tank was neglected (Fig. 2). The author conducted estimation based on two experiments (P1 and P2) with different operating parameters. The time of both experiment equal 71 hours.

3 Mathematical model

During estimation procedure author decided to use two numerical models of heat transport in borehole heat exchanger. The first one was a quasi 3D heat transport in BHE model [10,11] developed in ANSYS software and the second one was a MDF model presented in work [11].

3.1 The heat transport in the ground

Heat transport in the closest vicinity of a borehole heat exchanger is described by a function of spatial coordinates and time. This issue is also connected with the flow of fluid. It was assumed that the advection heat transport in the ground can be neglected (due to a lack of a water in sandbox). Also, evaporation and seepage of water has been neglected. In this case the heat transport in the rock mass was derived from Fourier's law and the conservation of energy, which is presented as partial differential equation of the transient heat conduction (Eq. 1).

$$\nabla^2 T_S = \frac{c_S \cdot \rho_S}{\lambda_S} \cdot \frac{\partial T_S}{\partial t} \tag{1}$$

3.2 Heat transport in the BHE using the q3D model

In this model the grout area is treated as a fully three-dimensional volume modelled by utilizing SOLID 70 type elements, whereas the fluid is modelled by one-dimensional FLUID 116 type elements in ANSYS software. Equation (Eq. 1) presents a partial differential equation of conduction in soil as well as in grout, and equations (Eq. 2) and (Eq. 3) present the heat transport process in a U-tube.

$$\rho_f c_f \cdot \frac{\partial T_i}{\partial t} + \rho_f c_f u \cdot \frac{\partial T_i}{\partial z} - \lambda_f \frac{\partial^2 T_i}{\partial z^2} - b_{ig}\left(T_g - T_i\right) = 0 \tag{2}$$

$$\rho_f c_f \cdot \frac{\partial T_0}{\partial t} + \rho_f c_f u \cdot \frac{\partial T_0}{\partial z} - \lambda_f \frac{\partial^2 T_0}{\partial z^2} - b_{og}\left(T_g - T_0\right) = 0 \tag{3}$$

Despite such dimensional simplification, real heat transfer within all the elements of the exchanger has been considered. Representing a fluid in the form of a one-dimensional finite element allows to maintain temperature variation along the vertical axis of this element while neglecting temperature variation in the radial direction. Heat fluxes that are normal to the contact surface of a U-tube pipe and the borehole along the vertical axis are fully considered. The quasi-3-D model was presented in detail in a paper [10].

3.3 Heat transport in the BHE using the MDF model

In this one-dimension finite element model, equations (Eq. 4 – 8) presented heat transport in the inlet pipe, outlet pipe and grout g1, g2, g3, respectively.

$$\rho_f c_f \cdot \frac{\partial T_i}{\partial t} + \rho_f c_f u \cdot \frac{\partial T_i}{\partial z} - \lambda_f \frac{\partial^2 T_i}{\partial z^2} - b_{ig1}\left(T_{g1} - T_i\right) - b_{ig3}\left(T_{g3} - T_i\right) = 0 \tag{4}$$

$$\rho_f c_f \cdot \frac{\partial T_o}{\partial t} + \rho_f c_f u \cdot \frac{\partial T_o}{\partial z} - \lambda_f \frac{\partial^2 T_o}{\partial z^2} - b_{og2}\left(T_{g2} - T_o\right) - b_{og3}\left(T_{g3} - T_o\right) = 0 \tag{5}$$

$$\rho_g c_g \cdot \frac{\partial T_{g1}}{\partial t} + \lambda_g \frac{\partial^2 T_{g1}}{\partial z^2} - b_{ig1}\left(T_{g1} - T_{g2}\right) - b_{g1g2}\left(T_{g1} - T_{g2}\right) - b_{g1g3}\left(T_{g1} - T_{g3}\right) = 0 \tag{6}$$

$$\rho_g c_g \cdot \frac{\partial T_{g2}}{\partial t} + \lambda_g \frac{\partial^2 T_{g3}}{\partial z^2} - b_{og2}\left(T_{g2} - T_o\right) - b_{g2g1}\left(T_{g2} - T_{g1}\right) - b_{g2g3}\left(T_{g2} - T_{g3}\right) = 0 \tag{7}$$

$$\rho_g c_g \cdot \frac{\partial T_{g3}}{\partial t} + \lambda_g \frac{\partial^2 T_{g3}}{\partial z^2} - b_{og3}\left(T_{g3} - T_o\right) - b_{ig3}\left(T_{g3} - T_i\right) - b_{g3g1}\left(T_{g3} - T_{g1}\right) - b_{g3g2}\left(T_{g3} - T_{g2}\right) = 0 \tag{8}$$

where :

- b – heat transfer coefficient,
- c – specific heat,
- T – temperature,
- t – time,
- u – fluid flow,
- λ – thermal conductivity,
- ρ – density,
- ∇ – Nabla.

Subscript :

- f – fluid,
- $g_1, g_2, g_3,$ – grouting,
- i – inlet pipe
- o – outlet pipe
- s – soil.

The proposed formulation allows to present the elements of a borehole as a 1D finite element that describes the pipes and grout. The presented model allows to maintain temperature variation along this element's vertical axis while neglecting the temperature variation in the radial direction. Heat fluxes normal to the contact surface of the borehole and rock mass along the vertical axis are fully considered.

3.4 Initial and boundary conditions

In order to solve the presented partial differential equations require assuming the initial and boundary conditions which were measured during the experiment test. It was assumed that the initial condition was connected with the influence of an undisturbed environment in the sandbox. As the initial condition, in the whole computational domain, the average ground temperature $T_{p1,s,g,f}(t=0)=294.5$ K for experiment P1 and $T_{p2,s,g,f}(t=0)=296.95$ K for experiment P2, was assumed. To simulate external sandbox conditions, for the all face surface of the computational domain (Fig. 2), the convection boundary condition was assumed as in equation (Eq. 9), where $T_{a,P1,P2}$=296.75 K is the three day average air temperature in the laboratory hall.

$$-\lambda_S \frac{\partial T_S}{\partial t} = b_a \left(T_s - T_{a,P1,P2} \right) \tag{9}$$

he assumed temperature at the inlet to the borehole heat exchanger U-tube was equal to the temperature at the inlet of the U-tube during the experiment test $T_{i,inlet}(t)=T_{inlet,P1,P2}(t)$. During measurements the volume flow rate was constant $\dot{V}_{P1} = 3.7 \cdot 10^{-5} \cdot m^3 \cdot s^{-1}$ and $\dot{V}_{P2} = 3.01 \cdot 10^{-5} \cdot m^3 \cdot s^{-1}$. As boundary condition in numerical model the calculated fluid velocity was equal $u_{P1} = 0.19\ m \cdot s^{-1}$ and $u_{P1} = 0.154\ m \cdot s^{-1}$. The analysis covered the same time as experiment (71 hours). In table (Tab. 1) the assumed values of specific thermo-physical constants of water, grout and sand are presented. The sand and grout thermal conductivity and specific heat are estimated therefore, they take values according to design experiment (Tab. 3). The variation range (Tab. 2) of estimated parameters was assumed according literature data.

Table 1. The assumed thermo-physical properties

Description	Water for P1	Water for P2	BHE Al pipe	Sand	Grouting	U-pipe copper
Density –ρ,kg,m^{-3}	986	977	2700	1735	1477	-
Specific heat – c,J·(kg·K)$^{-1}$	4184	4195	910	$c_s=x_2$	$c_g=x_4$	-
Thermal conductivity – λ,W·(m·K)$^{-1}$	0.58		200	$\lambda_S=x_1$	$\lambda_g=x_3$	391
Dynamic viscosity – μ, Pa·s	5.77·10^{-4}	4.86·10^{-4}	-	-	-	-

Table 2. The assumed thermo-physical properties

Estimated parameter	Min. value	Max. value
Soil thermal conductivity, $\lambda_S=x_1$	0.15	0.25
Soil specific heat, $c_s=x_2$	750	950
Grout thermal conductivity, $\lambda_g=x_3$	1.5	2.5
Grout specific heat, $c_g=x_4$	1000	1800

3.5 Grid independent test

Author did a grid independence test to ensure that the results would converge (Fig. 5). Solving problem is transient, so grid independence was checked for several time steps. The numerical simulation was done with 255 000 finite elements for q3D model and 130 000 elements for MDF model.

The Crank–Nicolson algorithm was used to solve the partial differential equations and an adaptive transient time step interval was assumed.

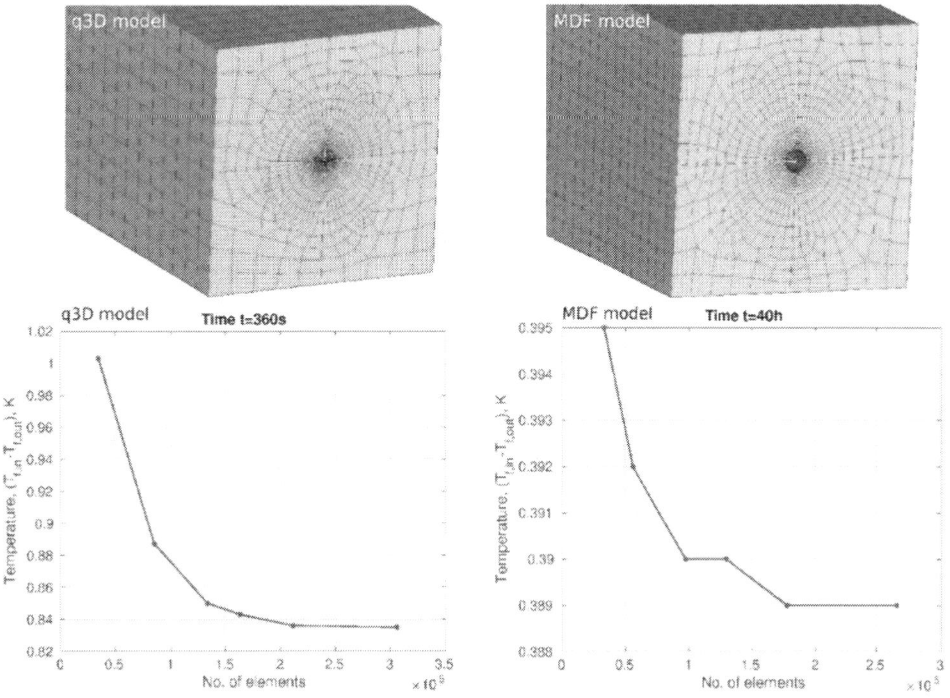

Fig. 5. Assumed grid and grid independence test

4 Results of the parameters estimation

The number of numerical experiments were reduced by using DOE. An extended central composite face centered (CCF) design in which factors adopt values on five levels were used. For four factors x_1, x_2, x_3 and x_4 there is a need to conduct 49 numerical experiments. The calculations were performed using "ZEUS" supercomputer at ACK Cyfronet Krakow. The table 3 presents the plan of the conducted simulations with the input and output parameters. For the output parameters, the author assumed two parameters described by the equations (Eq. 10 – 12). $Y1_{P1}$ and $Y1_{P2}$ described the average discrepancy in a time range $t \in \langle 0; 8h \rangle$ and $Y2_{P1}$ and $Y2_{P2}$ in a time range $t \in \left(8; 71h \right\rangle$.

$$B(t) = \frac{\left(T_{inlet}^{\exp}(t) - T_{outlet}^{\exp}(t)\right) - \left(T_{inlet}^{sim}(t) - T_{outlet}^{sim}(t)\right)}{\left(T_{inlet}^{\exp}(t) - T_{outlet}^{\exp}(t)\right)} \cdot 100\% \tag{10}$$

$$Y1 = \frac{\sum_{i=1}^{n} B(t_i)}{n} \quad for\ t \in \langle 0; 8h \rangle \tag{11}$$

$$Y2 = \frac{\sum_{i=1}^{n} B(t_i)}{n} \quad for\ t \in \left(8; 71h \right\rangle \tag{12}$$

Table 3. Results of conducted numerical experiment

No.	Input parameters				Output Parameters – q3D model				Output Parameters – MDF model			
	x_1	x_2	x_3	x_4	$Y1_{P1}$	$Y2_{P1}$	$Y1_{P2}$	$Y2_{P2}$	$Y1_{P1}$	$Y2_{P1}$	$Y1_{P2}$	$Y2_{P2}$
1	0.2	850	2	1400	1.85	7.75	7.73	12.17	1.79	7.74	9.62	13.69
2	0.15	850	2	1400	12.99	11.89	18.62	28.51	12.83	12.00	20.18	29.83
3	0.175	850	2	1400	6.93	2.12	13.01	20.19	6.79	2.16	14.75	21.61
4	0.25	850	2	1400	10.54	26.04	2.70	3.16	10.28	26.07	1.33	1.45
5	0.225	850	2	1400	5.21	17.04	2.75	4.39	5.04	17.07	4.76	6.03
6	0.2	750	2	1400	3.96	4.86	10.32	14.32	3.90	4.87	12.21	15.80
7	0.2	800	2	1400	2.69	6.33	9.00	13.22	2.61	6.34	10.89	14.73
8	0.2	950	2	1400	2.44	10.43	5.37	10.16	2.37	10.43	7.20	11.72
9	0.2	900	2	1400	1.75	9.11	6.52	11.15	1.70	9.11	8.39	12.69
10	0.2	850	1.5	1400	2.47	6.70	9.00	13.00	3.04	6.44	11.82	14.72
11	0.2	850	1.75	1400	2.03	7.30	8.19	12.53	2.08	7.18	10.58	14.13
12	0.2	850	2.5	1400	1.93	8.39	7.21	11.65	1.92	8.55	8.34	13.06
13	0.2	850	2.25	1400	1.84	8.11	7.43	11.89	1.79	8.20	8.87	13.34
14	0.2	850	2	1000	5.05	7.34	11.60	12.43	4.98	7.33	13.64	13.96
15	0.2	850	2	1200	3.26	7.55	9.58	12.30	3.22	7.54	11.63	13.82
16	0.2	850	2	1800	3.93	8.18	5.67	11.89	3.70	8.17	5.91	13.41
17	0.2	850	2	1600	1.86	7.96	6.33	12.03	1.71	7.96	7.61	13.55
18	0.15	750	1.5	1000	20.66	15.39	26.06	31.14	20.84	15.67	28.06	32.56
19	0.175	800	1.75	1200	10.58	3.77	16.65	21.60	10.67	3.91	18.77	23.07
20	0.25	750	1.5	1000	5.30	20.91	6.43	0.97	5.48	20.60	9.40	2.66
21	0.225	800	1.75	1200	3.13	14.77	6.54	6.07	3.26	14.66	9.11	7.79
22	0.15	950	1.5	1000	15.76	10.78	21.59	27.62	15.98	11.06	23.74	29.09
23	0.175	900	1.75	1200	8.00	1.77	14.25	19.67	8.07	1.85	16.43	21.18
24	0.25	950	1.5	1000	8.16	27.15	3.15	3.90	8.11	26.77	4.29	1.82
25	0.225	900	1.75	1200	4.28	17.75	3.89	3.89	4.35	17.62	6.55	5.65
26	0.15	750	2.5	1000	19.13	14.35	24.50	30.30	18.81	14.37	25.76	31.54
27	0.175	800	2.25	1200	9.81	3.16	15.82	21.08	9.61	3.14	17.38	22.43
28	0.25	750	2.5	1000	6.02	23.18	3.50	1.26	6.19	23.44	5.34	0.77
29	0.225	800	2.25	1200	3.50	15.73	5.40	5.31	3.62	15.85	7.27	6.86
30	0.15	950	2.5	1000	14.14	9.63	19.84	26.71	13.77	9.63	21.12	27.98
31	0.175	900	2.25	1200	7.22	1.51	13.39	19.14	6.99	1.51	14.95	20.52
32	0.25	950	2.5	1000	10.89	29.61	3.72	5.89	11.21	29.88	3.10	4.26
33	0.225	900	2.25	1200	5.32	18.73	2.72	3.10	5.49	18.85	4.60	4.67
34	0.15	750	1.5	1800	12.27	14.56	17.81	30.60	12.52	14.82	19.97	32.01
35	0.175	800	1.75	1600	6.78	3.36	12.78	21.33	6.71	3.49	14.74	22.79
36	0.25	750	1.5	1800	9.20	21.76	1.86	0.58	8.17	21.44	1.74	2.09
37	0.225	800	1.75	1600	4.90	15.20	3.11	5.80	4.39	15.09	5.17	7.52
38	0.15	950	1.5	1800	8.01	9.92	13.71	27.07	7.82	10.20	15.72	28.54
39	0.175	900	1.75	1600	4.47	1.43	10.52	19.40	4.29	1.51	12.42	20.90
40	0.25	950	1.5	1800	15.10	28.00	7.19	4.46	13.87	27.63	4.22	2.38
41	0.225	900	1.75	1600	7.82	18.17	1.48	3.61	7.25	18.05	2.63	5.37
42	0.15	750	2.5	1800	11.90	13.53	17.47	29.78	11.65	13.54	17.94	31.01
43	0.175	800	2.25	1600	6.40	2.76	12.36	20.81	6.15	2.75	13.29	22.17
44	0.25	750	2.5	1800	13.02	24.00	5.31	1.73	13.08	24.26	2.72	0.61
45	0.225	800	2.25	1600	6.56	16.14	2.80	5.05	6.54	16.26	3.35	6.58
46	0.15	950	2.5	1800	7.86	8.80	13.50	26.18	7.50	8.81	13.73	27.45
47	0.175	900	2.25	1600	4.20	1.28	10.15	18.87	3.88	1.27	10.92	20.25
48	0.25	950	2.5	1800	19.41	30.45	11.11	6.43	19.44	30.70	8.34	4.80
49	0.225	900	2.25	1600	9.58	19.15	2.22	2.83	9.55	19.28	1.23	4.40

The response points (Tab. 3) were approximated using the Kriging algorithm, which leads to the designation of the multidimensional response surface (Fig. 6 – 8). To verify the response surface the author checked the solution at six randomly selected points, of response space, and calculated the coefficient of determination (R^2). For q3D model R^2 equal: $R^2_{Y1_{P1}} = 0.92$, $R^2_{Y1_{P2}} = 0.94$, $R^2_{Y2_{P1}} = 1$, $R^2_{Y2_{P2}} = 1$. For MDF model R^2 equal: $R^2_{Y1_{P1}} = 0.92$, $R^2_{Y1_{P2}} = 1$, $R^2_{Y2_{P1}} = 1$, $R^2_{Y2_{P2}} = 1$.

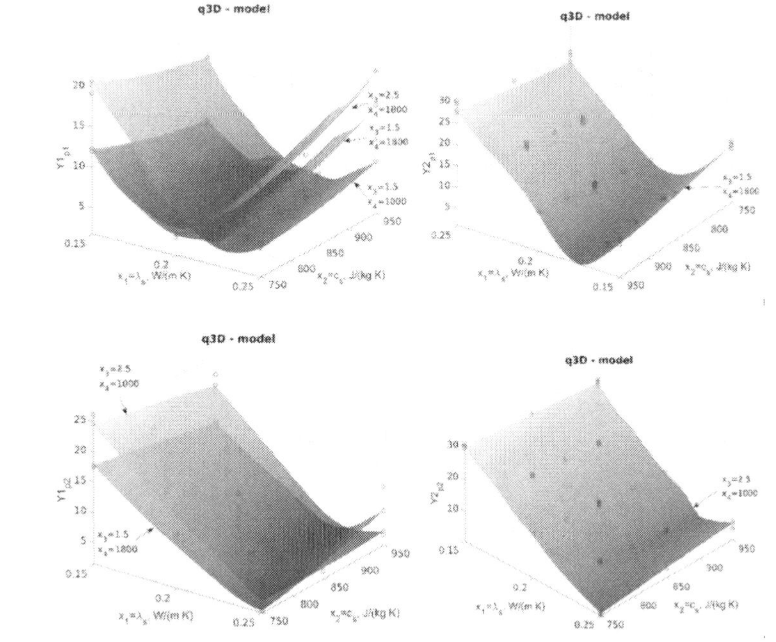

Fig. 6. The response surface for q3D - model

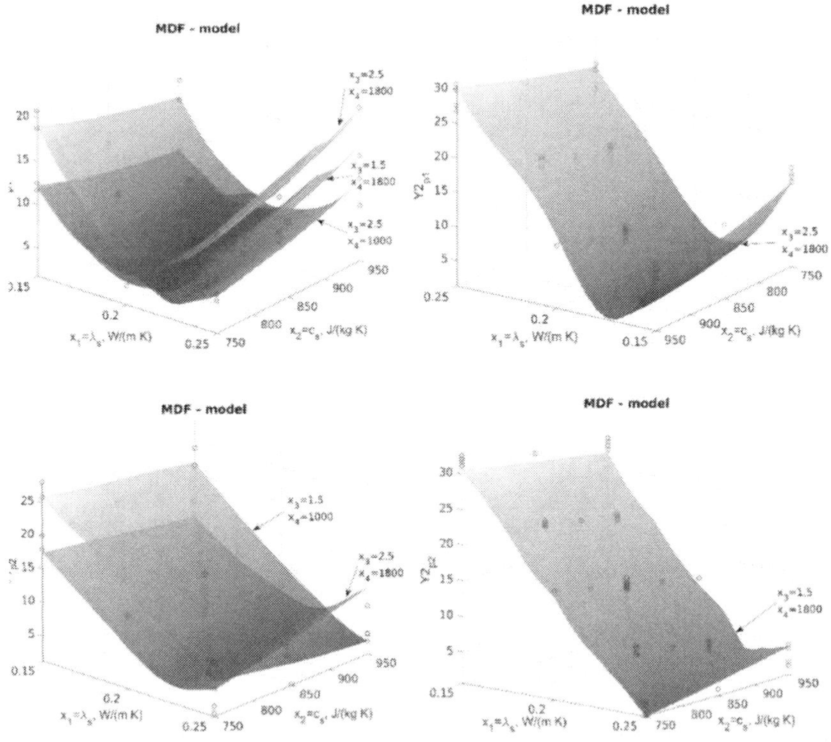

Fig. 7. The response surface for MDF - model

In figures 6 and 7 the presented multidimensional response surface was obtained from the design parameters, x_1, x_2 and the output parameters $Y1_{P1}$, $Y2_{P1}$, $Y1_{P2}$, $Y2_{P2}$ for q3D and MDF model, respectively. The response surface depends on two other design parameter, x_3 and x_4 which are equal $x_3 = [1.5, 2.5]$, $x_4 = [1000, 1800]$. The discrepancy, varies between 2% and 30%.

In figure 8 the presented multidimensional response surface was obtained from the design parameters, x_3, x_4 and the output parameters $Y1_{P1}$ for q3D and MDF models. The response surface depends on two other design parameter, x_1 and x_2 which are equal $x_1 = [0.15, 0.2\ 0.25]$, $x_2 = [750, 950]$. The discrepancy, varies between 2% and 20%.

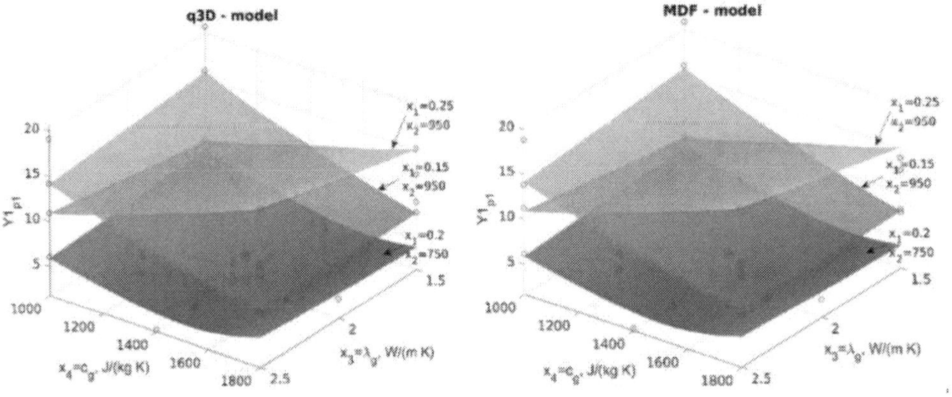

Fig. 8. The response surface for q3D and MDF models

To estimate the parameters the concept of Pareto dominance in multiple-objective optimization on predicted response surface was used. The objective of optimization was to minimize the output parameters ($Y1_{P1}$, $Y2_{P1}$, $Y1_{P2}$, $Y2_{P2}$) for both models. To conduct optimization procedure the screening algorithm in ANSYS 15 was used. The 1000 samples was set to generate and three candidate points was generated by optimization algorithm. In figures 9 and 10 the trade-off charts which shows the Pareto-dominant solutions was presented.

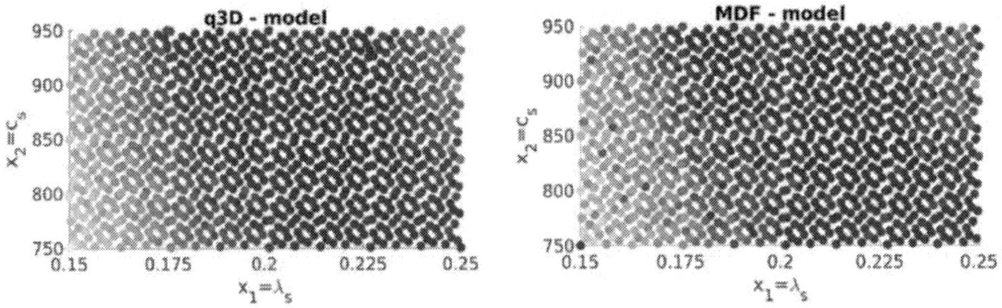

Fig. 9. The trade-off charts for design parameters and q3D and MDF models, feasible points marked as dark blue

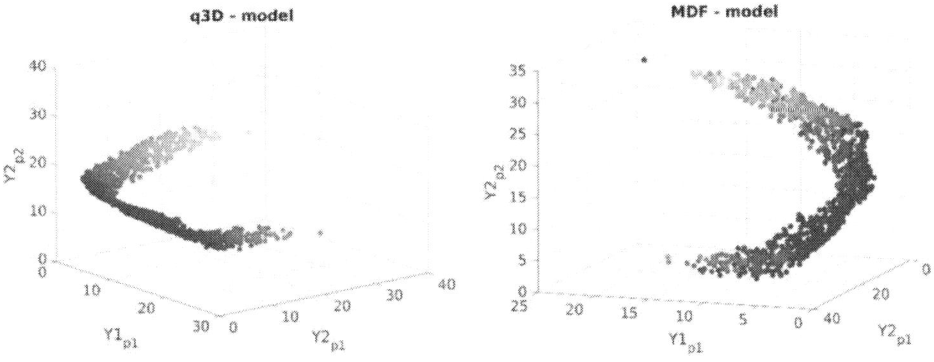

Fig. 10. The trade-off charts for output parameters and q3D and MDF models, feasible points marked as dark blue

In table 4 the best solution of estimated parameters was show and the output parameters ($Y1_{P1}$, $Y2_{P1}$, $Y1_{P2}$, $Y2_{P2}$) for estimated values and both models were presented.

Table 4. The estimated parameters value and discrepancies

Estimated parameter	Value	Output parameters	q3D – model	MDF – model
Soil thermal conductivity, $\lambda_S=x_1$	0.204	$Y1_{P1}$	3,86	3,24
Soil specific heat, $c_s=x_2$	944	$Y2_{P1}$	11,41	11,21
Grout thermal conductivity, $\lambda_g=x_3$	1.64	$Y1_{P2}$	3,95	6,19
Grout specific heat, $c_g=x_4$	1588	$Y2_{P2}$	9,35	11,06

The estimated parameters were verified by experiment data and simulation with different operating parameters than previous tests (P1 and P2). The discrepancies were calculated according equations (Eq. 11 – 12) and equal $Y_1 = 3.87$, $Y_2 = 1.8$ for q3D model and $Y_1 = 4.16$, $Y_2 = 1.62$ for MDF model. In figure 11 the outlet and inlet fluid temperature differences for experiment and both model with estimated parameters were shown. The discrepancy according equation (Eq. 10) was also presented (Fig. 11).

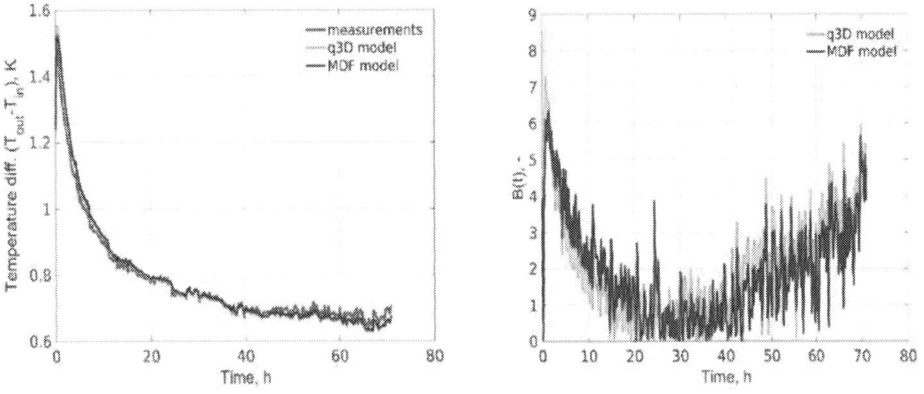

Fig. 11. The outlet and inlet fluid temperature differences (left) and discrepancy according eq. (10) (right)

The average discrepancy based on verification measurement for both heat transport models are below 5 %.

5 Conclusion

In this work, the estimation of the ground and BHE thermal properties has been investigated. Two numerical model of heat transfer in BHE were used to conduct estimation procedure. The work begins with a general determination of the balance equation for the flow and transport of heat within each element of the small laboratory sandbox. The BHE exchanger is modelled as a quasi-three dimensional and as a one-dimensional finite element with multiple degrees of freedom. The DOE is then used to obtain a response surface and conduct the multiple-objective optimization analysis. From the simulations mentioned above, the following conclusions were obtained:

- Based on the DOE, RSM and multiple-objective optimization the material properties of BHE and soil was determined. For both models the discrepancies Y1P1 are below 4 %, Y2P1 are below 11.5 %, Y1P2 are below 6.5 % and Y2P2 are below 11.5 %.
- For the heat transport models, the thermal properties variations with temperature and moisture content were neglected therefore, the calculated discrepancies below 11.5 % are considered satisfactory.
- The same optimal parameters values was determined based on two different heat transport models.
- The average discrepancies below 5 % based on the verification measurements with different operating parameters was obtained.
- The DOE, RSM and multiple-objective optimization as estimation procedure is an alternative to TRT.

As a further direction of research, it is proposed to conduct estimation procedure for BHE in full scale and compare estimated parameters with laboratory measurements.

This research was supported in part by PLGrid Infrastructure.

References

1. Aranzabal N., Martos J., Montero., et al.: Extraction of thermal characteristics of surrounding geological layers of a geothermal heat exchanger by 3D numerical simulations, Appl Therm Eng., **99**, pp 92-102., (2016)
2. Beier R.A., Smith M.D., Spitler J.D.: Reference data sets for vertical borehole ground heat exchanger models and thermal response test analysis, Geothermics, **40**(1), pp 79-85, (2011)
3. Beier R.A., Spitler J.D.: Weighted average of inlet and outlet temperatures in borehole heat exchangers, Appl Energy, **174**, pp 118-129, (2016)
4. Choi W., Ooka R.: Effect of natural convection on thermal response test conducted in saturated porous formation: Comparison of gravel-backfilled and cement-grouted borehole heat exchangers, Renew Energy, **96**, pp 891-903, (2016)
5. Christodoulides P., Florides G., Pouloupatis P.: A practical method for computing the thermal properties of a Ground Heat Exchanger, Renew Energy, **94**, pp 81-89, (2016)
6. Li M., Lai A.C.K.: Review of analytical models for heat transfer by vertical ground heat exchangers (GHEs): A perspective of time and space scales. Appl Energy, **151**, pp 178-191, (2015)
7. Mogensen P.: Fluid to duct wall heat transfer in duct system heat storages. *In: International conference on subsurface heat storage in theory and practice, Stockholm, Sweden*: Swedish Council for Building Research, pp 652–57, (1983)

8. Priarone A., Fossa M.: Temperature response factors at different boundary conditions for modelling the single borehole heat exchanger. Appl Therm Eng., **103**, pp 934-944., (2016)
9. Spitler J.D., Gehlin S.E.A.: Thermal response testing for ground source heat pump systems - An historical review, Renew Sustain Energy Rev., **50**, pp 1125-1137, (2015)
10. Wołoszyn J., Gołaś A.: Experimental verification and programming development of a new MDF borehole heat exchanger numerical model, Geothermics, **59**, pp 67-76, (2016)
11. Wołoszyn J., Gołaś A.: Modelling of a borehole heat exchanger using a finite element with multiple degrees of freedom, Geothermics, **47**, pp 13-26, (2013)
12. Yavuzturk C., Spitler J.D., Rees S.J.: A transient two-dimensional finite volume model for the simulation of vertical U-tube ground heat exchangers, ASHRAE Trans, **105**, pp 465–74, (1999)
13. Yavuzturk C. Modeling of vertical ground loop heat exchangers for ground source heat pump systems. Stillwater, Oklahoma: Oklahoma State University, (1999)

Evaluation of long-term operation of combined system for heating the building and preparation of domestic hot water

Dawid Taler[1,*], *Rafał* Pitry[1], and *Jan* Taler[1]

[1]Cracow University of Technology, Faculty of Environmental Engineering , Warszawska 24, 30-001 Krakow, Poland
[2]Thermo-Instal Company, Bałandy 4D/8, 32-600 Oświęcim, Poland

Abstract. The paper presents, the results of research on the operation and energy efficiency of a 186 kW gas-fired condensing boiler operating in a hybrid heat source system. The boiler co-operates with an 81.1 kW (electric) brine-to-water compressor heat pump, a 27.4 kW air-to-water heat pump and 6 flat solar collectors. A local, built-in, hybrid heat source is located in a public building and is intended to satisfy the building needs. The study was conducted over a period of 1 year – from 1 September 2014 to 31 August 2015. The gas-fired boiler operates in the heating buffer system all year round. The boiler performance is characterized both in the winter and in the summer season, in terms of the amount of heat produced and the heating power. The calculations results of the heat generation efficiency obtained in the measuring period are also presented

1 Introduction

In recent years, fossil fuel shortage and environmental pollution have resulted in the rapid use of renewable energy sources. Hybrid systems of supplying residential buildings with electricity, heat and domestic hot water that combine photovoltaic cells, solar collectors, ground and air heat pumps and gas boilers quickly develop. Combined heating, cooling and power systems including domestic hot water preparation have been the subject of numerous publications in recent years. The feasibility and affordability of hybrid photovoltaic-thermal (PV-T) collectors coupled with thermal or electrical solar heating and cooling systems (absorption chillers or heat pumps) were studied in [1], also using thermal energy storage. The proposed systems were analyzed in 4-5 person households with a 100 m^2 floor area and 50 m^2 rooftop area suitable for installation of solar collectors. Ten selected locations with different climatic conditions in Europe were studied. A least-cost investment methodology that optimizes the hybrid power-residential heating system, including thermal energy storage, was developed in [2]. Different configurations were considered: gas boiler-electric resistance heater, heat pump-gas boiler, and heat pump-electric resistance heater. The hybrid ground heat exchangers of ground-coupled heat pump systems using vertical, as well as horizontal heat exchangers, were studied in [3]. The optimum load ratio of horizontal to vertical ground

* Corresponding author: dtaler@pk.edu.pl

heat exchangers was found. The performance of a heat pump with hybrid heat source including solar and air energy using different refrigerants was simulated in [4]. Based on a lumped parameter model of the heat pump, the characteristics of the heat pump as the power consumption and coefficient of performance (COP) were established for three refrigerants R22, R134a, R744, and CO_2. There are no significant differences in the performance of the heat pump used three different refrigerants. The heat pump with CO_2 as the working fluid has a low COP, but it has a good low-temperature performance. The efficiency of new power sources (co-generation, tri-generation systems, fuel cells, photovoltaic systems) can be enhanced using heat pipe heat exchangers and solid sorption heat pumps [5]. In this paper, the results of experimental investigations of a hybrid heating and domestic hot water preparation system in a public building are presented. The wide variety of micro and mini heat sources currently offered by manufacturers makes it possible to design solutions not only with a single energy-generating device but also with more complex hybrid heat source systems including two or even more heat generators. Due to technical requirements, e.g., high supply temperature, or economic limitations, e.g., low cost of purchase, renewable heat sources, such as heat pumps or solar collectors, often need a traditional heat generator such as a gas-fired boiler. Each stage of development of this type of solutions – investment, design, construction or operation – is highly complex, and the comprehensive approach to the problem should not be based on a one-off, fragmentary and out-of-context analysis of the device, but on a comprehensive analysis of all aspects of the hybrid solution operation. The considerations presented herein are limited to the gas-fired boiler assessment and significance in a hybrid heat source in relation to the amount of generated thermal energy, the share of thermal output, working time and heat generation efficiency. Previous analyses of hybrid source systems for heating and preparing domestic hot water were carried out using various computer simulation programs. In this paper, the actual experimental results of one-year operation of hybrid source installations were presented.

2 Characteristics of investigated hybrid heat source

The aim of the analysis is a gas-fired boiler operating in a local hybrid heat source system located in the modern building of the Headquarters of the State Fire Service in *Oświęcim*, *Małopolska* province (voivodeship). The building and the installations were put in use in September 2011. The headquarters building is supplied with thermal energy by a local hybrid heat source, the center of which is a buffer tank. The hydronic system flowchart is shown in Fig. 1. The heat source includes:

- **KG**: a floor-standing Vitocrossal 200 CM2 186 kW gas-fired condensing boiler with a compact modulating MatriX burner,
- **PC1**: a Thermalia 90 81.1 kW (B0W35), 109.3 kW (W10W35) brine-to-water compression (ground source) heat pump made by the Hoval company, with a lower heat source in the form of 18-meter deep vertical brine probes,
- **PC2**: a Belaria 27 27.4 kW (A2W35) air-to-water compression heat pump made by the Hoval company,
- **KS**: Vitosol 100 SV1 flat solar collectors made by the Viessmann company, with thegross surface area of 6 x 2.51 m2.

The heat source operates for the heating, ventilation and domestic hot water preparation. Due to the different temperature characteristics, the heat-receiving systems are divided into two groups. The first group comprises low-temperature heating circuits (60/45°C) with low-temperature parameters (NP) 1-3 (marked in Fig. 1 as R1). The total power of the receivers is 129.2 kW. The second group includes heating circuits with numbers 4-7 and domestic hot water installation. This group is characterized by high-temperature parameters (WP) (marked in Fig. 1 as R2), with the total power of 231.7 kW and a design temperature of 80/60°C.

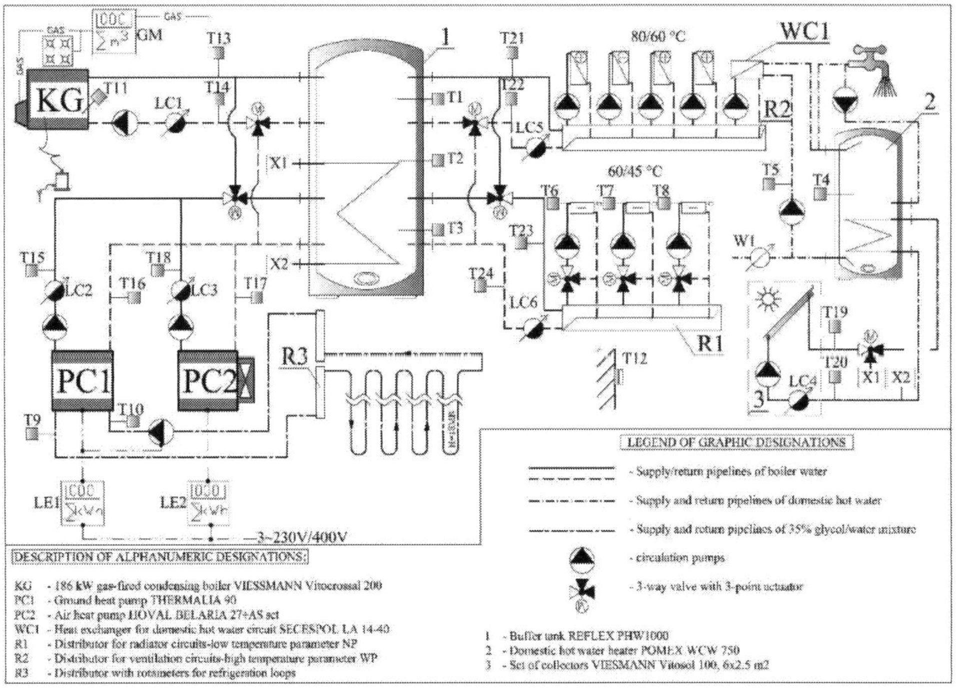

Fig. 1. Technological flowchart of the hybrid heat source including the measuring elements. Basic marking adopted in the measuring system: LC1÷LC6 – heat meters, LE1, LE2 – electricity meters, W1 – domestic hot water meter, GM – bellows gas meter, T1÷T22 – temperature transducers, the other markings are in the figure

The building of the Headquarters of the Local Government Fire Station is located in climatic zone III. The facility is composed of three units performing different functions. The building total cubic volume is 18432.15 m^3. The research was conducted from 1 September 2014 to 31 August 2015.

3 Test facility and measurements results

A new system of control, measurement, and recording of the entire heat source necessary operating parameters was designed for the needs of the research works. The measuring system flowchart is presented in Fig. 1. The technological system was equipped with water flow and heat meters, meters of the electric power network parameters and a *masterPLC*-type [6] data acquisition system made by the Altel company. Combustion analysis was performed to assess the gas-fired boiler combustion quality. The technological system and the measuring system are controlled by a master PLC controller.

4 Thermal energy production

During the measuring year (Fig. 2) the most significant amounts of thermal energy were generated by gas-fired boiler KG – 470.1 GJ/year (51.11%), brine-to-water heat pump PC1 – 395.7 GJ/year (43.02%), air-to-water heat pump PC2 – 35.8 GJ/year (3.89%), and solar collectors KS – 18.2 GJ/year (1.98%). The amount of thermal energy produced in the heating season is many times larger compared to the heat generated in the summer season: 878.2 GJ and 41.6 GJ, respectively. For this reason, the percentage share of energy generated by individual devices in the heating season (Fig. 3) is close to the annual value. In the summer season, the opposite is the case (Fig. 4) – the most significant amounts of thermal energy were generated by heat pump PC2 – 25.8 GJ/summer season (62.02%), solar collectors KS – 10.8 GJ/summer season (25.96%) and gas-fired boiler KG – 5.0 GJ/summer season (12.02%). During the summer, heat pump PC1 did not operate.

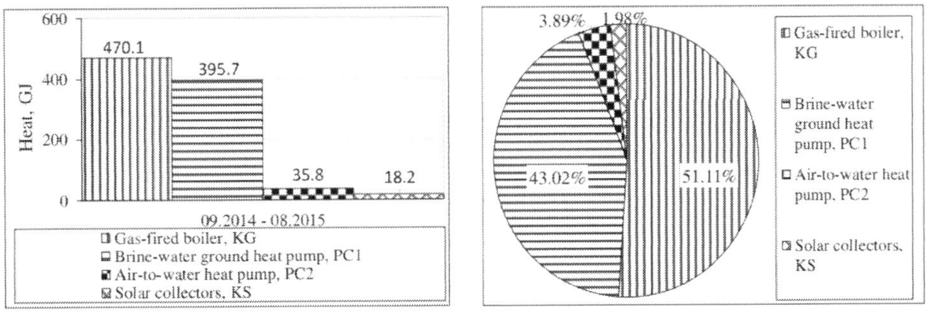

Fig. 2. Results of measurements of thermal energy production by individual generating devices in theentire measuring year by amount and percentage

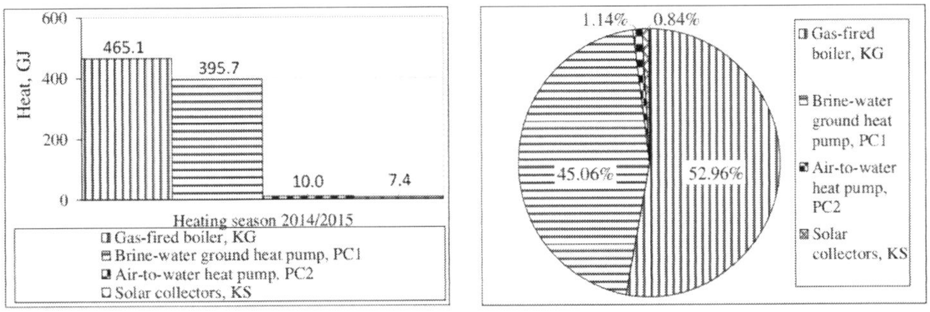

Fig. 3. Results of measurements of thermal energy production by individual generating devices in the heating season by amount and percentage

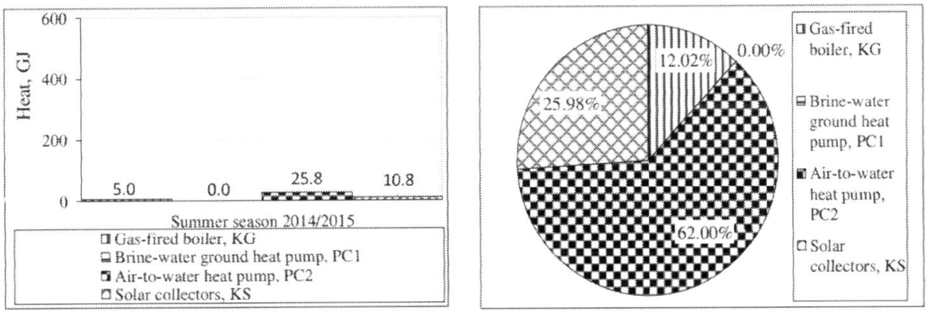

Fig. 4. Results of measurements of thermal energy production by individual generating devicesin the summer season by amount and percentage

5 Planning of diagnostic experiments

During the heating season, the gas-fired boiler operated with an average thermal power of 119.1 kW with a standard deviation of 42.6 kW. For the summer season, the thermal power and the standard deviation values were 95.3 kW and 47.3 kW, respectively, which in the whole year under consideration gives the boiler operation with the average thermal power of 118.8 kW with a standard deviation of 42.8 kW. The typical values for the winter (heating) and the summer seasons and the entire measuring year are 157.0 kW, 131.0 kW, and 157.0kW, respectively. The thermal power typical values are close to the gas-fired boiler maximum value of 186 kW. The charts are illustrating monthly thermal power values (Fig. 5 and Fig. 6) point to the gas-fired boiler cyclic operation ensuring an appropriate level of temperature in the buffer tank. Thermal power is supplied to the buffer tank in a wide range from 0 kW to 375 kW, which momentarily exceeds the boiler maximum power. This is the effect of the system being cooled down and the occurrence of a difference between the water temperature at the outlet and the inlet of the boiler higher than design temperature difference equal to 20K. Table 1 presents the monthly and annual characteristics of the gas-fired boiler output.

Table 1. Results of measurements of the gas-fired boiler (KG) thermal output

Period of analysis	Maximum output, kW	Mean output, kW	Period of analysis	Maximum output, kW	Mean output, kW
September 2014	152.0	84.0	May 2015	208.0	112.8
October 2014	279.0	85.5	June 2015	203.0	123.4
November 2014	247.0	87.3	July 2015	157.0	83.7
December 2014	329.0	123.8	August 2015	156.0	85.0
January 2015	343.0	122.4			
February 2015	375.0	120.6	Entire heating season	375.0	119.1
March 2015	319.0	153.9	Summer season	203.0	95.3
April 2015	341.0	149.7	Year	375.0	118.8

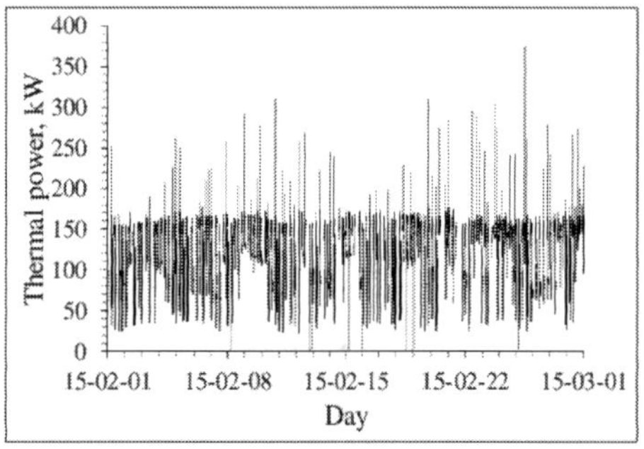

Fig. 5. Results of the gas-fired boiler (KG) thermalpower measurements in February 2015

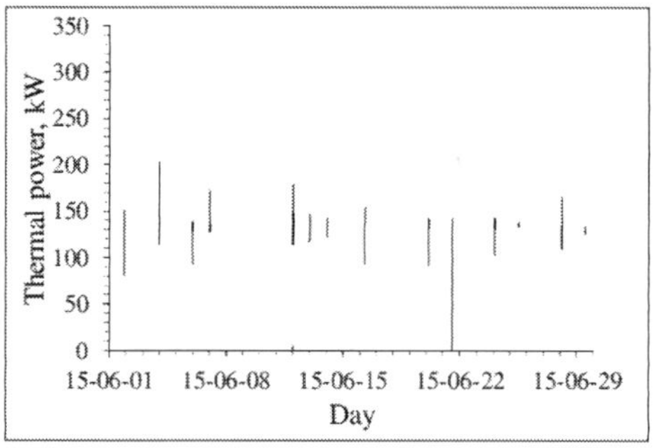

Fig. 6. Results of the gas-fired boiler (KG) thermalpower measurements in June 2015

6 Working time characteristics

The activation of each energy-generating device is related to transferring thermal energy to the buffer tank. An exception is the activation of the gas-fired boiler burner where thermal energy cannot be transferred unless the boiler circulation pump is activated (Fig. 1, where the pump is marked as P2). The longest working time is found for the brine-to-water heat pump PC1 – 1875.1 hours, 1103.6 of which was operation at the first degree of the power output and 771.5 hours – at the second. The next longest working times are for the solar collectors – 1341.4 hours, the gas-fired boiler (KG) – 1085.6 and the air-to-water heat pump

PC2 – 321.1 hours. As presented in Table 2, the annual working time of an individual device does not depend on the device installed power. The gas-fired boiler (KG) overall working time during the heating season – 1072.2 h – is almost 75 times longer than the working time measured in the summer season – 14.4 h. The average duration of the gas-fired boiler operation cycle in the entire year is 0.8 h.

The shortest – 0.1 h – was recorded in June and July 2015, and the longest – 1.5 h – in January 2015. 1411 starts were measured when the boiler joined in to operate within the heating system. The calculation results indicate that during the year the boiler operated using the phenomenon of water vapor condensation in flue gases for 283.5 hours, which makes up 26.1% of the entire operation during the year. The boiler total working time per month ranged from 3.1 to 226.3 hours. Table 2 below presents the calculation results obtained based on the working time of the gas-fired boiler.

Table 2. Results of measurements and calculations of the gas-fired boiler (KG) working time

Analysis period	Number of boiler starts	Total working time (number of boiler pump starts)	The average length of boiler operation	Operating time of the boiler with a percentage share of water steam condensation	
	-	*H*	*h*	*H*	%
Sept 2014	18	3.2	0.2	1.8	54.4
Oct 2014	130	83.9	0.6	15.9	19.0
Nov 2014	288	157.8	0.5	46.8	29.6
Dec 2014	164	203.2	1.2	22.9	11.3
Jan 2015	150	226.3	1.5	104.1	46.0
Feb 2015	137	195.7	1.4	21.1	10.8
Mar 2015	193	107.6	0.6	58.3	54.2
Apr 2015	171	79.6	0.5	44.2	55.6
May 2015	65	16.1	0.2	7.8	48.6
Jun 2015	40	3.1	0.1	2.2	72.8
Jul 2015	39	4.7	0.1	2.2	47.0
Aug 2015	16	4.3	0.3	2.0	45.9
Heating season	1296	1071.2	0.8	233.6	21.8
Summer season	115	14.4	0.1	49.9	-
One year period	1411	1085.6	0.8	283.5	26.1

The shortest – 0.1 h – was recorded in June and July 2015, and the longest – 1.5 h – in January 2015. 1411 starts were measured when the boiler joined in to operate within the heating system. The calculation results indicate that during the year the boiler operated using the phenomenon of water vapor condensation in flue gases for 283.5 hours, which makes up 26.1% of the entire operation during the year. The boiler total working time per month ranged from 3.1 to 226.3 hours.

7 Heat generation efficiency characteristics

The efficiency of thermal energy generation by the gas-fired boiler (KG) during the measuring period was calculated as the ratio between the thermal energy measured by heat meter LC1 and the measured amount of natural gas-fired in the boiler multiplied by the gas lower heating value:

$$\varepsilon_{WY,KG} = \frac{E_{LC1}}{V_{KG} \cdot W_{d_g}} \tag{1}$$

where :

E_{LC1} – thermal energy generated by the gas-fired boiler (KG) and recorded by heat meter LC1 in GJ,

V_{KG} – the boiler recorded natural gas consumption – gas meter (GM) reading reduced by gas consumption for cooking in m^3,

W_{d_g} – a lower heating value the gas fuel in GJ/m^3.

Based on the measuring data obtained in the measuring year under analysis, the heat generation efficiency calculated for the gas-fired boiler (KG) is $\varepsilon_{WY,KG}$=0.96 per year, for the heating season – $\varepsilon_{WY,KG}$=0.96 and for the summer season – $\varepsilon_{WY,KG}$=0.60. The arithmetic mean calculated from the monthly efficiency values is 0.82, and is therefore not comparable to the annual results of calculation of generation efficiency – $\varepsilon_{WY,KG}$. An analysis of the measurement data indicates that the monthly heat generation efficiency was the highest in December 2014 – $\varepsilon_{WY,KG}$ =0.99. The lowest efficiency value was found for the device in September 2015 – $\varepsilon_{WY,KG}$.=0.49. A rising trend can be observed for the changes in efficiency from October to April. The boiler momentary efficiency for the burner minimum and maximum power was 98.4% and 97.6%, respectively.

8 Conclusions

Based on the measurements and calculations carried out, the following conclusions can be drown:

- The amount of thermal energy produced by a given energy-generating device is not proportional to its installed power but depends substantially on the device working time within the system. The installed power of the gas-fired boiler KG is 186 kW, but its percentage share of the annual thermal energy production is 51.11%. For the 81.1 kW heat pump PC1, the percentage share of the overall annual thermal energy generation is 43.02%. These results are also confirmed by the simulation of the hydronic system operation using the Polysun program.
- Energy-generating devices operate cyclically. For this reason, the recorded values of power generated due to the operation of the devices varied in a wide range, including values exceeding the device rated power. This phenomenon is related to the fact that the system is considerably cooled at the start-up of the energy-generating device.
- The gas-fired boiler (KG) operation characteristics in the winter and the summer season differ substantially. In the winter (heating) season, the boiler is activated fast and dynamically with hot water transferred into the buffer system. In the summer season, long periods occur, even up to a few days, when water remains cool in the boiler water region.
- The heat generation efficiency of the gas-fired boiler (KG) averaged per year is $\varepsilon_{WY,KG}$=0.96. In the heating season, the efficiency value is almost equal to the annual average, but it drops significantly in the summer season to $\varepsilon_{WY,KG}$=0.60.
- In a hybrid system, heat generation efficiency is strongly dependent on specific system parameters. Such relations should be identified, e.g., by computer simulations in advance and can be a starting point for design and optimization of complex hybrid systems in which conventional and renewable heat sources are used.

References

1. Ramos A., Chatzopoulou M. A., Guarracino I., Freeman J., Markides Ch. N., Hybrid photovoltaic-thermal solar systems for combined heating, cooling and power provision in the urban environment, Energy Conversion, and Management, **150**, pp. 838-850, (2017)
2. Heinen S., Burke D., O'Malley M., Electricity, gas, heat integration via residential hybrid heating technologies – An investment model assessment, Energy, **109**, pp. 906-919, (2016)
3. Wei K., Li W., Li J., Wang Y., Zhang L., Study on a design method for hybrid ground heat exchangers of ground-coupled heat pump system, Int. J. of Refrigeration, **76**, pp. 394-405, (2017)
4. Li S., Li S., Zhang X. Comparison analysis of different refrigerants in solar-air hybrid heat source heat pump water heater, Int. J. of Refrigeration, **57**, pp. 138-146, (2015)
5. Vasiliev L.L., Kakaç S. Eds. Heat Pipes and Solid Sorption Transformations: Fundamentals and Practical Applications, CRC Press, Taylor & Francis Group, Boca Raton, Florida, USA, (2013)

Evaluation of the properties of coal bottom ash, as regards the possibilities of using it in biological land reclamation

Małgorzata Śliwka[1,*], *Waldemar* Kępys[1], and *Małgorzata* Pawul[1]

[1]AGH University of Science And Technology, Faculty Of Mining And Geoengineering, Department Of Environmental Engineering And Mineral Processing, Mickiewicza 30 Av. 30-065 Cracow, Poland

Abstract. The use of waste in land reclamation projects or road works is a generally applied method of waste recycling since coal bottom ash can be used in plant substrate. This paper presents the results of research on the physical and chemical properties and the toxicity of waste originating from power stations, in the form of coal bottom ash collected from pulverised-fuel and fluidised-bed boilers. To evaluate the eco-toxicological properties of waste, a series of plant growing tests were conducted, with the use of selected plant species, as well as germination tests in water extracts of waste. The latter were intended to determine whether coal bottom ash displayed cytostatic activity, while, in the case of pot experiments, the purpose was to determine the germination rates and growths of both above-ground and underground parts of plants. The test results indicated that the influence of the tested coal bottom ash on the plant development and growth depended on physical and chemical properties of that waste. The reactions of plants were also changing depending on the proportion of waste in soil.

1 Introduction

The production of electrical and thermal energy from coal is associated with the generation of various types of waste that is generally called a side product of fuel burning. Such products include bottom ashes, fly ashes, fly and bottom ash mixes, and the residues of exhaust-gas purification processes. Legal requirements demand that the quantities of waste should be constantly reduced, while the remaining waste should be either recycled or finally disposed of. What is decisive for the selection of the management method of the side products of fuel burning is primarily their physical and chemical properties. Those properties depend on a number of factors, e.g. type of coal, co-burning of biomass or other waste fuels, type of furnace, or exhaust-gas treatment methods, as described in many publications [1, 2, 3, 4, 5]. Generally, depending on the type of furnace or boiler, the coal burning processes are conducted at various temperatures [6]. The pulverised-fuel boilers operate at 1,200-1,400°C, while the fluidised-bed ones require a much lower temperature of 850°C. The temperature affects essentially both ash generation process and properties. The majority of the minerals contained in coal are melted in a pulverised-fuel boiler, while small particles of the liquid

* Corresponding author: sliwka@agh.edu.pl

phase are vitrified when removed from the boiler, together with exhaust gases. Consequently, spherical vitreous grains of fly ashes are generated. Besides the vitreous phase, ashes contain mullite, quartz, and hematite. The ashes collected from pulverised-fuel boilers hardly contain desulphurisation products because that process is mostly conducted by the wet method. A different situation occurs in the case of fluidised-bed boilers because lime sorbents are placed in such boilers to bind sulphur oxides. Consequently, the exhaust-gas desulphurisation products, mainly in the form of anhydrite, with the residues of unreacted sorbent and free CaO, remain in bottom and fly ashes. Since coal burning in fluidised-bed boilers occurs in the temperatures of ca. 850°C, the minerals contained in coal are not melted but only subjected to certain thermal transformations: silt minerals (e.g. kaolinite and illite) undergo dehydroxylation, while carbonates (calcite and dolomite) undergo decarbonatisation. The physical and chemical properties of waste belong to the main criteria of the determination of the possibilities and directions of the further use of a given type of waste. Such properties decide about the influence of the given waste on human health and the natural environment condition, as well as the utility values associated with the selected direction of waste application. Consequently, owing to ash properties, the ashes originating from pulverised-fuel boilers are used in the highest proportion. However, owing to the diversity and changeability of their properties (mainly chemical ones), the ashes originating from fluidised-bed boilers are used on a limited scale and usually disposed of.

Ashes are most often used as components of cement or concrete, building materials in road construction, reclamation of destroyed land, land surface paving, mining technologies, construction and remodelling of embankments etc. [4, 7, 8, 9]. One of the new directions in using ashes is land reclamation, relying on either complete or partial recreation of the soil-generating layer, or as a fertiliser. Ashes can become part of the soil backbone, or the material providing mechanical properties to the developing soil. In addition, ashes can become a fertilising material, or a component of fertiliser, ensuring physical, chemical, and biotic properties for the soil-generating backbone [10, 11, 12, 13]. Waste applied as substrate for plants should represent specific features that are essential for the plant growth and safe for the soil and water environment. The application of ashes for such purposes requires conducting prior chemical tests of pollution leaching, chemical content, as well as trials of the how ashes influence plant growth and development. This paper presents the results of research conducted on test plants with substrates made of bottom ashes originating from coal burning in fluidised-bed and pulverised-fuel boilers.

2 Materials and methods

Two types of bottom ashes, originating from bituminous coal burning, were used in our research: ashes from a fluidised-bed boiler (marked BAF) and bottom ashes from a pulverised-fuel boiler (marked BAP). In the first stage of our tests, the physical and chemical properties of ashes were established, being essential from the viewpoint of plants. The grain content was determined by sieve analysis and real density was determined by the pycnometric helium technique, using a Micrometrics Multivolume Pycnometer. The chemical composition was determined by using mass spectrometer with inductively coupled plasma mass spectrometry (ICP-MS) and inductively coupled plasma atomic emission spectroscopy (ICP-AES), a leaching test was conducted according to Standard EN 12457-2. The distilled water was mixed with the waste in the proportion of 1:10. The pH index was established for aqueous extract, as well as for the pollution content, with the use of ICP-AES and ICP-MS. Chloride was analysed by using the Volhard titration method.

The phytotoxic properties of bottom ashes were identified on the basis of phytotoxicity tests, in reference to *Lepidium sativum*, while the growth tests in reference to both *Lepidium sativum* and *Sinapis alba* [14, 15]. The phytotoxicity evaluation, in respect of the tested waste, consisted in the determination how waste influenced test-plant germination and growth. A standard phytotoxicity test was conducted, with the use of water extract from waste applied to *Lepidium sativum*. For that purpose, water extract was prepared of the tested waste (standard procedure), followed by making a number of extract dissolutions with distilled water, with the concentrations of 12.5%, 25%, 50%, and 100%. Petri dishes, lined with filter paper, were filled with 5 ml of solution each (in three repetitions) and planted with 10 seeds *(Lepidium sativum)* each. Control specimens were prepared in a similar way, however, without adding waste extract. All the specimens were left until germination occurred. After 72 hours, the numbers of germinated seeds were counted, while the length of plants' sprouts and roots were measured.

Next, the pot experiments were conducted. The test plants were cultivated on universal substrate (pH 5.5), mixed with tested waste (in the proportions of 20% and 40% vol.). Additionally, bottom ash collected from a fluidised-bed boiler was mixed, in the same proportions, with soil prepared of universal substrate (70% vol.) and sand (30%) because of its binding properties. Reference or control specimens were also prepared, without using waste addition. 10 test seeds were planted in each pot. Each pot specimen was prepared in four repetitions. Plants were grown in controlled laboratory conditions (in respect of humidity and access to light). The plant germination and early growth observations were conducted during fourteen days. A single-factor statistical analysis of variance (ANOVA) and post-hoc Tukey tests were applied to evaluate how substrates influenced the plant development, using the Statistica software for that purpose.

3 Results

The grain content of both types of bottom ashes is presented in Fig. 1. When one compares the test results, one can notice clear differences in the grain content. The bottom ash originating from a fluidised-bed boiler contained grains of up to 8 mm in diameter, while that from a pulverised-fuel boiler contained much larger grains, reaching up to 20 mm. The specific density of BAF amounted to 2.68 g/cm^3, while that of BAP was much lower, reaching 2.27 g/cm^3.

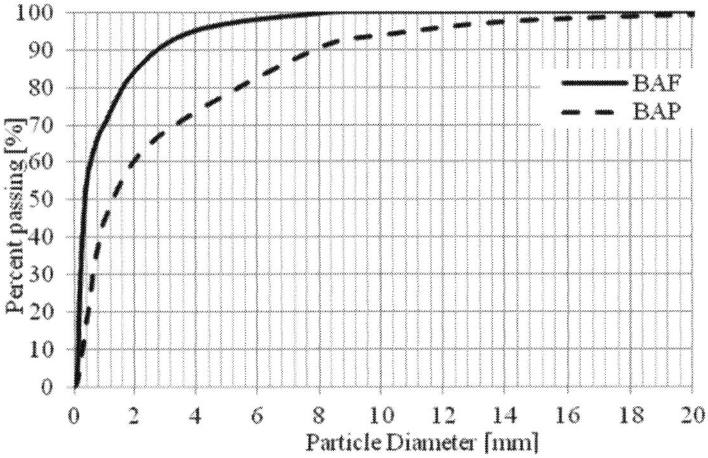

Fig. 1. *The grain content of the bottom ash originating from a fluidised-bed (BAF) and pulverised-fuel (BAP) boilers*

Chemical composition is presented in Table 1. The differences in compound content were associated with the origin of bottom ashes. In the case of bottom ash originating from a pulverised-fuel boiler (BAP), SiO_2 (78.3% wt.) was its main component, with much less of Al_2O_3 (7.9% wt.) and Fe_2O_3 (6.2% wt.). The chemical composition of BAF was rather different, with 49.9% wt. of SiO_2, 15.7% wt. of Al_2O_3, and 3.4% wt. of Fe_2O_3. One could notice the influence of the exhaust-gas desulphurisation process conducted in a fluidised-bed boiler, as the CaO content amounted to nearly 20% wt. and that of SO_3 to 7.5% wt. in BAF. The heavy-metal contents were higher in BAF than in BAP.

Water eluate reaction of BAF was alkaline and it amounted to 12.25, while that of BAP was neutral (pH = 7.06). The leaching rate of heavy metals and chlorides was very low in both types of ashes, although that of sulphates amounted to 555 mg/dm^3 in BAF. In the case of BAP, sulphate leaching rate was much lower and it amounted to 23.1 mg/dm^3. The sources of the origin of bottom ashes, associated with specific chemical compositions and pH reactions essentially influenced the plant development, as presented below.

Table 1. Chemical properties of bottom ashes

Symbol of ash	Chemical composition [% mass]									
	P_2O_5	Mn_2O_5	SiO_2	Al_2O_3	Fe_2O_3	CaO	MgO	K_2O	Na_2O	SO_3
BAF	0.1399	0.0383	49.9000	15.7331	3.4260	19.9045	1.1149	1.4460	0.4149	7.5897
BAP	0.1901	0.0375	78.3000	7.9575	6.2593	1.2902	0.6582	0.1999	0.1817	0.3559
	Trace element content [mg/kg]									
	As	Cd	Cr	Cu	Hg	Ni	Pb	Sr	Ti	Zn
BAF	17.84	0.54	77.40	25.92	0.09	38.61	59.74	137.72	164.21	167.04
BAP	7.18	0.16	10.29	8.15	0.03	4.85	1.75	88.13	86.10	123.66
	Concentration in leachates [mg/dm^3]									
	pH	Cd	Cr	Cu	Hg	Ni	Pb	Zn	SO_4	Cl
BAF	12.25	0.0004	0.0004	0.0004	0.0004	0.0030	0.0004	< 0.002	555.70	3.30
BAP	7.06	0.0007	0.0060	0.0066	0.0001	0.0058	0.0002	< 0.002	23.10	5.60

The phytotoxicity tests of the ashes indicated a beneficial influence of BAP on the development of test plants' rhisosphere, in reference to *Lepidium sativum* (Fig. 2). Based on a single-factor statistical analysis of variance (ANOVA) and post-hoc Tukey tests, it was found that the differences in the root length were statistically significant in the water solutions of the ash, made with the concentrations of 25% and 50%. However, no statistically significant differences were found in root lengths, in the case of the water solutions prepared of BAF.

The addition of bottom ash from pulverised-fuel and fluidised-bed boilers to substrates differently influenced the test plant's sprout and root growth (Figs. 3 and 4). To confirm the influence of substrate on the plant growth, a single-factor statistical analysis of variance (ANOVA) and post-hoc Tukey tests were conducted.

The pot experiments indicated a non-beneficial influence of the bottom ashes originating from a fluidised-bed boiler on test plant germination and development. The addition of ash caused suppression or delay of germination of both species planted in specimens, with the 40% ash content in substrates (Figs. 5 and 6). The addition of 20% of ash to substrate caused a delay in test plant germination. In both test groups (with 20 and 40% of ash added to substrate), strong growth inhibition was observed in the plant root and sprout zones (Figs. 5 and 6).

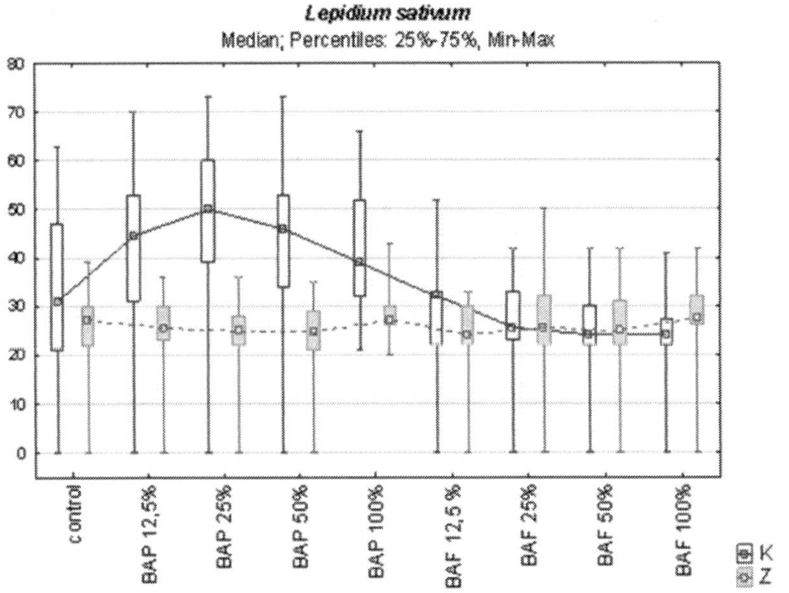

Fig. 2. *Sprout (Z) and root (K) lengths of Lepidium sativum: germination tests in waste water solutions*

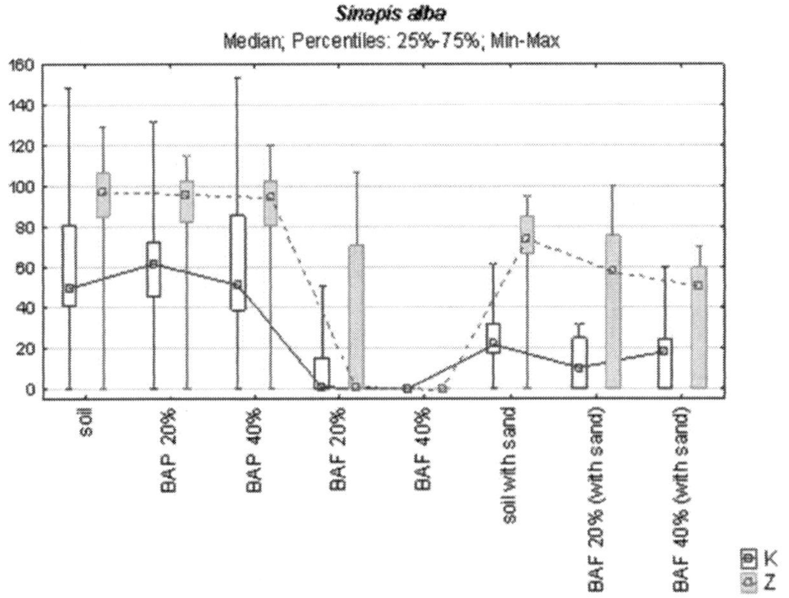

Fig. 3. *The Sinapis alba sprout and root lengths, with various substrates*

It was caused by the binding properties of ashes after hard coal burning in a fluidised-bed boiler and higher water demand of ashes from fluidized bed boilers, related to the shape of grains and lack of glassy phase. In addition, BAF contain a significant amount of CaO that absorbs moisture.

Upon statistical analysis completion, it was found that the inhibition of the plants' root and sprout growths were significant, in comparison to control specimens. It was also concluded that the differences between the sprout lengths of the test plants growing on the substrates with the addition of 20 and 40% of ash were also significant. A higher proportion of ash in substrate inhibited sprout growth. That phenomenon was not observed in roots.

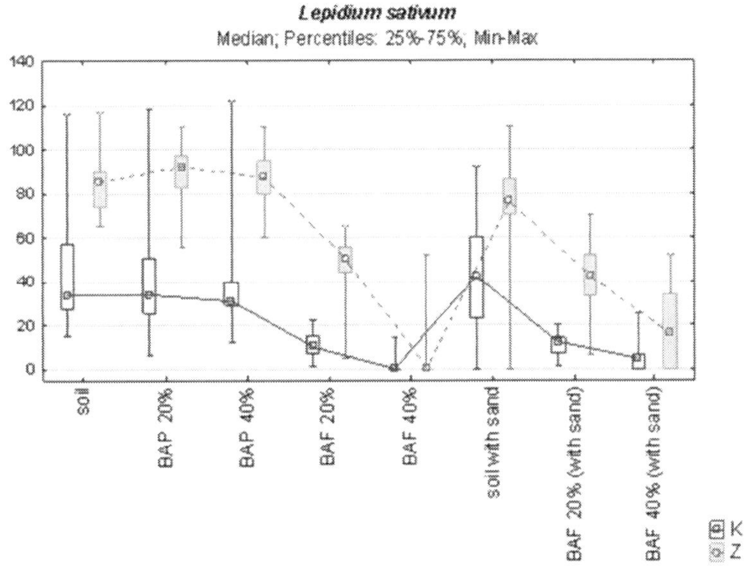

Fig. 4. *The Lepidium sativum sprout and root lengths, with various substrates*

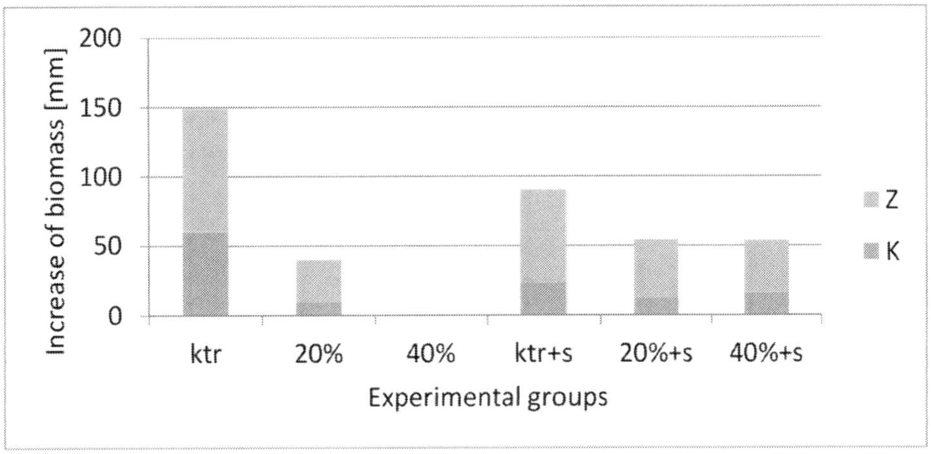

Fig. 5. *The Sinapis alba biomass growth in the experimental specimens with the admixture of BAF waste, with the proportions of 20 and 40% in substrate and the admixture of BAF in substrate with sand (20 and 40%, respectively) (Z – sprouts, R – roots, ktr - control group, s – with sand)*

However, the addition of sand to the BAF containing substrates caused a reduction of the binding properties of waste, and that further affected increased plant germination and positively influenced the sprout length in both test-plant species, at a 40% proportion of waste in substrate, in comparison to analogous specimens, without sand admixture. In comparison to control specimens (with sand admixture), the *Lepidium sativum* sprout and root grows and *Sinapis alba* root growths were smaller in the experimental specimens with the addition of sand. Those differences were statistically significant.

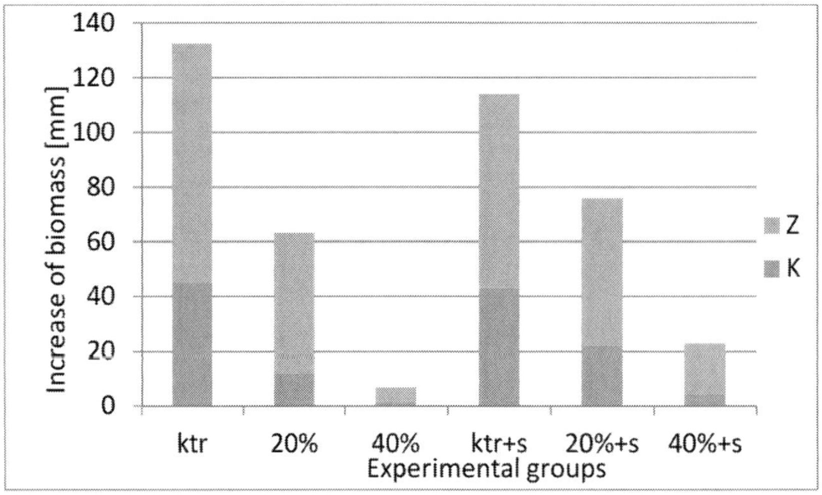

Fig. 6. *The Lepidium sativum biomass growth in the experimental specimens with the admixture of BAF waste, with the proportions of 20 and 40% in substrate and the admixture of BAF in substrate with sand (20 and 40%, respectively)*

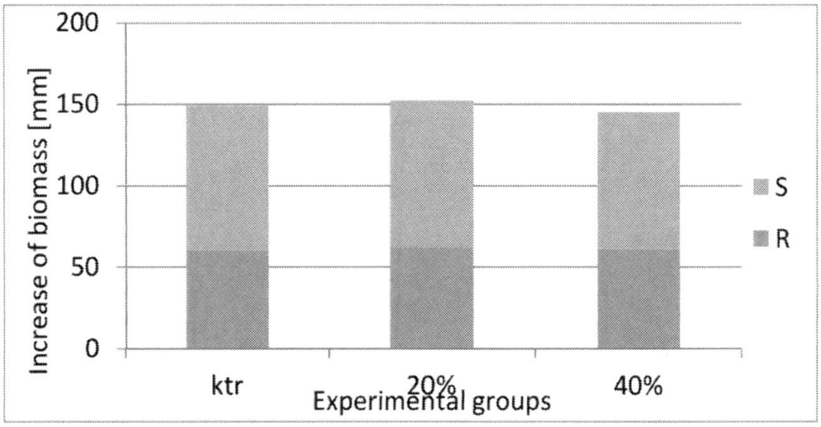

Fig. 7. *The Sinapis alba biomass growth in the experimental specimens with the admixture of BAP waste, with the proportions of 20 and 40% in substrate*

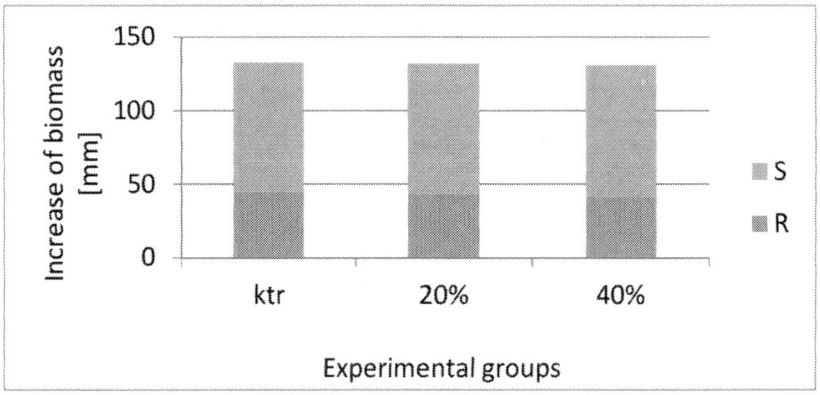

Fig. 8. *The Lepidium sativum biomass growth in the experimental specimens with the admixture of BAP waste, with the proportions of 20 and 40% in substrate*

The addition of BAP waste to cultivation substrates did not influence significantly the test plants' growth or development. The plants similarly tolerated the addition of 20 or 40% of waste (Figs. 7 and 8). Our research results indicated a possibility of using the tested types of waste in nature. Taking into account, however, the grain content of waste, one may also consider conducting experiments on a mixture of BAP waste with heavy soils (clayey or silt soils). Owing to the identified lack of waste toxicity in respect of test plants, content of components and granulometric composition, BAP waste can positively affect the properties of heavy soils, with low macro-component content.

4 Conclusions

This paper presents the results of the experiments concerning the evaluation of the properties of bottom ashes originating from coal burning in fluidised-bed and pulverised-fuel boilers, with the intention to use ashes in nature. Based on our tests, no toxicity was identified in bottom ashes from a pulverised-fuel boiler, in respect to test plants. When testing the influence of water solutions of the two types of ashes on the plant germination capability, stimulating activity was observed with the concentrations of 25 and 50%. At the same time, the pot experiments did not indicate any significant differences in sprout or root growths of *Sinapis alba* and *Lepidium sativum*, in the specimens with the admixtures of those types of waste, as compared to control specimens. Our test results indicated a possibility of application of the tested bottom ashes from pulverised-fuel boilers in nature. Besides, it seems to be justified to continue our experiments, with higher ash content in experimental substrates and in mixtures with heavy soils (clayey or silt soils). Owing to the identified lack of waste toxicity in respect of test plants, macro-component content, and granulometric composition, BAP waste can positively affect the properties of heavy soils, with low macro-component content. Our tests conducted on water extracts of the bottom ashes originating from fluidised-bed boilers did not indicate any ash toxicity, despite high sulphate content and high pH. Our pot experiments identified inhibiting influence of wastes on plant germination and growth. Such ashes caused substrate solidification and that suppressed plant germination in the time that was specific for each of the test species. The plants germinated with a delay and their condition was visibly poorer in comparison to that of control specimens. The addition of sand to substrates caused higher sprout growths at 40% vol. proportion of the bottom ash originating from a fluidised-bed boiler in substrate, however, growths were significantly lower than those of the control specimens. A high sulphate content and the binding properties of ashes suggested that it was not possible to apply those ashes in nature, especially directly as additives to soil or land.

This study was conducted under statutory research of the Department of Environmental Engineering and Mineral Raw Material Processing (Project No. 11.11.100.482), Faculty of Mining and Geoengineering, AGH University of Science and Technology in Kraków, Poland.

References

1. Ahmaruzzaman M., A review on the utilization of fly ash, Progress in Energy and Combustion Science, **36**, pp. 327–363, (2010)
2. Uliasz-Bocheńczyk A., Mazurkiewicz M., Mokrzycki E. Fly ash from energy production – a waste, by product and raw material. Gospodarka Surowcami Mineralnymi - Mineral Resources Management. **31**, pp. 139–150, (2015)
3. Śliwka M. Pawul M. Kępys W. , Pomykała R., A Waste management options for the combustion by-products in the context of the retardation of soil resources' depletion. Journal of Ecological, **18**, 5, pp. 216–225, (2017)

4. Galos K., Gawlicki M., Hycnar E., Lewicka E., Nieć Marek., Ratajczak T., Szlugaj J., Wyszomirski P., Raw materials from mineral waste. Ed. IGSMIE PAN, Krakow., (2009)
5. Grammelis P., Skodras G., Kakaras E., Effects of biomass co-firing with coal on ash properties. Part I: Characterisation and PSD. Fuel, **85**, pp. 2310–2315, (2006)
6. Vassilev S.V., Vassileva Ch.G., Karayigit A.I., Bulut Y., Alastuey A., Querol X. Phase–mineral and chemical composition of composite samplesfrom feed coals, bottom ashes and fly ashes at the Soma power station, Turkey. International Journal of Coal Geology **61**, pp. 35– 63, (2005)
7. Kępys W., Pomykała R., Research into the sefulness of Ash from the Co-Combustion of Lignite and Biomass in Mining Technologies. Polish Journal of Environmental Studies. **23**, 4, pp. 1381-1384, (2014)
8. Singh M., Siddique R., Effect of coal bottom ash as partial replacement of sand on properties of concrete. Resources, Conservation and Recycling, **72**, pp. 20– 32, (2013)
9. Andrade L.B., Rocha J.C., Cheriaf M., Influence of coal bottom ash as fine aggregate on fresh properties of concrete. Construction and Building Materials, **23**, pp. 609–614, (2009)
10. Śliwka M., Uliasz-Bocheńczyk A., Pawul., An appraisal of the properties of bottom waste obtained from bio-mass congestion to estimate the ways of its environmental use. Polish Journal of Chemical Technology, **19**, 2, pp. 33-37, (2017)
11. Antonkiewicz J.,. The use of bottom ash to bind of heavy metals in soil. Ochrona Środowiska i Zasobów Naturalnych, **41**, pp. 398 – 405, (2009)
12. Bielińska E. J., Baran St., Stankowski S., Ocena przydatności popiołów fluidalnych z węgla kamiennego do celów rolniczych, Inżynieria rolnicza, **6**, 15, pp. 7-15, (2009)
13. Czech T., Gambuś F. Wieczorek J., Assessment of chemical composition of waste materials from hard coal burning in view of their agricultural and environmental applications. Inżynieria Ekologiczna, **34**, pp. 89-95, (2013)
14. Baran A. Śliwka M. Lis M., Selected properties of flotation tailings wastes deposited in the Gilów and Żelazny Most waste reservoirs regarding their potential environmental management. Archives of Mining Sciences, **58**, 3, pp. 969–978, (2013)
15. Traczewska T. M., Biologiczne metody oceny skażenia środowiska. Wrocław. Oficyna Wydawnicza Politechniki Wrocławskiej, (2011)

E3S Web of Conferences **46**, 00020 (2018) https://doi.org/10.1051/e3sconf/20184600020
3rd International Conference on Energy and Environmental Protection

Experiment designing in operating investigations of subassembly of the marine energy systems

Jan Monieta[1,*]

[1]Maritime University of Szczecin, Wały Chrobrego 1-2, 70-500 Szczecin, Poland

Abstract. To choose the receive locations of diagnostic signals, reliability and preliminary investigations were applied. The functional systems and the most unreliable subassemblies were determined. The studies were conducted for the auxiliary marine engines with the same rated speed and fueled with same distillation and residual fuels. In the article a choice of the ship's object, which has a significant impact on reliability, operating costs and the functions performed, is presented. Reliability, economic and risk of exploitation investigations were carried out, which allowed selecting the facility with the highest failure, costs of adverse events and threat to the natural environment. The research covered the objects of a ship owner operating mainly bulk cargo vessels for mass transport and their essential assemblies – combustion engines.In order to obtain a reasonable amount of information on the state of the examined combustion engine, the following experimental planning methods were used: experiment type, essential elements, location of measurement sensors, measured quantities, test conditions, measurement frequencies, measurement data analysis and inference methods, etc.The reliability indicators and preliminary diagnostic tests were used to select the diagnostic signal reception points. Functional systems and the most unreliable subassemblies were defined in this way. The tests were carried out for auxiliary marine engines with the same rotational speed and fueled by distillation and residual fuels. The tests were conducted for several types of combustion engines.

1 Introduction

Floating objects are required to minimize threats to people and the environment, the highest efficiency, quality, minimum operating costs, safety, high reliability, etc. [1, 10]. As a result, the complexity of these objects, the power of the propulsion systems, while striving to reduce the size, increases. Therefore, the demands for people involved in the design, manufacture and operation of ship machinery and processes performed by them are increasing. Thus, the works are carried out to improve reliability, efficiency and reduce harmful pollution, vibrations and noise of floating objects [12].

* Corresponding author: j.monieta@am.szczecin.pl

Designing of diagnostic experiments in relation to marine combustion engines was carried, among others, by the authors of the works [3, 9, 10, 12, 14]. Used diagnostic procedures for planning experiments included selection: research facilities, the technical characteristics of the object, of the diagnostic parameters and locations of observation and diagnostic symptoms. To include the reasonable amount of information on the state of the internal combustion engine test, use the methods of experiment planning to make the choice: the type of experiment, the sensors locations, measured values, test conditions, frequency of measurements and methods of reasoning.

2 Planning of diagnostic experiments

A properly planned and carried out experiment is the starting point for further organized activities, the result of which should be a diagnosis of the object's condition. The basic element of the experiment is measurement, which is the process of receiving and processing information about the value in order to obtain a quantitative result for transmission in space and time through registrations, mathematical development or application to control [15]. One of the measures of informativeness is the quotient of a number of measurements necessary to implement a given program to the number of measurements of the complete program. The development of the theory of experiment was determined by the following facts [15]:

- the combinations number of values exceeds the possibility of conducting tests,
- not all factors cannot be tested separately,
- are factors whose values cannot determined or are unmeasurable and unknown.

The experiment theory is aimed at improving the efficiency of scientific research, as a quotient of the quantity and quality of information to the costs and time research incurred [15]. For complex objects, it is required to include n factors and measurements:

$$m = m_1 m_2 \dots m_i \dots m_n \tag{1}$$

where :

m – number of measurements,
m_i – i-th factor,
n – n-th factor.

A need to determine the functions of many variables in the form of:

$$Y = F(x_1, x_2, \dots, x_i) \tag{2}$$

where :

m – vector output quantitiess,
F – function,
x_i – input quantity.

A feature of some classical methods is factor analysis, as estimation of the frequency of events subjected to testing, randomization, sequential inference, selection of information from studies, statistical analysis, etc. [7, 8, 15]. The possibilities of isolating dominant factors were also addressed in order to avoid overly complex models.

3 State of knowledge

The authors of the works, among others, designed diagnostic experiments in relation to internal combustion engines [2, 9, 12, 14, 15]. The applied procedures of planning diagnostic experiments included the selection of: research objects, features of the technical state of the object, diagnostic parameters and observation sites as well as diagnostic symptoms. In order to take into account a reasonable amount of information on the condition of the combustion engine under test, the following experimental planning methods shall be selected: type of experiment, location of measurement sensors, measured quantities, test conditions, measurement frequency and inference methods. Diagnostic actions result in a large set of diagnostic parameters, so there is a need to select relative attributes.

Selection of engine's parameters and conditions of taking measurements or registering vibration signal are usually based on references, researcher's experience or intuition. The article presents exemplary vibroacoustic diagnostics of combustion engine [13]. Selection of test points of vibration on the basis of impact tests' results was suggested.

For a more efficient and environmentally friendly ship operation, a number of measures have been proposed [5]: the introduction of the electronically controlled versions of marine engines, in which the fuel injection and outlet valve opening/closing timings can be controlled and the engine can operate in various modes with high efficiency and low emission, the application of the outlet gas bypass, the usage of turbochargers with variable geometry turbines and the installation of waste heat.

In cases where a set of parameters is numerous and the values of these parameters are similar, there the authors proposed a methodology for classification of acquired diagnostic parameters. This method can be used to build a diagnostic parameter ranking that divides the set of diagnostic variables into stimulators, dissimulators and nominators. Input quantities limited to three elements, i.e. engine shaft speed, engine torque and fuel injection advance angle, were taken into account. Range of output values: carbon dioxide condensation in the outlet manifold, hydrocarbon condensation in the outlet manifold, nitrogen oxide condensation in the outlet manifold, outlet gas temperature and excess air ratio.

4 Results and analysis

A motor vessel is a complex system, which is why it is important to choose an object for diagnostic tests. The system is the set of elements called objects, related to each other and the relationships between elements *R*:

$$O = \{O_1,\ O_2, \dots, O_n,\ R_1,\ R_2, \dots, R_m\} \tag{3}$$

where :

O_l – Element of the system) O,i=1, 2, …, n,

R_j – relationships between elements of the system, j = 1, 2, ..., m.

The significance of the subsystem can be determined by the set of criteria and is the function of the degree of fulfillment of the conditions set by C_{ri} criteria:

$$I = f(C_{r1}, C_{r2}, \dots, C_{ri}, \dots,\ C_{rn}) \tag{4}$$

where :

C_{ri} – criteria 1, 2, ..., n.

In the bulk carrier you can also selected the important elements, among others, on the basis of the frequency and costs of adverse events [4, 10]. For the selected populations of ships the measure of those losses *L* for a period $(\tau, \tau + \Delta\tau)$ can be described by the relationship [10]:

$$L(\tau, \tau + \Delta\tau) = \sum_{i=1}^{n(\tau,\tau+\Delta\tau)} L_i \tag{5}$$

where :

L_i – collective loss of the i-th of the event in the period $(\tau, \tau + \Delta\tau)$

n – 1, 2, ..., ∞ is the number of adverse events in this period.

Collective losses can be presented in relative formulate L_{ci} and expressed as a percentage:

$$L_{ci} = \frac{L_i}{L(\tau,\tau+\Delta\tau)} 100\% \tag{6}$$

The measure used to assess the safety is the risk of losses of the object. For to the risk assessment was adopted the qualitative and quantitative safety criteria [11]:

$$R = f \cdot E \tag{7}$$

where :

R – risk,
f – frequency of occurrence of an accident sea,
E severity of the effect of case.

The risk of the accident is defined as the product of the probability of the risks and consequences of the accident:

$$R_a = P_i \cdot C_i \tag{8}$$

where :

R_a – Risk of an accident,
P_i – Probality of the risks,
C_i consequences of the accident.

The greatest losses were caused by fires in marine propulsion units, which described adverse events with significant losses affecting the safety of the crew and the environment. The evaluations show that neither the quantitative nor the economic analyses are satisfactory, but the risk analysis.

The research covered a period of 12 years in which 1040 adverse events were analyzed. This is a sufficient sample to conclude on the state of the population. Maritime ship to transport of bulk cargo was divided into a deck and machinery equipment, and functional systems. It was decided to decomposition the ships taking into account the literature and book of Safety Management Systems of the company. Separated systems have been allocated to deck and engine room crew [10].

The unreliability function, i.e. the probability of damage to the functional system (element) for a fixed instant is [9]:

$$P^*(\tau) = \frac{m(\tau)}{n} \tag{9}$$

where :

$m(\tau)$ – number of objects damaged in the time interval <0,τ>,
n – number of examined objects.

Fig. 1 the results of reliability tests for 3 tested engines of the same type and fueled with residual fuels are presented. The results also depend on the assumed of engine damage and the boundary in the serving systems. For tests carried out for internal combustion engines driving generators fueled distillation fuels, the most unreliable system was the feed system [9], and for the same type of engine, but adapted to residual fuel combustion, the lubrication system turned out to be the most unreliable. Frequent works on oil are replacement from contamination with combustion products. Feed and lubrication systems also the most unreliable components of the injector and oil filter have the significant impact on energy effects and the environment.

The usefulness of various signals and measures is presented in various works [11]. This work presents the design of an experiment for vibration acceleration and fuel pressure signals in injection subsystem. Because the feed system turned out to be the most frequently failing engine system, the influence of the signal reception site on the level of measures in various domains was studied.

Fig. 2 presents the impact of the place of mounting the vibration acceleration sensor on the signal courses. The signals from various measuring points are synchronized.

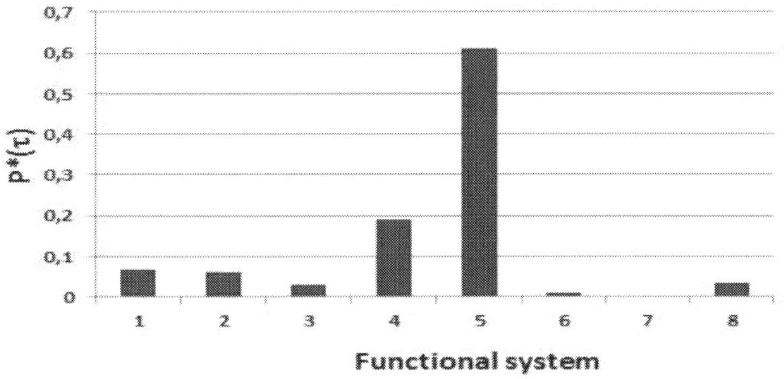

Fig. 1. Diagram of the probability of the unserviceableness state of functional systems of marine engines fuelled by residual fuels: 1 – block, 2 – piston-crank, 3 – exchange, 4 – starting air, 5 – feed, 6 – lubricating, 7 – cooling, 8 – control

Fig. 3 presents the results of research on the influence of the vibration acceleration signal direction on the values of amplitude estimates and kurtosis for 4 tested cylinders of combustion engine, at the same engine load. The highest values of the amplitude estimate were obtained for the sensor with signals received along the injection pipe.

The engine operating conditions are determined by: the ambient conditions, the control of the engine by the operator and the moment of resistance depending on the character of work performed by the engine. The engine work is characterized by set **Z** that is a sum of the sets of the engine conditions and engine operating states [2]:

$$Z = W \cup T_s = \{Z_1, Z_2, \dots, Z_{in}\} \tag{10}$$

where :

Z – set,
W – engine conditions,
T_s – engine operating states.

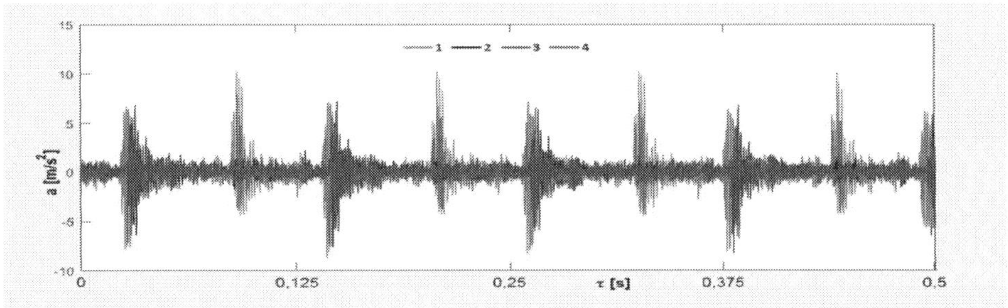

Fig. 2. Impact of the location of the vibration acceleration sensor on the course of the vibration acceleration signal: 1÷4 – sensor attachment points

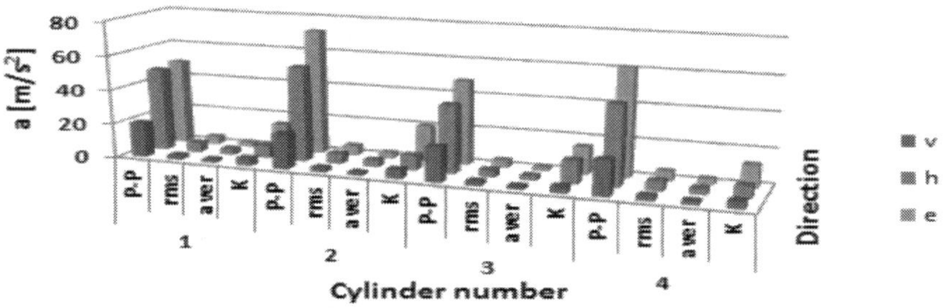

Fig. 3. The influence of the vibration acceleration signal reception direction on the values of amplitude estimates and kurtosis for 4 tested cylinders of one engine: P-P – peak to peak, rms – root-mean-square value, aver – average value, K – kurtosis, v – vertically, h – horizontally, e – along

5 Selection of signals and diagnostic parameters

The reduction of the signal vector was done in preliminary tests, from which the useful one was selected. The basis for the selection was the correlation matrix for the selection of parameters independent of each other, and at the same time having very close links to the independent variable. It consists in determining the correlation coefficient between the values of parameters obtained in subsequent measurements and independent variables (engine load) according to the formula:

$$k = \frac{\sum_{i=1}^{n}(x_i - \bar{x})(y_i - \bar{y})}{\sqrt{\sum_{i=1}^{n}(x_i - \bar{x})^2 \sum_{i=1}^{n}(y_i - \bar{y})^2}} \tag{11}$$

where :

- i – the sequence number of the input or output quantities,
- x_i – another value of the input size,
- $\bar{x}$ – average value of the input volume in the set of observations,
- y_i – successive value of a given diagnostic parameter,
- $\bar{y}$ – average value of the given diagnostic parameter.

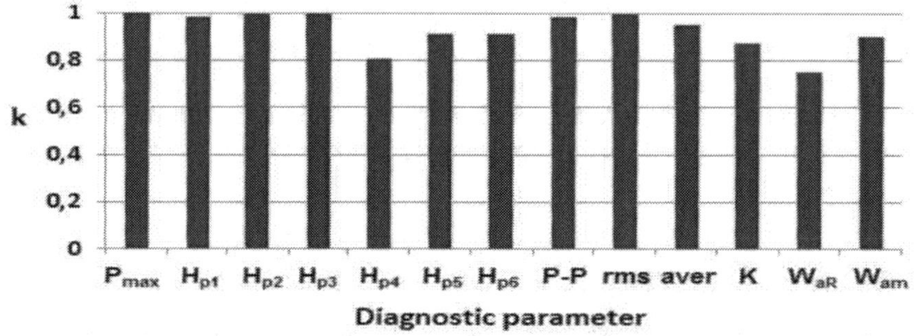

Fig. 4. Values of the correlation factor between exemplary diagnostic parameters and relative engine load: Pmax –the maximum value of the pressure course in the injection subsystem, Hp1÷Hp6 – amplitude components of the pressure spectrum in the injection subsystem, WaR – wavelet range, Wam – wavelet max

Values of the correlation coefficient for assessment of the usefulness of exemplary diagnostic parameters in time, amplitude, frequency and wavelet domain indicated the Fig. 4. The values of the correlation coefficient in the scope of very strict compounds (over 0.9) were considered as useful. Most parameters meet this criterion.

6 Conclusion

The article presents the criteria for selecting the test object on the example of the marine vessels. The analysis of adverse events in sea-going vessels should take into account both the number of damages, the economic technical aspects and the risk. The selected objects of sea-going ships are combustion engines and their elements are injectors.

Methods for selection of measurement points using the criterion of reliability and preliminary test results are presented. The input quantities in the preliminary diagnostic tests of the marine combustion engine are: relative load of the engine (measured setpoint, of fuel dose, etc.), in the basic research the technical condition of the engine. In the conducted research, a static, determined plan was used.

The possible solutions for the reduction of measurement sizes and methods of data analysis have been pointed out.

In order to obtain the reasonable amount of information about the state of the tested combustion engine, it was selected by means of experimental design methods: experiment type, relevant elements, location of measurement sensors, measured quantities, test conditions, measurement frequencies, measurement data analysis and inference methods, etc.

This research outcome has been achieved under the research project "Ecological and economic aspects of the operation of selected elements of marine power systems" No. 1/S/IESO/2014 financed from a subsidy of the Ministry of Science and Higher Education for statutory activities.

References

1. Blasco J., Duran-Grados V., Hampel M., Moreno-Gutierrez J. Towards an integrated environmental risk assessment of emissions from ships' propulsion systems. Environ International, USA, **66**. Pp. 44–47, (2014)
2. Chłopek Z. Some remarks on engines testing in dynamic states. Combustion Engines, Poland, pp. 60–71, (2010)
3. Girtler, J., Kuszmider, S. & Plewiński, L.: Selected issues of the operation of ships in the safety aspect of the navigation, Maritime University of Szczecin, Szczecin (in Polish), (2003)
4. Grzywaczewski Z.: Reliability of ships, WEMA Machine Industry Publishers, Warsaw (in Polish), (1988)
5. Guan C., Theotokatos G., Chen H. Analysis of two stroke marine diesel engine operation including turbocharger cut-out by using a zero-dimensional model, Energies, Switzerland, **8**, pp. 5738–5764, (2015)
6. Kontovas, C.A., Psaraftis, H.N. Formal safety assessment: a critical review, Marine Technology, USA, **46**, 1, pp. 45–59, (2009)
7. Korzyński M.: Experiment methodology, WNT, Warsaw (in Polish), (2013)
8. Leszek W.: Selected methodical issues of empirical research. Library of Exploitation Problems, Radom (in Polish), (2006).
9. Monieta J.: Choice of most deceptive functional system and sub-assembly of marine diesel engines. Increasing the Work Efficiency of Energy Systems. International Collection of Scientific Works. Kaliningrad, Russia, pp. 141-146, (2002)

10. Monieta J. Estimation of undesirable events in operation of bulk carriers. Journal of KONES Powertrain and Transport, Poland, **20**, 2, pp. 291-297, (2013)
11. Monieta J. The application in-cylinder pressure course in different domains for diagnostics of medium speed marine diesel engines in operation conditions. Combustion Engines, Poland, **3,** 154, pp. 104, cd. pp. 153-160, (2013)
12. Tarełko W.: Methodology of designing exploitation properties of complex technical objects, Library of Problems of Exploitation, Gdynia – Radom (2011)
13. Tomaszewski F., Szymanski G.M. Frequency Analysis of vibrations of the internal combustion engine components in the diagnosis of engine processes. The Archives of Transport, Poland, **24**, pp. 117 –125, (2012)
14. Zadrąg R. Kniaziewicz T. Utilization of the zero notarizations method for the building of a ranking for diagnostic marine engine parameters. Combustion Engines, Poland, **171**, 4, pp. 44-50, (2017)
15. Żółtowski B., Cempel C.: Engineering of diagnostics machines. Polish Society of Technical Diagnostics, Warsaw, Bydgoszcz, Radom (in Polish), (2004)

Experimental research on the domestic ORC micro power plant with a commercial biomass boiler

Tomasz Kaczmarczyk[1,*], *Grzegorz* Żywica[1], and *Eugeniusz* Inhatowicz[1]

[1]Polish Academy of Sciences, Institute of Fluid Flow Machinery, Department of Turbine Dynamics and Diagnostics, Generała Józefa Fiszera 14, 80-231 Gdańsk, Poland

Abstract. The purpose of the work was to experimentally evaluate the operation of the domestic ORC micro power plant that uses a commercial biomass boiler fueled with wood pellets. The boiler, with a maximum thermal power output of $45kW_t$, uses a heating jacket and thermal oil as a working medium. The prototypical domestic ORC micro power plant was equipped with a multistage radial-flow microturbine that can generate electricity ($2.5kW_e$ at a rotational speed of 24,000 rpm). The microturbine is a key component of the turbogenerator, which was manufactured in oil-free technology. The turbogenerator's high-speed bearings are lubricated with the low-boiling medium's vapour. The HFE7100 fluid was used as a working medium in the ORC installation. The paper discusses the thermodynamic conditions to be met for effective operation of the boiler and the results of experimental research. The operating characteristics of the ORC installation and the biomass boiler were presented. Problems that occurred while testing the micro-cogeneration power plant with the boiler and their impact on the electric and thermal efficiency of the cycle were discussed.

1 Introduction

In the European Union countries, the development of modern technologies that can help in reducing global climate changes and CO_2 emissions has been observed for the last dozen or so years. The data provided in [1,2] show that in recent years in Poland, hard coal consumption by households and small enterprises accounts for over 30% of the total domestic consumption. It is estimated that approximately 50% of the world's energy production is wasted in the form of heat released into the atmosphere, due to the limitations of energy conversion processes [3]. Waste heat can come from various sources, including industrial and domestic installations [4,5]. By using heat recovery systems, it is possible to harvest energy not only from commercial energy systems but also from natural renewable energy resources. Therefore, it is reasonable to consistently introduce new energy technologies and smart energy management systems [6] in order to increase the efficiency and profitability of existing and newly built energy system installations [7]. There are

* Corresponding author: tkaczmarczyk@imp.gda.pl

promising signs of progress in the field of distributed power generation, which enables high-efficiency generation of electricity, heat or cold. The use of micro-CHP systems can meet the demand for thermal and electric energy in single-family houses in which the demand for electricity does not exceed 10 kW_e [8,9]. Hence the constant search for expansion machines that can be used in CHP systems [10-12]. Commercial solutions with electric power capacities below 10 kW_e are virtually not available [13], though their potential use in energy-efficient buildings (e.g. public utility buildings or single-family and multi-family houses) appears to be unquestionable. It should be noted, however, that noise is the second largest environmental problem in the European Union, after air pollution [14]. Therefore, new CHP ORC energy installations should be characterised not only by high energy efficiency and operational reliability, low noise and vibration levels but should also have no negative impact on the environment [15].

This article presents the results of research on a small cogeneration system equipped with a biomass boiler. Such installations, with a power output capacity adapted to use them in single-family houses, are not yet available on the market, and their further development will certainly contribute to the implementation of assumptions of the new energy policy of the European Union.

2 Domestic cogeneration system

The domestic cogeneration system (Fig. 1) comprises the following components: ORC installation (1), biomass boiler (2) and the heat removal system (fan cooler). A radial-flow microturbine (3), specially designed for this purpose, was used as an expansion machine in the ORC installation. The microturbine (3) is a four-stage device, with the nominal rotational speed of 24,000 rpm at which it can produce 2.5 kW_e of electric power. It is noteworthy that the micro-turbogenerator (3) is an oil-free machine, equipped with gas bearings lubricated with the low-boiling medium's vapour. The correct operation of the cogeneration system is supervised by the control and conditioning/receiving system for the produced electric energy (4). The system guards against an excessive increase of the microturbine's rotational speed and the electrical operating parameters (current and voltage).

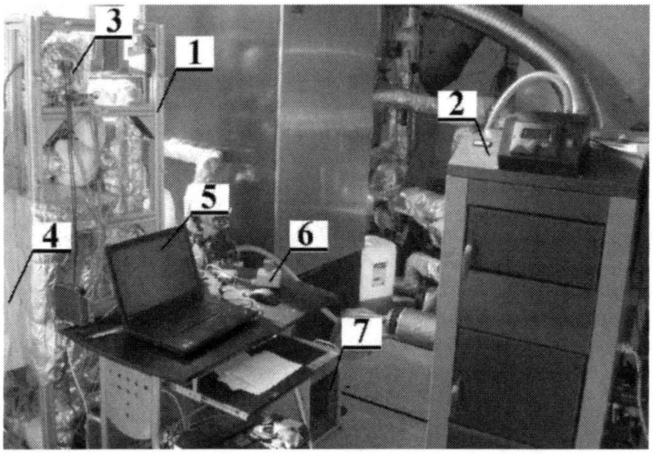

Fig. 1. The photo of the test rig which was taken during tests: 1- ORC installation, 2 – biomass boiler, 3 – radial-flow microturbine, 4 – conditioning/reception/control system for the produced electric energy, 5- control panel, 6 – emergency shutdown button of the ORC system, 7 – electricity receiver (R type)

In addition, the level of vibration of the microturbine is monitored, and if the permissible vibration level is exceeded, the machine can be stopped using the emergency shutdown button (6). The cogeneration system can work in automatic mode or in manual control mode using the user's desktop (5). As the system uses a three-phase high-speed synchronous (4-pole) electric generator, it was necessary to process the electrical signal. For this purpose, a rectifier system was specially designed with the energy consumption system that uses electric power produced by the microturbine (3). The generated electricity is consumed by a set of DC bulbs with a total electric power of 1 kW_e and also by an electric heater (7) with a rated power of 5 kW_e. The cycle uses the organic working fluid sold under the trade name HFE7100. This medium is incombustible, inexplosive, nontoxic and environmentally friendly as its ODP (Ozone Depletion Potential) coefficient is zero. The ORC (1) system consists of the following components: microturbine (3), control system (4), evaporator and condenser (that are plate heat exchangers), regenerator (shell-and-tube heat exchanger) and circulation pump. The boiler (2) with a thermal power of 45 kW_t, equipped with a heating jacket, was used as a heat source. The thermal oil (Iterm 6Mb) whose maximum operating temperature is 280°C was employed in the system as a heat carrier. The boiler (2) was equipped with a fuel feeder. During the cogeneration system tests, the biomass-fired boiler used pellets made of softwood (pine and spruce) with was a diameter of 6 mm and a calorific value of 5 kWh/kg. It should be mentioned that the humidity rate of the biomass was not measured.

3 Results of the experimental research

The research of the biomass boiler coupled with the ORC installation was of a long-term nature and lasted continuously for three days. The selected results of the research are presented in the following part of the paper. During measurements, it was found that the maximum temperature of the thermal oil at the boiler outlet and at the evaporator inlet was about 200°C. Using a thermal imaging camera (FLIR E50), it was estimated that the temperature of the flue gas in the flue was approx. 170°C (Fig. 2) and was slightly lower than the temperature of the lid (172°C) that fixed the screw-conveyor feeder and the fan that supplies boiler with air (Fig. 3).

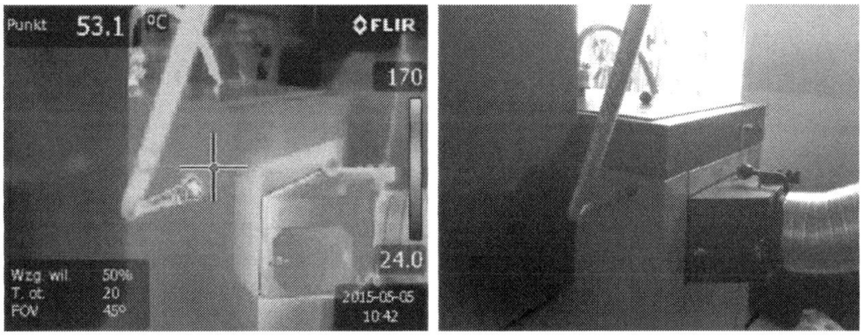

Fig. 2. The thermal image and photo of the biomass boiler's flue

Fig. 2 shows that the surface temperature of the boiler cover plate was approximately 53°C. In order to determine the electrical efficiency of the ORC cogeneration system, the tests were carried out in such a way that the electric power produced by the micro-turbogenerator was constant. An exemplary graph depicting electric power that was produced by the cogeneration system during the tests (over a period of 8,000 seconds) is shown in Fig. 4.

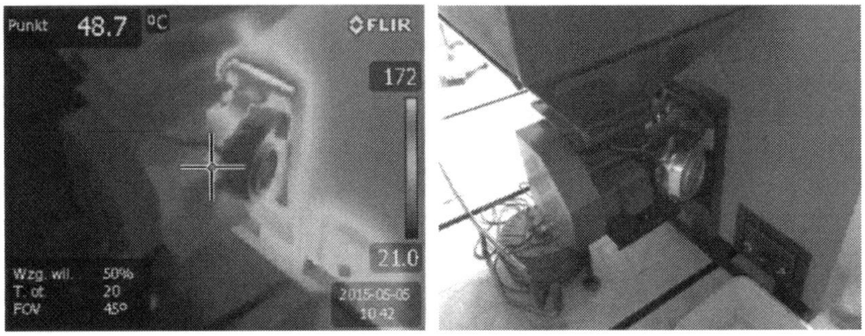

Fig. 3. The thermal image and photo of the system for attaching the feeder and the air fan to the biomass boiler

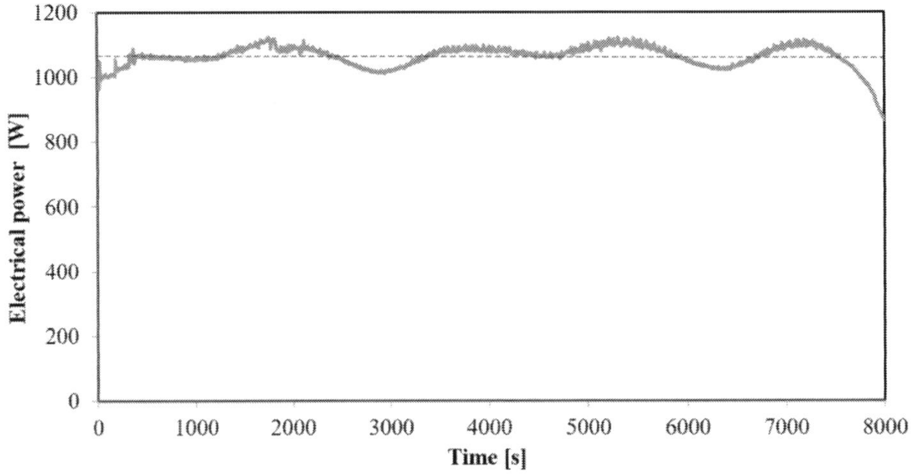

Fig. 4. Electrical power generated by the cogeneration system vs. time

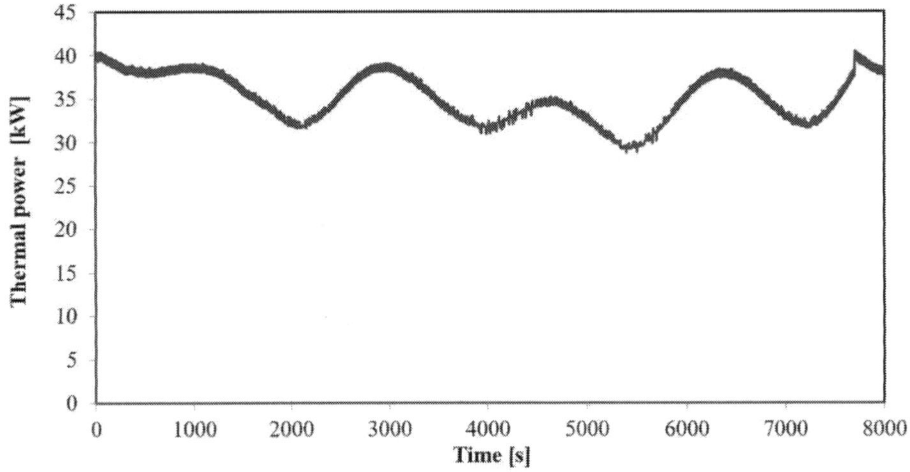

Fig. 5. Thermal power of the biomass boiler vs. time (registered during testing of the ORC cogeneration system)

The average electric power of the biomass-fired boiler, measured when it operated in the ORC installation, was 1,100 W_e. The thermal power of the biomass boiler was presented in Fig. 5. The thermal power of the boiler varied from 29.5 kW_t to 40.6 kW_t, and the average value was about 36.7 kW_t.

There are two reasons for the fluctuations in the thermal power level: 1. the high thermal inertia of the boiler as there were about 200 litres of thermal oil in the boiler and in the heating cycle (i.e. in the evaporator and piping together with its fittings); 2. the feeder and the air fan worked in cycles—resulting from the PID controller settings done by the manufacturer of the boiler; other thermophysical properties of the oil compared to water. Moreover, heating the thermal oil can cause an excessive and uneven increase in the oil temperature inside the boiler's heating jacket, which in turn causes incorrect operation of the temperature controller, which is intended for use in water cycles. The fluctuations in the thermal power of the boiler caused changes in the electrical efficiency of the organic Rankine cycle, which was shown in Fig. 6.

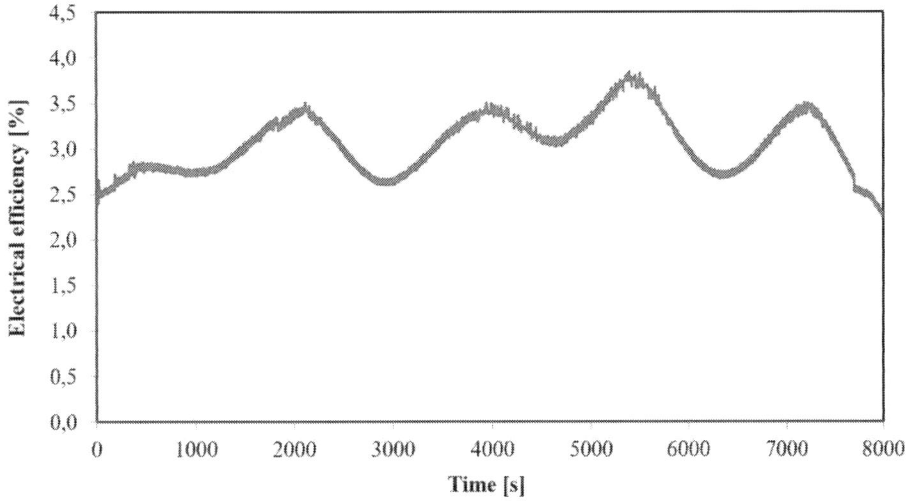

Fig. 6. Electrical efficiency of the ORC cogeneration boiler vs. time

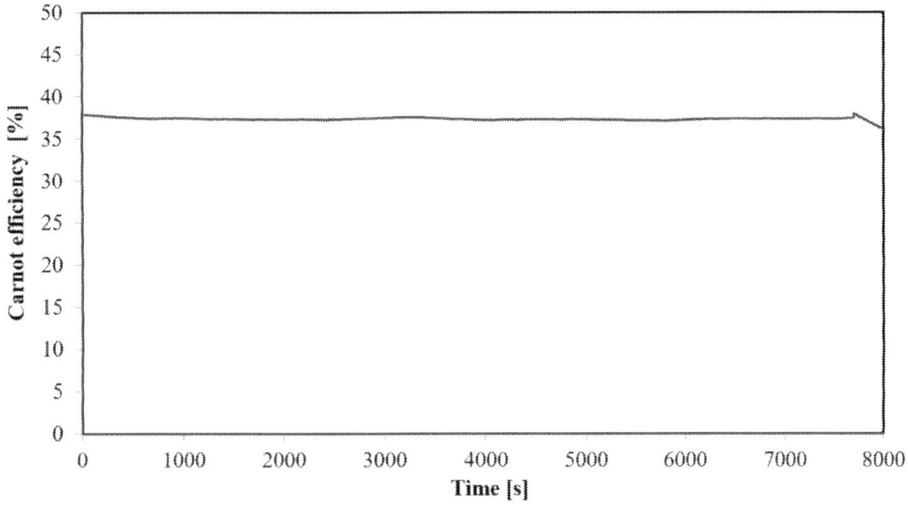

Fig. 7. Carnot efficiency of the ORC cogeneration system vs. time

The result of the incorrect operation of the control system and fluctuations in the thermal power of the boiler was the low electrical efficiency of the ORC, which ranged from 2.5% to 3.9%. The changes in the temperature of the HFE7100 medium resulted in changes in the temperature of the coolant (a solution of 40% ethylene glycol). As the matter of fact, it caused the Carnot efficiency of the cycle to be virtually constant during the research and its average value was approx. 37.2% (Fig. 7).

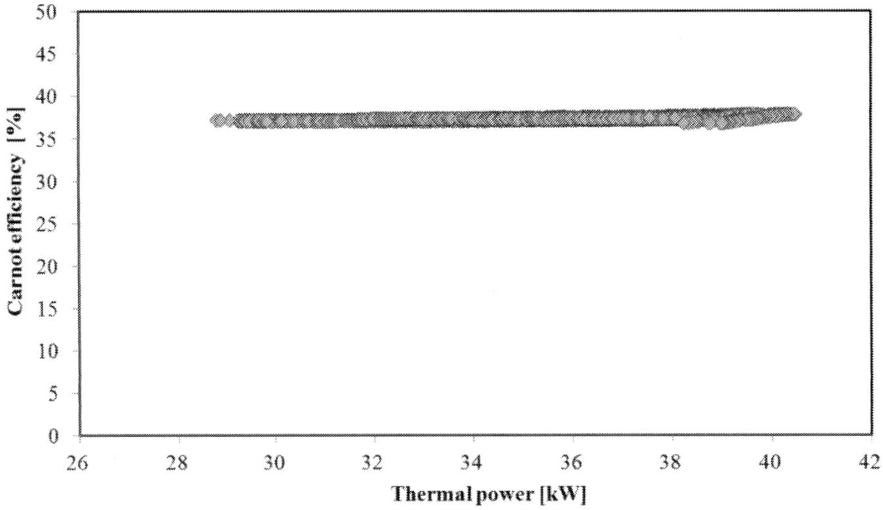

Fig. 8. Carnot efficiency of the ORC cogeneration system vs. thermal power of the boiler

According to the above statement, Carnot efficiency was constant and was not dependent on the thermal power of the biomass boiler, as shown in Fig. 8.

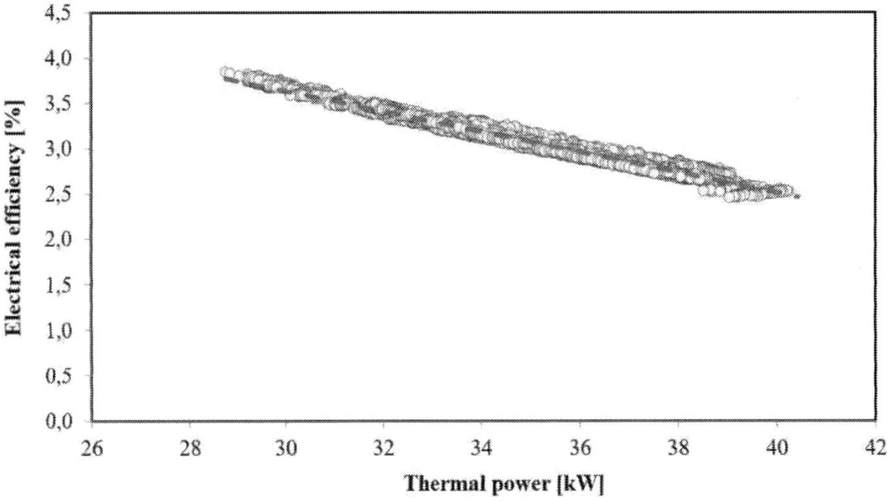

Fig. 9. Electrical efficiency of the ORC cogeneration system vs. thermal power of the boiler

Furthermore, it was found that the electrical efficiency of the ORC decreases as the thermal power of the boiler rises, which is depicted in Fig. 9.

It should be borne in mind that the microturbine worked practically at constant parameters (i.e. temperature, inlet pressure and flow rate of the working medium). What is more, the heat produced by the boiler was not effectively used by the microturbine but was directed to the cooling system, which affected the ORC system's electrical efficiency. On the other hand, the boiler operated at nominal parameters, striving to get the highest possible performance. It was found that biomass boilers operating in ORC systems should be selected according to nominal parameters of the expansion machine—in this case, the vapour microturbine. The oversized boiler caused significant heat waste that contributed to the low efficiency of the cogeneration cycle. Due to the very intense operation of vapour microturbines, boilers' thermal inertia should be as low as possible in order to ensure the optimal regime. Classic boilers with a heating jacket are characterised by high thermal inertia, which impairs the ability to control and regulate the system's operation. Besides, the control and regulation system for the boiler's components (i.e. feeder and air fan) should be properly calibrated depending on the type of working medium used (e.g. thermal oil). It should be borne in mind that the instability of parameters and any disturbances in the boiler's operation have a direct impact on the operation of the entire ORC cogeneration system.

4 Conclusion

After the tests of the 45 kW_t biomass boiler coupled with during the ORC installation with an electric power of 2.5 kW_e, it was determined that the too high nominal power of the boiler caused a non-optimal operation of the entire cogeneration installation. It was also the reason for high heat losses and the low electrical efficiency of the ORC. The maximum electric efficiency of the ORC was about 3.9% and it decreased with an increase in the thermal power of the boiler. The maximum electric efficiency of the ORC was about 3.9% and it decreased with an increase in the thermal power of the boiler. It was found that, regardless of the boiler's thermal power level, the Carnot efficiency was constant with a value of approx. 37.2%. Moreover, it was stated that the control and regulation system for the subassemblies of the biomass boiler should be individually adapted to the working medium (heat carrier). An essential parameter affecting the operation of the entire cogeneration installation is the boiler's thermal inertia, which should be as small as possible (due to the intense operation of expansion machines). It would ensure an optimal operation in a wider power range.

References

1. Consumption of Fuels and Energy Carriers in 2013, Central Statistical Office (GUS), Warsaw 2014, in Polish.
2. Consumption of Fuels and Energy Carriers in 2014, Central Statistical Office (GUS), Warsaw 2015, in Polish.
3. Rahbar K., Mahmoud S., Al-Dadah R.K., Moazami N., Mirhadizadeh S.A., Review of organic Rankine cycle for small-scale applications, Energy Conversion and Management, **134**, pp. 135-155, (2017)
4. Yaxuan X., Shuo A., Peng X., Yulong D., Chuan L., Qunli Z., Hongbing Ch., A novel expander-depending natural gas pressure regulation configuration: Performance analysis, Applied Energy, **220**, pp. 21-35, (2018)
5. Zhu K., Li X., Campana P.E., Li H., Yan J., Techno-economic feasibility of integrating energy storage systems in refrigerated warehouses, Applied Energy, **216**, pp. 348–357, (2018)
6. Lund H., Werner S., Wiltshire R., Svendsen S., Thorsen J.E., Hvelplund F., Mathiesen B.V., 4th Generation District Heating (4GDH): Integrating smart thermal grids into future sustainable energy systems, Energy, **111**, pp. 923 – 932, (2016)

7. Frigo S., Gabbrielli R., Puccini M., Seggiani M., Vitolo S., Small-Scale Wood-Fuelled CHP Plants: a Comparative Evaluation of the Available Technologies, Chemical Engineering Transactions, **37**, pp. 847-852, (2014)
8. Liu H., Shao Y., Li J., A biomass-fired micro-scale CHP system with organic Rankine cycle (ORC) – Thermodynamic modelling studies, Biomass and Bioenergy, **35**, pp. 3985-3994, (2011)
9. Onovwiona H.I., Ugursal V.I., Residential cogeneration systems: review of the current technology, Renewable and Sustainable Energy Reviews, **10**, pp. 389-431, (2006)
10. Zywica G., Kaczmarczyk T.Z., Ihnatowicz E., A review of expanders for power generation in small-scale organic Rankine cycle systems: Performance and operational aspects, *Proceedings of the Institution of Mechanical Engineers, Part A: Journal of Power and Energy*, **230**, 7, pp. 669-684), (2016)
11. Kaczmarczyk T.Z., Ihnatowicz E., Żywica G., Kiciński J., Experimental investigation of the ORC system in a cogenerative domestic power plant with a scroll expanders, Open Engineering, **5**, pp. 411–420, (2015)
12. Rak J., Błasiak P., Kolasiński P., Influence of the Applied Working Fluid and the Arrangement of the Steering Edges on Multi-Vane Expander Performance in Micro ORC System, Energies, **11,** 892, pp. 1-16, (2018)
13. Landelle A., Tauveron N., Haberschill P., Revellin R., Colasson S., Performance Evaluation and Comparison of Experimental Organic Rankine Cycle Prototypes from Published Data, Energy Procedia, **105**, pp. 1706-1711, (2017)
14. Leśnikowska–Matusiak I., Wnuk A., The impact of transportation noise on the acoustic environment of man, Transport Samochodowy, **3**, pp. 37-63, (2014)
15. Kaczmarczyk T.Z., Żywica G., Ihnatowicz E., Vibroacoustic diagnostics of a radial microturbine and a scroll expander operating in the organic Rankine cycle installation, Journal of Vibroengineering, **18**, 6, pp. 4130-4147, (2016)

Financing pro-ecological investments in the Częstochowa municipal enterprise

Magdalena Wrońska[1,2,*], *Anna* Kwarciak-Kozłowska[1], *Katarzyna* Długosz[1], and *Monika* Galwa-Widera[2]

[1]Czestochowa Municipal Enterprise, Konwaliowa 1 Sobuczyna, 42-263 Wrzosowa, Poland

[2]Czestochowa University of Technology, Faculty of Infrastructure and Environment, Dąbrowskiego 73, 42-201 Czestochowa, Poland

Abstract. The paper presents the opportunities for funding ecological investments with the example of the Czestochowa Municipal Enterprise. Modernizations of facilities and technologies are usually very capital-intensive investments. High costs of operation and changes introduced to programs of waste management have stimulated creation of various forms of support for the enterprises, including companies involved in municipal waste management. Both Poland and the European Union provide financial support to enterprises for ecological and innovative purposes. For nearly 15 years, both national and international programs have offered their support for changes in the system of waste management. The Czestochowa Municipal Enterprise has been successful in meeting legal requirements. Subsidies from the EU operational programs (nearly a third of investments in 2004-2014) allowed for the achievement of a high level of investments. One should emphasize performance of such tasks as municipal waste sorting plant, composting facility, station of bulky waste disassembly, station of construction debris fragmentation or municipal point of dangerous waste collection. 200 points of selective waste collection have been created in the community and educational campaigns were performed concerning waste sorting. Each of these units represents a necessary component of modern and ecological waste management. Nowadays, waste landfills represent modern enterprises with huge economic potential, whereas investments and subsidies allow for the development and building infrastructure of the municipal industry which is essential in development of cities and villages.

1 Introduction

The statistical Pole produces about 303 kg of waste per year (GUS 2016 data), which means that Poland is the only one of the largest waste producers in the EU. Paradoxically, looking at the level of waste generated by the inhabitants of the western EU countries, who produce about 200kg more waste, it can be concluded that in Poland waste is not a problem and its proper management. This disproportion may result from differences in the level of economic development and the fact that only 80% of Polish households are included in the

* Corresponding author: magdalena.wronska90@gmail.com

system of organized waste collection. According to the Eurostat data for the EU countries, only 38% of waste reaches the landfills, whereas the rest is burnt, recycled or composted. Unfortunately, most of the waste in Poland is stored using conventional methods [1, 2]. The Act of 14 December 2012 on waste (Journal of Laws 2013, Pos. 21 with further amendments) imposes the obligation of updating voivodship waste management plans with the development of investment plans. Their aim is to indicate the infrastructure necessary for the achievement of the conformance to the directives of the European Union (e.g. 2008/98/EC and 1999/31/EC) concerning municipal waste management, including implementation of hierarchy of the methods to manage the waste and reach the required levels of preparation for recycling and reusing the waste and limitation of storage of biodegradable municipal waste [1-4]. This imposes the obligation of continuous modernization and extension on the waste management enterprises. The expenditures for ecological investments often exceed the annual budgets of the beneficiaries. Therefore, the investments in modern ecological technologies would be impossible without various forms of support, both from national funds, EU funds and foreign subventions i.e. the Norwegian Financial Mechanisms or the European Economic Area Financial Mechanism [3].

1.1. Waste management in the Silesian Voivodship in Poland

The key objective of the National Waste Management Plan (Kpgo2022) is to determine waste management policy consistent with hierarchy of waste management and the polluter pays principle. The main objectives include e.g.:

- reduction in waste generation,
- reduction in the amount of biodegradable municipal waste disposed in landfills so that not more than 35% of the mass of such waste will be stored in 2020 compared to 1995.
- striving for reduction in the among of waste stored,
- achievement of the required level of recycling and preparation for the renewed use of paper, metal, plastics and glass from the stream of municipal waste.
- increased contribution of the waste subjected to recycling processes [2, 5].

The main environmental problems of the Silesian Voivodship include e.g. poor air quality, insufficient quality of surface and underground waters, soil pollution, burden of the communication noise and poorly organized waste management. The scale of phenomena which have a negative effect in the natural environment in community of the Silesian Voivodship is incomparably higher than in other country regions. The Silesian Voivodship is the 14th biggest voivodship in Poland and the 2nd in terms of population. The voivodship is characterized by the highest population density (372 people per square metre, with the country's mean of 123 people per square metre) [2, 5]. It should be noted that the consequence of the achievement of the high level of urbanization and industrial development is irreversible transformations in the natural environment [2, 5]. The Silesian Voivodship Council (council is a provincial-level elected assembly) of 24 April 2017 concerning the adoption of the Waste Management Plan for 2016-2022 divided the Silesian Voivodship into three municipal waste management regions (RGOK). Four RGOKs were present in previous years. The plan defines activities according to the hierarchy of managing the waste to achieve the objectives determined in the National Waste Management Plan [2]. The Czestochowa Municipal Enterprise is located in the area of the waste management region I. Based on the waste management plan for the Silesian Voivodship, Table 1 presents the municipal waste mass, including biodegradable waste mass for individual regions of municipal waste management.

Table 1. Vehicles dimensions

Specification	Region I	Region II	Region III	Region IV	region: śląskie voivodeship
Population, [Os]	705 194	2 021 124	1 004 219	762 056	4 492 593
Mass of municipal waste, [Mg]	204 174	688 930	377 737	227 314	1 498 156
Mass of municipal biodegradation solid waste, [Mg]	101 115	347 289	188 288	112 943	749 636
Generic production index, [kg/M/year]	290	341	376	298	334

1.2. Sources of financing waste management

Sources of financing pro-ecological investments in waste management can be divided into three groups (table 2):

- co-financing from Polish funds,
- co-financing from European Union funds,
- co-financing from foreign funds.

The Operational Programme: Infrastructure and the Environment represents an important instrument of implementation of the new Lisbon strategy. Expenditure on priority objectives of the EU accounts for 67.89% of total expenditure from EU funds. Total amount of financial resources involved in the Operational Programme: Infrastructure and the Environment for the years 2007-2013 was 37.7 billion EUR, of which the EU contribution was 28.3 billion euro, whereas the national contribution was 9.4 billion euro. The most capital-intensive environmental investments in the area of water, wastewater and waste management could be made with the EU subsidies reaching 85% [1].

Table 2. Sources of financing waste management [7-10]

Sources of funding	Target, mission	The method of co-financing
National funds		
Funds for environmental protection and water management • National Fund for Environmental Protection and Water Management (existing since 1989), • 16 provincial funds for environmental protection and water management,	Supporting environmental actions in the main priorities: - Protection and sustainable management of water resources - Rational waste management and protection of the earth's surface - Atmosphere protection - Protection of biodiversity and ecosystem services	Subsidies and loans on preferential terms
BOŚ (Bank of Environmental Protection)	The goals are closely related to NFOSiGW operational programs	Preferential loans granted on the basis of the NFOŚiGW
EU funds		
Operational Program Infrastructure and Environment	- Decrease in the emissivity of the economy	Refunds (up to 100% of costs incurred) or

Sources of funding	**Target, mission**	**The method of co-financing**
	- Environmental protection, including adaptation to climate change - Improving energy security - Development of low-emission public transport in cities - Development of rail transport in Poland	advance payments
Operational Program Innovative Economy	-Increasing the innovativeness of enterprises - Increased competitiveness of Polish science - Increasing the role of science in economic development - Increasing the share of innovative products of the Polish economy in the international market - Creating sustainable and better jobs - Increase in the use of information and communication technologies in the economy	Refunds or advances
Intelligent Development Program	- Support for conducting B & R works by enterprises - Supporting the environment and the potential of enterprises to conduct B & R & I activity - Increasing the intensity of cooperation within the national innovation system - Innovation support in enterprises - Increasing the scientific and research potential	Refunds or advances
Regional Operational Programs	- counteracting and minimizing the effects of extreme climatic events, - early response systems in case of catastrophic events, - projects related to waste management, - projects in the field of wastewater management and water supply, - nature protection, especially in protected areas.	Loans and subsidies
Foreign funds		
Norwegian Financial Mechanism And Financial Mechanism of the European Economic Area	- biodiversity and environmental monitoring - energy saving and renewable sources	Subsidies

Sources of funding	Target, mission	The method of co-financing
	- research and scholarships - innovations in the field of green technologies in addition, general programs such as - civil society - social and regional development - cultural heritage - decent work and tripartite dialogue - Schengen and internal affairs	
	Other	
Commercial bank loans	Any investments to improve or change technology, structure, etc.	Credits and loans
Own funds of enterprises	Any investments to improve or change technology, structure, etc.	Funds from commercial activity - profits
Municipal budgets	Pro-ecological investments within the commune	Financing or co-financing of investments or taking up shares in investments as indirect actions
Private investors	Ecological and innovative activities	Co-financing, co-financing, sponsorship,

1.3.The ecological investments in the area of the Czestochowa Municipal Enterprise

Czestochowa Municipal Enterprise. in Sobuczyna implemented the investments connected with building the waste management plant within the Integrated Operational Programme for Regional Development. Its net cost was 20,0 million PLN. With additional funding, the following investments were made:

- Municipal waste sorting plant: (current capacity: 95 000 Mg/year). Municipal waste sorting plant - the basic task of the sorting plant is separation from the entire stream as much as possible of waste, re-usable. The remaining part, unsuitable for recovery, is transferred for further processing in case of high energy calorific value and in the case of low calorific value of depositing on the component inventory.
- Waste composting plant in the open-prism system (current capacity 40 100 Mg /year) Composting is an aerobic treatment of waste, the process takes place in open, aerated piles. In the aerobic prisms, composting occurs on the surfaces of moist and aerated components due to the participation of microorganisms, such as bacteria and fungi. Intensive growth of thermophilic bacteria and high temperature (>70°C) cause the destruction of parasitic and pathogenic organisms and plant seeds, and as a result, the emergence of compost. Compostable waste is directed to the composting plant: the "bio" fraction from municipal waste, so-called waste. green (leaves, grass), waste trees and shrubs with clipping, other wooden waste, organic and green waste from marketplaces.
- The demolition station for large-size waste (current capacity of 25 000 Mg/year), such as: furniture, doors, etc. - both waste separated on a sorting line from municipal waste

and waste collected selectively. On the square, they are dismantled and recovered raw materials that can be re-used (including plastic, metal, wood).

- Station for grinding construction debris (current issue of 15 000 Mg/ year)Station for shredding renovation and construction waste. There are such waste as: rubble, waste from building renovation, window joinery, etc. The recovery process of the waste provided includes segregation, disassembly, debris crushing, which is later used as a foundation for technological roads.
- Municipal hazardous waste collection point. Including hazardous waste storage. Mobile point of scraps of hazardous waste (MPZON) is a specialized car that hears the collection of hazardous waste from the city of Czestochowa. Collected wastes are temporarily stored in the area of ZZO in Sobuczyna and are then transferred for utilization or processing.

Funding from the Cohesion Fund within the Operational Programme: Infrastructure and Environment with net value of 17,8 million PLN (total value of the investment: 54,5 million PLN) allowed for technological extension of the Regional Installation of Municipal Waste Processing to meet the European standards. The physical execution status of the project was ca. 38.52%, whereas financial progress status was ca. 14.84%. The project included three investment tasks:

- Technical closing of the landfill cell No 1: the task was aimed to seal the top parts and slopes of the 16.8 ha cell that has been used since 1987 with 3 million Mg of waste and division with intake of leachate waters and rainfall waters. Reclamation of the landfill cell No. 1 fully protects the environment from the effect of waste deposited in this cell.
- Building the stage I of the landfill cell No. 2: the effect of this task was a new landfill cell with usable surface area of 2.6 ha with drainage system and synthetic sealing. The technologies used for the cell construction meet al the current standards and guarantee a full security of individual environmental components during the functional use.
- Building the composting facility using the closed technology: building the composting plant helped adjust the regional installation of waste management with European standards. This results in the implementation of modern technologies based on aerobic composting in reinforced concrete bioreactors, which allows for the use of the initial phase of intensive composting under hermetic and controlled conditions. The composting plant processes the subscreen fraction from the mixed municipal waste. The post-processing air is cleaned and, if monitoring detects exceeding the permissible parameters, the afterburning system will be switched on automatically. This technological process ensures a 100% air protection from possible pollutants.

The benefits of the extension of the Waste Management Plant included:

- reduction of harmful emissions to air, land, and waters,
- limitation of the sanitary threat and limitation of the substantial amount of liquor as a result of closure and reclamation of the landfill cell,
- reduction of greenhouse gas emissions as a result of limitation of landfilling of biodegradable waste,
- safe landfilling of waste which is impossible to be recycled (ballast) using a modern landfill cell equipped with facilities that protect all the components of the environment,
- returning of the degraded land (ca. 16.8 ha) to the environment, used previously for landfilling of waste by means of reclamation.

The ecological effects of the investment include:

- Achievement of the efficiency of the Waste Management Plant in Sobuczyna of up to 116 000 Mg/year.
- Limitation of the amount of waste discharged to the landfill site by 32% (from 68% in 2009 to 36% in 2015)

- Including 150 000 people in the municipal waste management system.

Another important ecological investment was limitation of the amount of waste deposited in the environment through implementation of the systematic solution of selective waste collection in the city of Czestochowa. This was possible due to the contract with the Marshall's Office in the Department of Regional Development in Katowice. The net value of the entire project was 5,8 million PLN. This involved the campaign aimed to promote the activities of selective waste collection. The educational campaign helped the community realize how important selective waste collection is and what measurable benefits it can produce to the natural environment. The project included three investment tasks:

- 200 local points of municipal selective waste collection were created. Each point was composed of 3 containers for paper, plastics/metal and glass.
- Five hook containers were purchased as local points of biodegradable waste collection.
- The Stationary Point of Selective Collection of Municipal Waste (SPSZOK) was designed and construed in the area of regional installation of waste management in Sobuczyna. This is a comprehensively equipped facility used to collect and temporary store municipal waste, dangerous waste, post-renovation waste and bulky wastes. Each inhabitant of Czestochowa can supply sorted waste to this facility, such as: post-renovation waste, metal, glass, bulky wastes, plastics, tyres, car windows, used electrical and electronic equipment, dangerous waste, foil and waste paper.

Table 3 presents the levels of municipal waste management achieved by the Community City of Czestochowa in 2017. These results were possible due to the ecological investments made in the area of the enterprise. According to the Silesian Voivodeship Waste Management Plan for 2016-2022, the needs connected with current waste management system in the Region I of the Silesian Voivodeship were diagnosed. They are connected with building and modernization of current installations for processing of green waste and other biodegradable waste. It is also necessary to increase capabilities of these installations as the deficit was observed already in 2016. The investment will be made within the project, consisting in the extension of the composting plant dedicated to green and biodegradable waste with capacity of 15 000 Mg/year.

Table 3. The levels for municipal waste management achieved by the City of Czestochowa in 2017 [10]

Type of management	**Acceptable level [%]**	**Level reached [%]**
recycling, preparation for reuse and recovery using other fraction methods: paper, metals, plastics, glas	20	24.58
recycling, preparation for reuse and recovery by other methods of non-hazardous construction and demolition waste	45	97.76
reduction of biodegradable municipal waste transferred to landfilling	**Acceptable level [%]**	**Level reached [%]**

2 Conclusions

The most important effect of the investments carried out was the extension of the inhabitants of the Northern Subregion of the Silesian Province to an organized waste collection system, thus improving the condition of the natural environment in Czestochowa and throughout the Czestochowa poviat.

Thanks to the implemented projects:

- the amount of waste sent to the landfill has been reduced. Currently, the landfill is directed primarily to the so-called post-process waste;
- separation in the process of sorting of biodegradable fraction directed to the stabilization process and the fraction containing secondary raw materials and the component for alternative fuel;
- reduction of biodegradable waste to the level required by law;
- separation from municipal waste: large-size waste along with its dismantling; construction waste with crumbling and preparation for re-use; hazardous waste to be disposed of for disposal.

The effect of the implementation of projects is primarily to increase the efficiency of municipal waste treatment and to reduce the inconvenience of the existence of Waste Management Plants for the surrounding population.

References

1. Program Operacyjny Infrastruktura i Środowisko w województwie śląskim, wrzesień 2013, https://www.wfosigw.katowice.pl/poiisfiles/biuletyn_POIiS_2013.pdf (2013)
2. Plan gospodarki odpadami dla województwa śląskiego na lata 2016-2022, https://bip.slaskie.pl/dokumenty/2016/11/02/1478090442.pdf
3. Jura J., Zasady finansowania przedsięwzięć proekologicznych, Budownictwo o zoptymalizowanym potencjale energetycznym, **1,** 15, pp 85-90, (2015)
4. Toruński J., Wyrębek H., Źródła finansowania inwestycji ekologicznych w Polsce, Zeszyty Naukowe Akademii Podlaskiej w Siedlcach, **82**, pp 43-63 , (2009)
5. Bień E., Źródła finansowania gospodarki odpadami w gminach województwa śląskiego, Inżynieria i Ochrona Środowiska, **13**, 1, pp. 5-15, (2010)
6. Opracowanie własne na podst. PGO Woj. Śląskiego na lata 2014-2020
7. Barczak A., Kowalewska E., Źródła finansowania zadań z zakresu ochrony środowiska w Polsce – przegląd stosowanych rozwiązań, Prawo Budżetowe Państwa i Samorządu, **1**, 20, pp. 37-58, (2014)
8. https://rpo.slaskie.pl/
9. https://www.pois.gov.pl/strony/o-programie/zasady/finansowanie/
10. http://www.mos.gov.pl/

Liquidation of ammonia clouds using a mobile turbine rescue and firefighting system

Małgorzata Majder–Łopatka[1,*], and *Tomasz* Węsierski[1]

[1]Szkoła Główna Służby Pożarniczej (Main School of Fire Service), ul. Juliusza Słowackiego 52/5, 01-629 Warszawa Poland

Abstract. Given its extensive industrial use and chemical properties ammonia is a source of serious hazards in the event of its uncontrolled releases. The mobile turbine rescue and fire-fighting system (MTSRG), equipped with a special SO3-W aircraft turbine with 10 kN power and water efficiency of 6 m^3/min, can be used to liquidate spatial gas clouds by diluting them with combustion gases and sorption on the surface of the produced water microdroplets. The publication presents calculations related to the effectiveness of the system during the uncontrolled release of ammonia from a pressure vessel. To estimate the performance of this system, use was made of water stream droplet size research, kinetics studies of ammonia sorption, SO3-W turbine parameters at a safe working distance, as well as results of ammonia concentration simulation using the ALOHA program. A combined simulation carried out for a rupture of DN100 pipe supplying liquid ammonia to spherical tank with a diameter of 16.6 m and filled in 70% showed that it is theoretically possible to reduce the ammonia concentration by as much as 95.75% while the vehicle is operating at a working distance of 60 meters from the rupture. The exhaust gases together with the entrained air contribute mainly to the reduction of the concentration of ammonia through the mechanism of dilution (91.6%). Water mist generated by the vehicle is theoretically responsible for the reduction of ammonia concentration by only 4.15%.

1 Introduction

Given its wide range of applications, ammonia may cause diverse chemical hazards [1-4]. The production of ammonia only in Poland in anhydrous form and in an aqueous solution (calculated as pure NH3) amounted in 2016 to 2623 and 96.4 thousand tones, respectively [5]. Consequently, the elimination of spatial contamination of ammonia has often been the object of scientific research [6-9]. The greatest hazards occur in case of uncontrolled release is ammonia in liquefied form. Due to the pressure in tanks containing liquid ammonia and the fact that ammonia in such a state is at a temperature much higher than the boiling point, its release quickly generates a significant cloud with corrosive and toxic properties. The spatial liquidation of ammonia clouds would be possible by using highly efficient technologies that provide significant amounts of water per unit of time in the form of a fine

* Corresponding author: mmajder@sgsp.edu.pl

spray. One of such technologies is the turbine rescue system that ejects water in the form of a fine spray generated by breaking of solid water jet by the exhaust stream of the turbine. Such a vehicle called MTSRG (Mobile turbine rescue and firefighting system) with a SO^{3-}W turbine of 10 kN thrust is presented on Figure 1. The technology enables the dissipation of 6 m^3 of water per minute, whereby the obtained water fine spray has an average drop diameter of 51 □m and covers an area of 800 m^2. Such a droplet size generated over a huge surface should ideally absorb vapours of dangerous substances that are soluble in water. An additional advantage is the huge amount of exhaust gases diluting the cloud and the ability of operating the system from a considerable distance.

Fig. 1. Mobile turbine rescue and firefighting system (MTSRG) equipped with SO3-W turbine having a 10 kN thrust

The contamination elimination method generally used by rescue units consisted in the first place in limiting the leakage and further spreading by applying water curtains. However, it did not allow spatial liquidation of chemical clouds. This was made possible by using high efficiency water monitors. However, the depth of penetration and the sorption efficiency is considerably smaller than the solution presented in this article.

2 Research methodology

In analyses of diverse possibilities of applying turbine extinguishing systems in the liquidation of ammonia chemical clouds, the following analytical methods, technical parameters and test results were used:

- estimation of the effects of uncontrolled outflow of ammonia in the computer program ALOHA 5.4.4,
- kinetic studies of reduction of ammonia concentration as a function of time through water curtains,
- testing of average droplet diameters using a Drop Spectrum Analyzer IPS (AWK) ,
- technical parameters of the SO_3.

An analysis was carried out for leakage of a ball-shaped pressure vessel with a diameter of 16.6 m and liquid ammonia filled in 70%. The release scenario assumes that a DN100 supply pipe becomes broken up as a result of a catastrophic event at the level of the tank bottom. According to assumptions the ambient temperature was 10°C, at average cloudiness.

The atmospheric stability class was defined as D. The storage temperature of liquefied ammonia in a spherical tank with incomplete cooling was set at 10°C (maximum permissible).

Studies related to kinetics of reducing ammonia concentration were performed at a test stand for kinetics of the effectiveness of water curtains as shown in Figure 2. The stand has a test chamber measuring 1.2 x 1.2 x 1.2 meters, a boom for mounting the spray nozzle, an ammonia supply system to the test chamber consisting of a heating bowl and a flask, as well as a supply line characterised by adequate mechanical strength and chemical resistance. In the course of the research, ammonia was obtained by desorption of NH_3 from 25% ammonia water. The stand was also equipped with a water supply system having a tank capacity of 0.25 m^3 enabling the setting and regulation of liquid flow parameters - pressure and efficiency, a drop tank and a gas analyzer (Photoionization detector MX 6 made by Industrial Scientific) to record the concentration of ammonia in the chamber as a function of time.

For needs of the tests used was made of three types of spray nozzles made by BETE Deutschland GmbH with the markings: TF 6 (spiral nozzle), CW-50 F (nozzle with full spraying cone), P48 (fog nozzle). For working pressures, they provided drops with mean surface diameters of 600 mm, 235 mm and 185 mm, respectively. The Drop Spectrum Analyzer was used to study average droplet diameters and pulverization spectra IPS (AWK) made by Zakład Elektronicznej Aparatury Pomiarowej KμK in Warsaw. The AWK analyzer makes it possible to determine the average diameter of droplets and the distribution of droplet diameters within the range from 0.5 to 3000 μm. The analyzer consists of a measuring probe, a stand, power cords, a signal converter and a PC (Fig. 3).

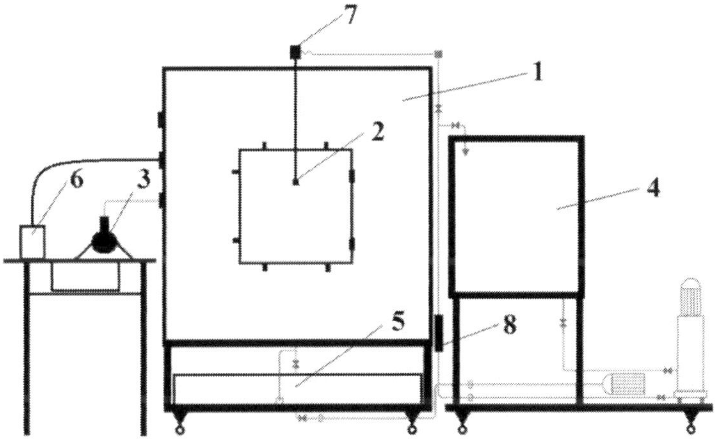

Fig. 2. Diagram of testing arrangement; 1 – research cell, 2 - spray nozzle, 3 - bowl heater with the flask, 4 - water tank, 5 – dumping tank, 6 - gas detector (PID), 7 - pressure sensor, 8 – flow sensor

The technical parameters of the SO3-W turbines were assumed as parameters of the turbine system taking into account a safe working distance of 60 m for rescuers with full protection. At the start position, the turbine air expense is 13.8 m^3/s [10]. Assuming a significant excess of air in relation to fuel and an increase in the temperature of combustion gases from 20°C to 650°C, the volume of exhaust gases escaping per unit of time is 46.7 m^3/s [11]. This is a volume of gases similar to that of Turbolöscher II for which the volume of generated exhaust gases at the outlet of the turbine is 50 m^3/s. Considering these parameters, the estimated volume of exhaust gases together with the air sucked in by the high-velocity exhaust gases at a working distance will be about 1850 m^3/s. The water capacity of the system was assumed at 6 m^3/min. The direction of the projection of the turbine water stream, which is consistent with the wind direction, was assumed as the direction perpendicular to the axis of the direction of the cloud's propagation.

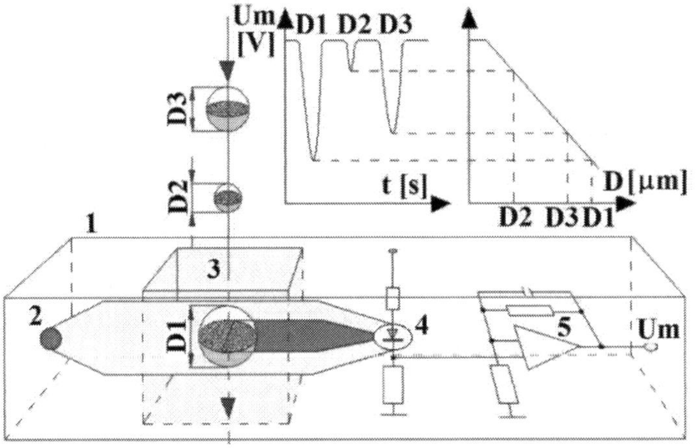

Fig. 3. Drops Spectrum Analyzer IPS: 1 - probe 2 - transmitter, 3 – measuring space , 4 -receiver, 5 – electronic arrangement, D1, D2, D3 - adequate size of droplets

3 Results and discussion

The simulation performed in the program ALOHA 5.4.4 indicates that for pressure vessel break described in the test methodology an outflow may be anticipated at a maximum intensity of 7690 kg / min. During the first hour, the intensity of the outflow will only slightly decrease. The maximum concentration of ammonia at a distance of X from the place of outflow oriented along the propagation of the cloud is shown on Figure 4. The time for the reaction for rescue teams is very short. For example, at a distance of 2 km, the maximum concentration value outside objects equalling to 1330 ppm would be reached practically after only 6 minutes (insert figure 4). This means that it would only be possible to use efficiently the turbine technology if the leakage occurred as a result of a secondary accident on the site of the plant where other activities with the use of a turbine vehicle would have been going on.

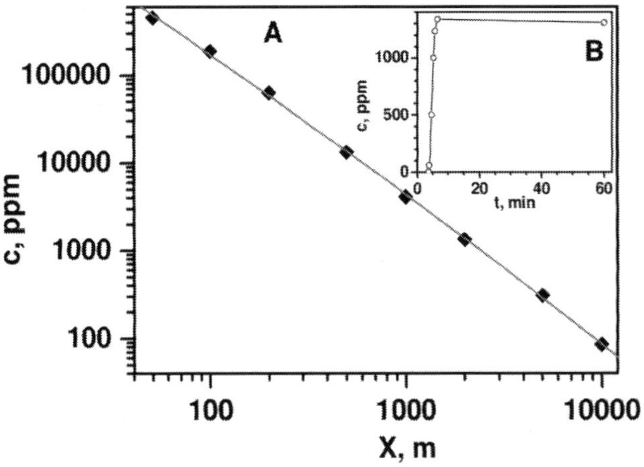

Fig. 4. A) Maximal concentration of ammonia at a distance X from the point of release for assumed simulation conditions B) Outdoor concentration of ammonia for assumed conditions 2 km from the point of outflow.

At the maximum flow intensity determined with the use of the program as being 7690 kg/min (assuming some cooling of steaming ammonia) approximately 170 m^3 of ammonia vapour could be generated in a second. Hence, effective application of the same exhaust gas stream with an intensity of 1850 m^3/s could theoretically reduce the concentration of ammonia to 8.4% of the initial value, if such a variant would be feasible taking into account security reasons. Consequently, the concentration reduction would be as high as 91.6%. It should be borne in mind that ammonia is a combustible substance and at a working distance of 60 meters one could expect ammonia concentrations of up to 360000 ppm when only the direction of the wind changes. As a result there would be a great risk of LEL exceeding 15% (150000 ppm) at the presence of a very large ignition source such as a turbine, despite being supplied with water. This would be feasible at the aforementioned change of wind direction at the moment of starting the turbine and being aware that water is fed from the top parallel to the exhaust gas stream. However, it should be clearly emphasized that during the turbine operation, a large volume of exhaust gases and their high velocity would not allow such a high concentration of ammonia in the vicinity of the turbine.

The reduction of ammonia concentration resulting from the action of dispersed water streams was estimated for the analyzed case by means of kinetic studies of the change in ammonia concentration in the chamber as a function of time.

Changing the concentration of ammonia in the test chamber at the time of supplying water streams through spray nozzles is similar to the exponential decay which can be described by the pseudo-first order kinetic equation given below:

$$-\frac{d[C]}{dt} = k_p[C] \tag{1}$$

where :

C – ammonia concentration In chamber [ppm],

k_p – apparent rate constant of the sorption process [s^{-1}].

The apparent absorption rate constant kp depends on parameters of the liquid flow and the degree of water spray fragmentation. The course of changes in ammonia concentration as a function of time in the test chamber for the tested nozzles was shown in Figure 5. The determined values of the apparent sorption rate constant for the tested nozzles at a water efficiency of 130 dm^3/h are shown in Table 1.

Table 1. The apparent velocity of ammonia absorption kp in the kinetic region of maximal sorption

Type of nozzle	Model	Pressure	Efficiency	Mean surface droplet diameter	k_p
		P [bar]	V [dm^3/h]	Ds [µm]	[s^{-1}]
Spiral nozzle	TF 6	0,5	130	600	0.0087
Nozzle with full sprinkling cone	CW-50 F	4	130	235	0.0142
Fog nozzle	P48	6,5	130	185	0.0164

An analysis of the relation between the reciprocal of the apparent ammonia sorption rate constant and the mean surface diameter of droplets allows achieving the equation $1/kp=0.127D_s + 38.92$ with a correlation of R2 = 0.997 (Fig. 6). On this basis, by extrapolating we can predict that for droplets $D_s = 51$ μm equal to the size of the mean surface diameter of the distributed water streams generated by MTSRG, corresponds to $k_p = 0.022\ s^{-1}$.

The range of effective operation of MTSRG is an ellipse having an area of approx. 800 m^2. With a water effectiveness of 6000 dm^3/min = 100 dm^3/s, this gives an average surface spray intensity of 0.125 dm^3/m^2s (450 dm^3/m^2h). The surface intensity of spraying in the test chamber with base dimensions of 1.2x1.2 meters and water capacity 130 dm^3/h is 0.025 dm^3/m^2s.

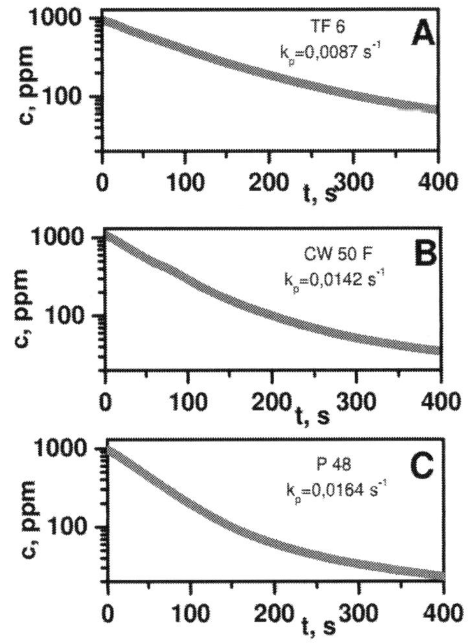

Fig. 5. Ammonia concentration vs time in research chamber for the TF 6, CW50-F and P 48 nozzle. Water efficiency V= 130dm3/h, water temperature T=20°C.

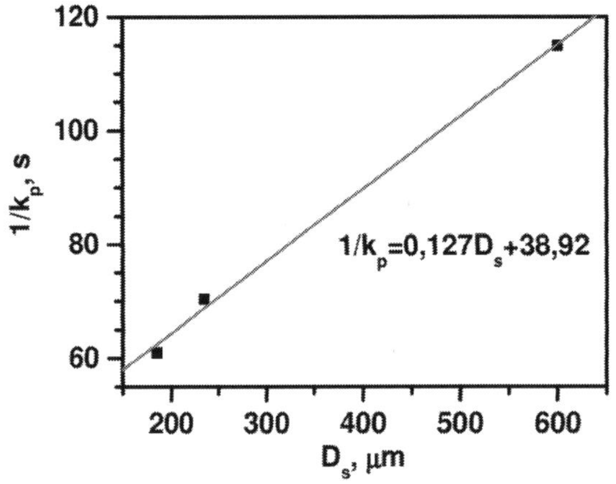

Fig. 6. Dependence of the reciprocal apparent rate constant of adsorption vs. mean surface diameter

To determine the value of the apparent rate constant of the sorption process k_p corresponding to the surface intensity of spraying specific for MTSRG, a study was made on the relationship between water efficiency and the value of the constant k_p (Fig. 7). The measurements were carried out for a full cone CW50-F nozzle. The obtained dependence is expressed by the equation $k_p = 8.78x10^{-5}V + 0.00287$. Hence the extrapolated k_p value corresponding to the intensity of MTSRG spraying is $k_p = 0.04238 \ s^{-1}$.

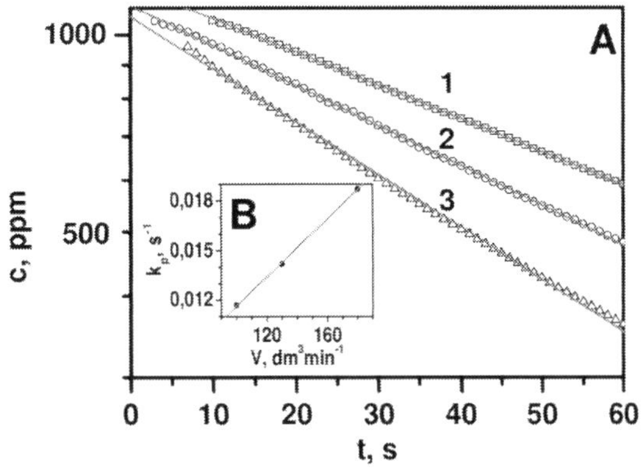

Fig. 7. A) Ammonia concentration vs time in kinetic range for the water efficiency. 1) 100 dm^3/h, 2) 130 dm^3/h, 3) 180 dm^3/h B) dependence between apparent rate constant of adsorption and water efficiency

Further calculations related to reducing the concentration by applying water currents can be made using the dependence of $lnC-lnC_0 = -kt$ and based on the presumption that at such a significant emission in the vicinity f the place of release one can expect practically pure ammonia. A time interval of 1 second was assumed for calculations. Hence the maximum degree of absorption by dispersed water spray for continuous ammonia outflow allow absorption in the first second of 4.15% of the ammonia released within the examined time interval in the space covered by the MTSRG operation.

For the analyzed very unfavourable leakage scenario involving a rupture of a DN100 pipe at the bottom of a 16.6 meter diameter spherical tank filled in 70%, the total concentration reduction through the MTSRG vehicle would therefore be approximately 95.75% at a working distance of 60 meters. This value could be theoretically even greater considering that the liquid phase that has not managed to evaporate during the release would be partly diluted by the already fallen water which could be in the first phase frozen. However, the amount of ammonia absorbed in this way is difficult to estimate. It should be borne in mind that although one volume of water is theoretically able to absorb 1176 volumes of ammonia gas in normal conditions, the surface of sorption and mixing (not taking into account small short-lived puddles of non-vaporized ammonia) would be very small in relation to the sorption surface of mist droplets.

4 Conclusion

The calculations showed that MTSRG furnished with a SO_3-W turbine having a 10 kN thrust and generating a volume of 1850 m^3/s of exhaust gas with air sucked into the stream from the environment, can provide an effective means for the liquidation of chemical contamination on a very large scale. Combined simulation carried out in the ALOHA 5.4.4 program of a rupture of a DN100 pipe supplying liquid ammonia to a spherical tank with a diameter of 16.6 m and filled in 70%, along with measurements of the mean surface diameter and measurements of the rate of ammonia sorption by water streams in the test chamber have shown that it is theoretically feasible to reduce the ammonia concentration by as much as 95.75% while the vehicle is operating at a working distance of 60 meters away from the rupture. The exhaust gases together with the entrained air contribute mainly to the reduction of the concentration of ammonia through the mechanism of dilution. This effect constitutes up to 91.6% of reduced concentration. Kinetic tests in the test chamber showed, however, that the water mist generated by the vehicle theoretically reduces ammonia concentration by only 4.15%. This indicates that water efficiency is greater than in the case of Turbolöscher II, despite the smaller amount of water, which is related to the smaller diameter of the droplets generated in the exhaust stream of the SO^{3}-W engine [6].

The research was carried out as part of the DOB-BIO6 / 06/11/2014 project financed by NCBiR "Mobile turbine rescue and firefighting system"

References

1. Keim M. E., Introduction to chemical disasters, [in:] Ciottone's Disaster Medicine (Second Edition), Elsevier, (2016)
2. Pritchard J.D., Compendium of chemical hazards ammonia, Health Protection Agency, London, (2011)
3. Amshel C.E., Fealk M.H., Phillips B.J., Caruso D.M., Anhydrous ammonia burns case report and review of the literature, Burns, **26**, 5, pp. 493-497, (2000)
4. Wróblewski D., Węsierski T, Retrospektywna analiza uwolnienia się amoniaku w Drzycimiu w kontekście faz przygotowania i reagowania modelu zarządzania kryzysowego, Przem. Chem., **90**, pp. 2100-2104, (2011)
5. Rocznik statystyczny przemysłu 2017, Główny Urząd Statystyczny, Warsaw ,(2017)
6. Węsierski T., Majder-Łopatka M., Wąsik W., Likwidacja skażeń przestrzennych amoniaku za pomocą turbinowych pojazdów gaśniczych, Przem. Chem., **96**, pp. 1080-1084, (2017)
7. Hua M., Xiaohui S., Juan Z., Xuhai P., Protective water curtain ammonia absorption efficiency enhancement by inorganic and surfactants additives, Process Safety and Environmental Protection, **116**, pp. 737-744, (2018)
8. Hua M., Qi M., Yue T., Pi X., Pan X., Jiang J., Experimental Research on Water Curtain Scavenging Ammonia Dispersion in Confined Space, Procedia Engineering, **211**, pp. 256-261, (2018)
9. Xiaohui S., Juan Z., Hua M., Pan X., Experimental research on decontamination effect of water curtain containing compound organic acids on the leakage of ammonia, Process Safety and Environmental Protection, **105**, pp. 250-261, (2017)
10. Silnik SO3-W. Opis silnika, Wytwórnia sprzętu kominikacyjnego PZL Rzeszów, First Edition, (1977)
11. Ruchała P, Dziubiński A., Silniki odrzutowe SO 1 i SO3, Polska Technika Lotnicza. Materiały historyczne, **66**, 1, pp. 1-8, (2011)

Modernity and ecology in the aspect of a packaging industry company management

Agnieszka Ociepka-Kubicka[1,*]

[1]Czestochowa University of Technology, Faculty of Management, Armii Krajowej 19B, 42-200 Częstochowa, Poland

Abstract. The article discusses the issue of innovativeness in the area of packaging with reference to the balanced development. It also presents the threats of the packaging industry to the natural environment and the methods and tools of their restriction. The importance of biodegradable packaging and packaging from recycled materials is stressed. Moreover, the problems of the burdensomeness of packaging products and the packaging manufacturers are emphasised. Modern enterprises must implement pro-ecological, innovative activities, both at the level of production and the company's management. The changing economic conditions and the social pressure result in the new company's management goals – alongside the production, financial and other aspects. These new goals refer to the widely understood environment, and human life and health. The analysis of the innovative and pro-ecological activities implementation was presented on the example of a modern, dynamically developing joint stock company, United Packaging S.A., producing packaging from corrugated board with the increased water resistance for food industry, household chemicals and electronic industry.

1 Introduction

The world packaging market, according to forecasts in the coming years, will develop very dynamically. At present, the packaging fulfils numerous functions such as storage and protection against pollution or damage, transportation, distribution and product advertising. Research by an international consultancy company, Smithers Pira, shows that by 2020 the world's packaging sector will grow by 3.5% per year, to reach 1 trillion US dollars in that year. According to the Polish Chamber of Packaging (PIO) forecasts, with a 4-5% growth rate of the Polish packaging market by 2020, it will reach a comparable level with the developed markets of Western Europe. Polish industry meets domestic needs and 15-20% of its output is directed at the foreign markets. Taking environmental issues into consideration in the rapidly growing packaging industry is one of its major challenges [1]. Extensive data presented in the Report indicate the world-wide tendencies to the increasing application of ecological packaging, in order to minimise the negative impact on the environment. Research show that the region in the Report referred to as Asia-Pacific, will become the leader of the ecological packaging market. It is expected that North America and Western and Eastern

* Corresponding author: agnieszkaociepa22@wp.pl

Europe will have a substantial deal in the ecological packaging market thanks to their rigorous governments' environment politics. Research show also that Japan will have higher and higher position on the ecological market due to its highest use per capita [2]. In the packaging process, both traditional and – more and more frequently – new raw materials, subsidiary materials and technologies are used. This evokes the waste generation, discharge of sewage into water and/or earth, and dust and gases into the atmosphere. Also the management of the packaging waste constitutes a problem. The article discusses the innovation issues in the packaging industry in terms of sustainability. The work consists of two parts. The purpose of the first part is to identify the environmental risks associated with the packaging industry and the possibilities of their reduction. The second part presents the modern, dynamically developing company, United Packaging S.A., which produces corrugated cardboard packaging with increased hydro resistance for the food industry, household chemistry and the electronics industry.

2 Innovation in the packaging industry and sustainable development

Packaging is one of the most important processes without which nowadays products would not be finished. Observing the principles of sustainable development, which proclaim the need of reconciliation of the economic development with the social expectations and the protection of the environment, is increasingly perceived as necessary. This awareness poses a difficult challenge for companies, as they have to follow the market requirements of their customers while taking into account the environmental requirements [3]. In addition, demographic and lifestyle changes, related among others to the aging of society, force the production of qualitatively innovative packaging. The ecological aspect of packaging must be taken into account not only at the manufacturing stage, but also for the use and disposal of packaging [4].

Table 1. Structure of the packaging market in Poland and Europe from the packaging material type point of view, 2020 (own study based on [5])

Fabric	Market share [%]	
	Poland	Europe
Plastics	38.0	38.0
Paper and cardboard	36.2	38.5
Metal	10.0	8.0
Glass	8.2	7.5
Wood and others	7.6	8.0

It should be emphasized that the introduction of new packaging to the market should be based on a comprehensive analysis of its impact on the environment at different stages of the life cycle. The impact of the packaging on the environment depends on the: design, sourcing, production, distribution, use and recycling. Research indicates that the packaging industry can be a serious threat to the natural environment, inter alia due to the:

- possibility of harmful substances' emissions into the atmosphere, soil and water during the production process,
- possibility of environmental pollution during the packaging waste deposition in landfills,
- potential presence in packaging materials of heavy metals such as lead, cadmium, mercury, chromium and other harmful substances that may be hazardous to health and life,
- environmental impact of packaging and packaging waste from the production and distribution, resulting from the lack of use of various methods of their disposal.

According to the report of DNB Bank Polska: "Packaging. Industry Report, 2017", the world packaging market is going to grow dynamically (around 6% in 2018) and to a great extent is dependent on the material used in the packaging production. The report characterises the packaging materials and presents the anticipated growth of particular packaging types in 2018.

Table 2. Anticipated growth of packaging sectors in 2018 on the basis of DNB Report (own study based on [5])

Packaging material	Short characteristics	Growth in 2018
Paper	The raw material is widely available, lightweight, paper packaging production is cheap, easily processed, printable, biodegradable and environmentally friendly. Its disadvantage is soaking which reduces the mechanical durability	2%
Glass	Very good properties for storing food and beverages (it is sterile and does not absorb moisture), in addition, similarly to paper, it is 100% recyclable. The disadvantage is often the need to additionally use protective packaging	3%
Metal	Mechanically strong, with adequate hardness which facilitates transport, higher packaging production costs as compared to paper, not as ductile as, for example, plastic. The disadvantage is its high thermal conductivity	5%
Plastic	It is light, durable, sterile, suitable for long-time food storage. Plastic is not biodegradable, it is the environment-polluting waste, its recycling is difficult and expensive	9-10%

Although the plastic is not biodegradable, is a waste polluting the environment and its recycling is difficult and costly, the greatest increase in packaging production is forecasted for plastic packaging (9-10 per annum). The use of plastic packaging is primarily determined by economic and practical factors. On the other hand, taking into account that the average time of decomposition of a plastic bottle is up to 450 years, recycling of such packaging becomes a serious challenge and a major risk for their development. For instance, France has introduced a law stating that by 2020 disposable plates and cutlery should only be made of biodegradable materials. The European Union plans to reduce the number of disposable bags by 80% within 15 years.

2.1 Possibilities to limit the negative impact of packaging on the environment

Packages must meet numerous requirements, e.g. technical, promotional, informative, ecological and economic, to fulfil their functions properly. Each one is important, but environmental requirements are, more and more frequently, playing a substantial role. According to estimates, annually we generate about 530 thousand tons of used paper and cardboard packaging, 466 thousand tons of glass packaging, 276 thousand tons of plastic packaging and laminates, as well as 130 thousand tons of metal packaging. The weight of used packaging that affects the environment per year per one inhabitant of Poland equals approximately 70 kg and is still much lower than in Western Europe. The world average is around 150 kg. In terms of value in the above-mentioned countries, the use of packaging per capita varies between EUR 300-340, with approximately EUR 212 in Poland. The packaging market in Poland is systematically growing. According to the forecast of the Polish Chamber of Packaging, with approximately 4-5% pace of economic development in 2014-2020, it will reach a comparable level with the markets of Western European countries. In accordance with the principles of sustainable development, the production of packaging should not cause negative changes in the environment. In order to achieve this principal goal the packaging industry should: be based on renewable natural resources, protect ecosystems, biodiversity, minimize harmful emissions. The packaging waste management is also a serious problem. In accordance with the waste management principles of the EU Waste Framework Directive (2008/98/EC), waste prevention is the primary priority in the waste management. The Directive of the European Parliament and of the Council on packaging and packaging waste (94/62/EC, as amended by Directive 2004/12/EC), also contains provisions for the prevention of packaging waste, especially those deposited in landfills. Under these rules, the re-use of packaging and the recovery and recycling of packaging waste is undoubtedly a good means of preventing waste. Research shows that consumers much more willingly purchase products in natural, ecological packaging [6]. In addition, modern food packaging is expected to extend the shelf life of the product. Advances in new food packaging technologies include delaying oxidation, preventing microbial development, preventing moisture, etc. New food packaging technologies beyond the primary function must increase the safety of storage and transport of food.[7]Therefore, when designing a new packaging, in order to reduce the burden on the environment, the following aspects are important: the production of reusable packaging, the manufacture of packaging materials from recyclable materials, the use of natural-origin materials in the production of packaging, the use of smaller quantities of materials for the packaging production, the production of packaging free from pollution, e.g. with heavy metals. At the design and production stage, what also should be taken into account is the following: the emission of gases from the production processes, greenhouse effect, the energy consumption of the production, the adaptation of packages to the requirements of their biodegradation, the adaptation of packages to the requirements of incineration of packaging waste with energy recovery. Recycling as the process of the waste materials reprocessing, for economic reasons, as well as environmental protection, is a very desirable solution. Recycling of packaging is of great importance in the recovery of metals, glass, waste paper and certain plastics. Organic recycling is the oxygen or anaerobic treatment of biodegradable packaging waste. This recycling is used for packages made of biodegradable polymers, e.g. synthetic polyesters or the ones of microbial origin, starch and cellulose. Waste generated from ecological packaging, as opposed to the plastic waste, is biodegraded. Therefore, this type of packaging is considered the most environmentally friendly. The materials based on paper are the most recyclable materials in the world: the paper and cardboard recycling level equals 70-80% in North America, and even 90% in Central Europe. Another way of re-using paper packaging is degradation (biodegradation, compost). Packaging may be considered as biodegradable if it is made from organic substances, such as cellulose or synthetic compounds, which may be easily decomposed by natural bacterial or fungal processes. The cellulose packaging biodegradation level depends on the substances used in the production process and on other materials used as surface treatment [6]. According to the European Standard EN 13432:2002, a biodegradable material should degrade at least 90%in 6 months, in an environment rich in carbon dioxide (UNI EN 13432:2002). On the other hand, according to the UNI EN 14995:2006, a material is "compostable" if it degrades at least 90%n biodegradable fragments of the dimension of 2mm, in contact with organic materials in a period of 3 months (UNI EN 14995:2006) [8, 9]. In spite of the vast activities in the field, many authors underline that the recycling of – among others – packaging waste, is insufficient and still remains a global issue [10]. Analysing the packaging life cycle, the total greenhouse gas index is considered, produced directly and indirectly at the various stages of the product life cycle, which is referred to as the carbon footprint. Taking into account the packaging material production stage, the materials derived from plant materials are more advantageous in terms of CO_2 emissions. This is due to the fact that plants assimilate CO_2 during photosynthesis. An example is a fully biodegradable polymer, such as polylactide. It is obtained from renewable raw materials such as maize. Another example is the use of synthetic petroleum based polymers for the production of packaging [11]. In the packaging industry, rapidly growing innovation is observed. Research is being done around the world to use smaller quantities of packaging materials and renewable sources [12]. More and more often, new composite materials based on natural fibers of plant origin, such as coconut, reed, wheat straw etc., are used for production [13-15]. Design labels sensitive to the changes in the internal state of the package, which changes through a visible colour, will alert the consumer about the freshness or reaching the expiry date of the product [16]. In order to minimize the impact on the environment, printing ink is improved. Water-based printing ink, containing no toxic chemicals is preferred. Research on methods and measures to prevent the migration of harmful substances in the form of sorption coatings, in packaging materials of cellulose origin has been carried out.

3 United Packaging SA as an example of modern and ecological company

United Packaging S.A is located at Skarbimierz in the Opole Province .The company acts in accordance with the principles of sustainable development. In this way, it approaches the use of natural resources since their only rational use leads to the natural environment protection. It also attains a positive effect on the natural environment through many actions, such as increasing the share of renewable energy sources, reducing the material consumption of production, as well as in the area of waste, water and sewage management. A solar farm of an area of approx. 700 m^2 has been built on the premises of the Production Plant, which consists of 290 VITOSOL 200-F SV2-type high-performance flat collectors supplied by Viessmann, along with a heat accumulation unit, a heat distribution unit and an emergency cool-down unit. The total annual heat yield by the solar farm resulting from the local insolation and the optical efficiency of the installed solar collectors exceeds 690 MWh. Considering the solar installation's thermal losses that have a considerable influence on the overall performance of the system, it can be stated that the Skarbimierz Solar Farm allows an annual saving of about 43.000 m^3 of gas. The total annual reduction of CO_2 emission exceeds 80 tons. The Company utilizes solar energy for heating up the production and storage and the welfare and office areas. Another action undertaken by the company within the framework of ecological activity is the reduction of the production material consumption by cutting down the raw material losses in the corrugated cardboard printing process. Moreover, a digital Quality Control System implemented in the corrugated cardboard printing technological line has reduced the initial loss level by another 30%. These losses are reduced by setting the individual colours of the print relative to one another and controlling the process in an automatic manner by a digital camera system. Reduction of the washing water consumption level is another pro-ecological action undertaken by the Company. It makes it possible to save the water, generate less waste water, so the expenditures incurred for the purchase and commissioning of this innovative (Eco-drive) system are paying off. The Company's "by-product" is waste water containing residues of paints that significantly affect the natural environment; therefore, we have also invested in the construction of the Company's mechanical and biological waste treatment plant. These are major pro-ecological projects that will also yield specific economic effects. The Company has a functioning Environmental Management System complying with the ISO 14001 standard. In addition, it has a Quality Management System in place complying with the ISO 9001 and BRC IOP standards for companies manufacturing packaging intended for contact with food.

3.1 The Company's innovative activity

From the outset, the Company has been focusing on innovation. As a matter of fact, it is constantly looking for state-of-the-art solutions it could implement. Such modern technologies are often linked with pro-ecological solutions. It has been searching for solutions which, on the one hand, would modernize and speed up the production process and, on the other hand, would initiate environmental protection activities. Research and development processes have been launched that encompass both the materials technology, and the unique design supporting the function of protecting products against moisture. In addition, by equipping the printing line with the innovative Eco-drive system, it is possible to reuse the last washing-stage water for pre-washing of the printing units in the next production cycle, as well as to use the digital Quality Control System in the corrugate cardboard printing technological line.

4 Conclusion

Fast expansion entails an increasing influence of businesses on the natural environment. Modern companies should be managed in a strategic, market-oriented and innovative manner. The world packaging market has been developing rapidly and the necessity of ecological packaging has been increasing. For their products to be competitive, they must care for the product quality, which in turn requires them to be open to change. Implementing ecological solutions in the packaging industry on a wide scale may help in reducing many key environmental problems, which include primarily the unfavourable climate change, the depletion of natural resources, the environmental pollution or the loss of biodiversity. Therefore, the success of a company in the contemporary economy is more and more often assessed from the ecoinnovation point of view. Ecoinnovative technologies or products enable a company to compete in the market, which is the indicator of its success. For this reason, managing modern companies must be based on thorough knowledge and competences that enable the creation of innovative solutions. A good example is the dynamically growing United Packaging SA company that manufactures corrugated cardboard packaging of enhanced hydro-resistance.

References

1. Koskela S., Dahlbo H., Judl J., Korhonen M.R., Niininen M. Reusable plastic crate or recyclable cardboard box? A comparison of two delivery systems. Journal of Cleaner Production, **69,** pp 83–90, (2014)
2. https://www.transparencymarketresearch.com/ecological-packaging-market.html (20.04.2018)
3. Heide M., Olsen S. Influence of packaging attributes on consumer evaluation of fresh cod. Food Quality and Preference, **60**, pp 9–18, (2017)
4. Wikström F., Williams H., Verghese K. Clune S. The influence of packaging attributes on consumer behaviour in food-packaging life cycle assessment studies - a neglected topic. Journal of Cleaner Production, **73**, pp 100-108, (2014)
5. Wasiak W., Przemysł i rynek opakowań w Polsce [Polish Packaging Industry and Market]. Opakowania [Packaging], pp 18-21, (2014)
6. Grundey D. Functionality of Product Packaging: Surveying Consumers' Attitude Towards Selected Cosmetic Brands. Economics & Sociology, **3**, pp 87-103, (2010)
7. Majid I., Nayik G., Dar S.M.,Nanda V. Novel food packaging technologies: Innovations and future prospective. Journal of the Saudi Society of Agricultural Sciences. https://doi.org/10.1016/j.jssas.2016.11.003, (2016)
8. Piselli A., Garbagnoli P., Alfieri I., Lorenzi A., Del Curto B. Natural based coatings for food paper packaging. International Journal of Design Sciences and Technology, **20**, pp 55-78, (2014)

9. Garbagnoli P., Altomare L., Nardo L.,Curto B. Development of active packaging for agricultural food. International Journal of Design Sciences and Technology, **20**, pp-79-95, (2014)
10. Tencati A., Pogutz S., Moda B., Brambilla M., Cacia C. Prevention policies addressing packaging and packaging waste: Some emerging trends. Waste Management, **56**, pp-35-45, (2016)
11. Liua B.,Xu H. Preparation and characterization of intelligent starch/PVA films for simultaneous colorimetric indicator and antimicrobial activity for food packaging applications. Carbohydrate Polymers, **157**, pp 842–849, (2017)
12. Krepker M., Shemesh R. Active food packaging films with synergistic antimicrobial activity. Food Control, **76**, pp 117-126, (2017)
13. PatríciaRomani V., Prentice-Hernández C. Active and sustainable materials from rice starch, fish protein and oregano essential oil for food packaging. Industrial Crops and Products, **97**, pp 268–274, (2017)
14. Pogrell H., Market development of bioplastics and latest biopackaging trends, *4th-Conference The future of biodegradable packaging, Warsaw*, (2011)
15. Da Cruz N.F., Simões P., Marques R.C. Costs and benefits of packaging waste recycling systems. Resources, Conservation and Recycling, **85**, pp 1–4, (2014)
16. Lee S., Yam K. L., Piergiovanni L. Food packaging science and technology. CRC Press, London (2008)

Numerical modeling of airflow over column of vehicles using Ansys® package

Władysław Hamiga[1,*] and *Wojciech* Ciesielka[1]

[1]AGH University of Science and Technology, Mickiewicza Av. 30 30-059 Cracow, Poland

Abstract. Growing needs in transportation determinate systems development which improve efficiency of travel and reduce harmful influence to environment. Computational Fluid Dynamic (CFD) uses numerical analysis to find optimal solutions in terms of chosen objective function. Time save and reduce costs for experiments on prototypes are one of the advantages of this method.
The aim of this research is to analyze airflow around different motor vehicles which are moving together in the same direction. To reduce fuel consumption and, at the same time, decrease negative influence to environment, the primary target was reducing total drag force during a ride. The vehicles were set in a column - one after another. In this work considered three types of vehicles: Car, Van and a Truck. Presented vehicles were organized into appropriate groups, creating different configuration. Additional parameter in simulation was distance between vehicles. Simulations of singular vehicles were also done. It allows to evaluate influence of moving vehicles in a column for generated drag force. Described traffic situation were modeled and numerically calculated using ANSYS® package. The purpose of this work was to assess the impact of the distance between vehicles, in a given configuration, for generated drag force.

1 Introduction

More and more studies concentrate on realistic situation on road, where information about aerodynamic characteristic such as drag and lift of singular vehicle is not enough. Researchers investigation going around issue of motion of group of vehicles traveling together (platoons). The advantages of forming vehicles in convoys is both. Increasing of fuel efficiency and improving driver comfort and safeness. Fuel efficiency increase because of reducing pressure and friction drag force, which is result of moving in environment of surrounding fluid (air). Reduce of air drag force is significant aspect in saving energy of moving vehicles. Estimating, the air drag is generating about 52% of total resistance in motion at 100km/h for large, heavy tracks [8]. For this reason aerodynamic characteristics of vehicles are very important.

Study in full-scale models are performed on real vehicles under normal operating conditions. However, road tests are sensitive to ambient conditions such as weather, or they

* Corresponding author: hamiga@agh.edu.pl

are disturbed by vehicles which are not participating in the study. Another way are tests realized in wind tunnel with controllable and known boundary conditions. Experimental study in wind tunnel giving good results but it is very expensive, if good representation of reality is necessary (moving track, rotating wheels, full-scale geometric model, shape's details). Therefore this research method is commonly use in validation process, where geometry and boundary condition are simplify and model can be made in scale. Data from experiment is useful to create numerical model, which is used in Computational Fluid Dynamic (CFD) analysis. Numerical calculations are beneficial for saving time and reducing cost of study.

The study of aerodynamic interaction between vehicles has been investigated for years. Experiments are different between each other in terms of the applied research method, vehicle geometry, and their configuration. Interaction of two cars moving on highway was analyzed by Watkins et al. [10]. Special consideration is emphasis on spacing between vehicles. The experimental studies have been conducted on two Ahmed bodies with backlight angle 30 degree. Experiment was executed in close-jet and fixed-ground type of wind tunnel. Authors prove drag force reduction for close spacing and significant changes in lift. Experimentally study of vehicle platooning in wind tunnel is hard because of huge size of model. To compensate model scaling and increase measurement accuracy, the wake generators was used by Hesham et al. [5]. Further CFD analysis was conducted using STAR-CCM+ package. Behavior of flow pattern around vehicle, during acceleration and overtaking maneuver, was described by Manimaran [6]. In simulations, geometric model of two cars is Ahmed body with backlight angle 25 degree. Author using k-ε model of turbulence in OpenFOAM software. Schito et al. [7] conducted his research in two stages. At first, CFD simulation was performed on Ahmed bodies with backlight angle of 0 and 30 degree. In second part the Ahmed bodies have been replaced by simplified vehicles shapes of compact car, sedan, van and a truck. In both cases, three different spacing was considered. Calculation of singular vehicle has been performed too.

The aim of this work is to reduce drag force of moving column of vehicles. Three type of vehicles was considered: Car, Van and a Truck. Decrease of acting drag force is effect of appropriate configuration and spacing. Appropriate drive in column will reduce fuel or electrical energy consumption, and will have beneficial effect on environment.

2 Methodology

In order to study aerodynamic parameters of moving column of vehicles, a series of CFD simulation has been conducted. Six configuration of cars was investigated. In each of configuration it was considered three type of vehicles. Spacing between vehicles is from 0.3 to 100 meters. The condition of simulation is set to steady-state. Therefore, fluid properties like velocity, pressure or mass flow rate do not change over time. The volume meshes were built using hexahedral cells. Elements around vehicle's geometric has been applied using inflation layer. Inflation layer is 20 cell height with corresponding y+ value near 35 [9]. As turbulence model, in simulation, was used k-ω Shear Stress Transport (SST). Output parameter of CFD simulation is drag force, acting on each of the vehicles individually. Forces were recalculate to dimensionless drag coefficients according to equation:

$$C_d = \frac{2F_d}{\rho A v^2} \tag{1}$$

where :

F_d – acting drag force [N],
ρ – air density [kg m^{-3}],
A – frontal area of the vehicle [m^{-2}],
v – relative fluid velocity [m s^{-1}].

Grid validation was realized comparing CFD simulation of airflow over Ahmed body, with experimental data from the literature [1], [4]. Ahmed body with backlight angle 25°, without stilts was considered. Based on the conducted simulation tests, which were compared with the data in the above-mentioned literature, it was shown, that the calculated drag coefficient error did not exceed 5%.

All of CFD analysis were conducted in ANSYS FLUENT, while mesh was prepared in ICEM software (Fig.1)

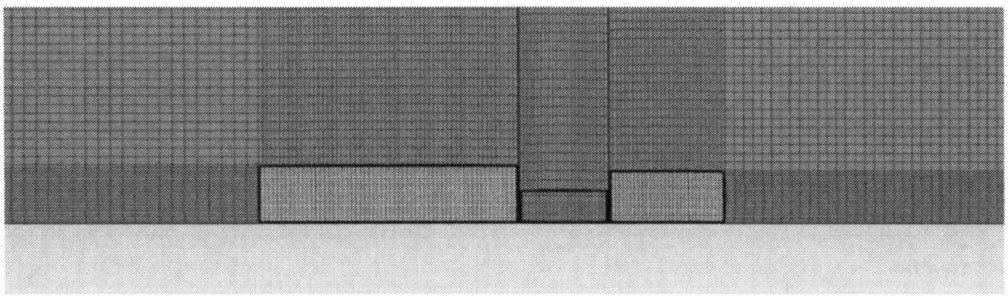

Fig. 1. Hexahedral mesh. Vehicle configuration: Truck-Car-Van and spacing 0.3m

2.1 Object of research

Geometry models of vehicles used in this study, are simplified to the shape of rectangular cuboid. Size of the vehicle was considered as main criterion. Therefore, differences between three types of vehicles are only in height, width and length. Individual dimensions are given in the Table 1.

Table 1. Vehicles dimensions

Vehilce	Height H [m]	Width W [m]	Length L [m]
Car	1.49	1.77	4.55
Van	2.5	2.05	6
Truck	2.8	2.48	14

Clearance between vehicles and the road is set to 0.15m.

2.2 Boundary conditions

Domain of CFD analysis is rectangular cuboid with dimension 20m x 20m x 125 – 325m (last dimension depends of spacing between vehicles). Geometric model of domain is shown at Fig.2. To reduce number of cells, symmetry condition is set up on one of the plane, so only half of model is considered. Velocity of fluid is define as 25m/s at the inlet. Road is set up as "moving wall" with absolute velocity 25m/s. Elements such as the surfaces closing the area of simulation (top and side), as well as the surfaces of the vehicles (vehicle 1, vehicle 2, vehicle 3) has been set as "walls" with zero normal velocity. The boundary condition at the outlet is define as the ambient pressure.

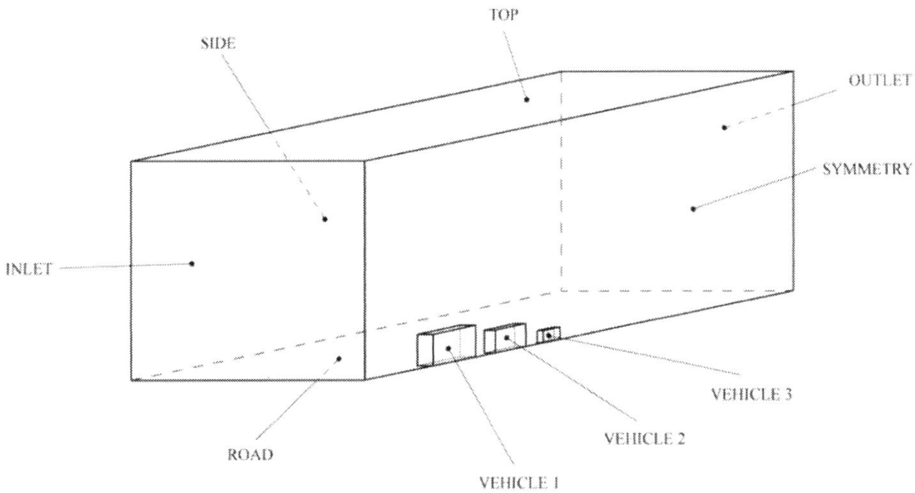

Fig. 2. Model description

3 Results and discussion

The study of aerodynamic interaction between vehicles has been investigated for six different configuration. The column of vehicles consisted of Car, Van and a Truck. Relationship between spacing and drag coefficient for all configurations is presented on charts (Fig.4 - Fig.9). Drag coefficient is calculated for each of the vehicle individually.

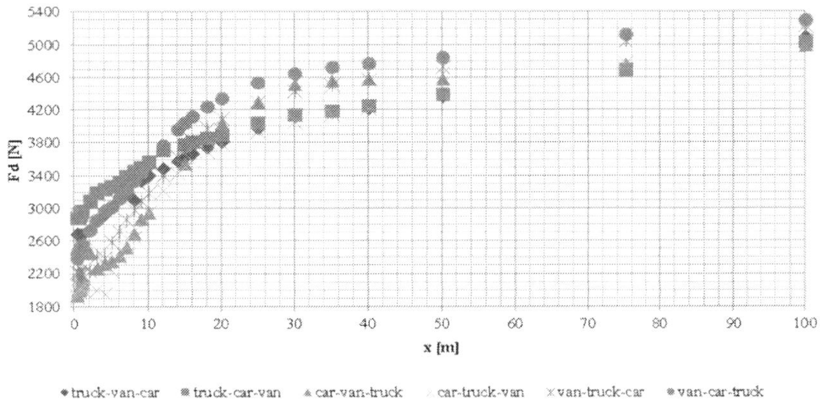

Fig. 3. Total drag force as a function of spacing for all configurations

Drag force presented on Fig.3 is the sum of the forces generated on individual vehicles (drag force of column). Drag force reduction is observable in all configurations. However, significant changes are only on second and third vehicle (Fig.4 - Fig.9). Reduction, in this case, in drag force reaching up to 66% for Car-Truck-Van configuration. It should be noted that distance of 0.3 meters is dangerously small, therefore study on higher vehicle spacing was conducted. For spacing 1 to 18 meters, the most beneficial configuration is with Car on lead. Above 18 meter, in configuration Car-Van-Truck, drag coefficient of Van and Truck is rapidly increasing (Fig.6). Decrease of drag coefficient is also visible for leading vehicles. However, it is only for spacing below 20 meter for Car (Fig.6, Fig.7) and below 12 meters for Truck (Fig.4, Fig.5). Configurations with Vans on lead do not show any benefits for them in drag coefficient reduction (Fig.8, Fig.9).

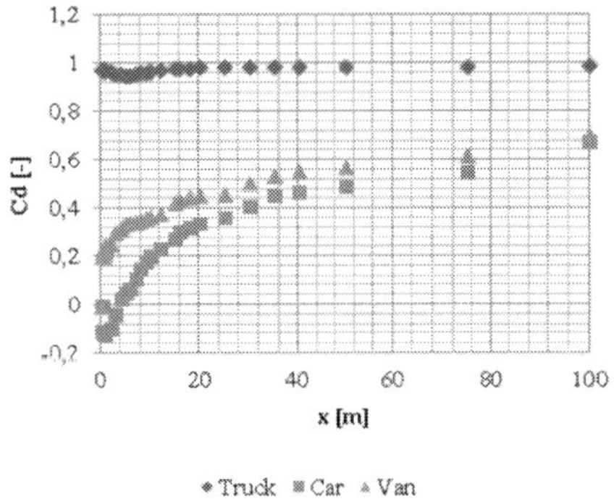

Fig. 4. *Drag coefficient as a function of spacing. Configuration: Truck-Car-Van*

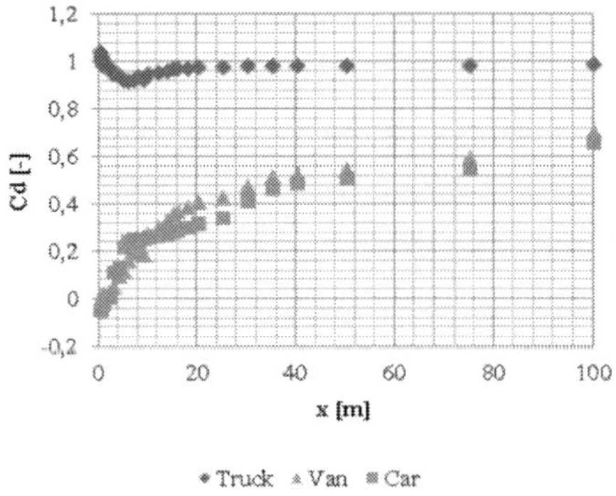

Fig.5. *Drag coefficient as a function of spacing. Configuration: Truck-Van-Car*

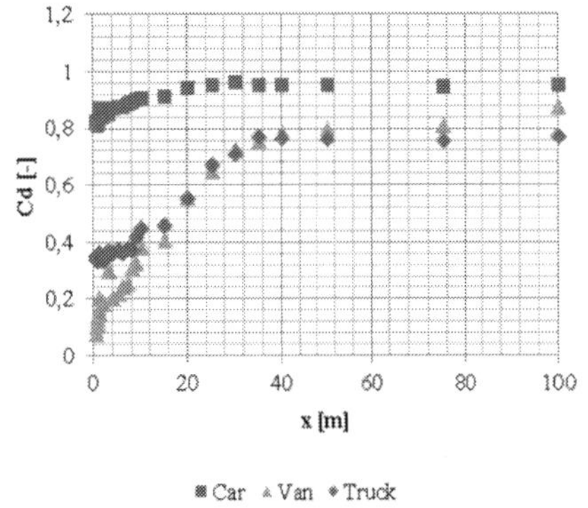

Fig. 6. *Drag coefficient as a function of spacing. Configuration: Car-Van-Truck*

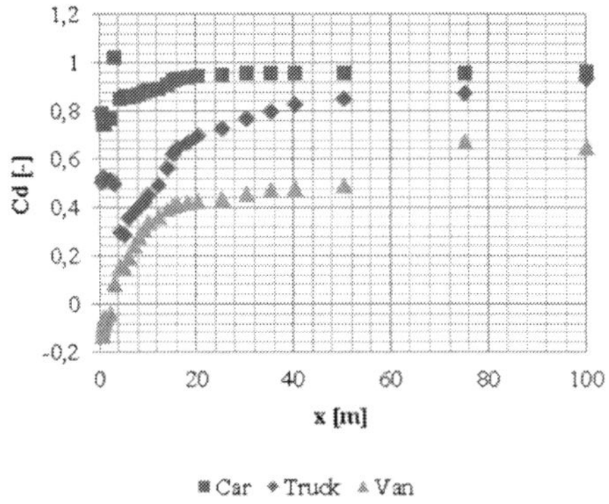

Fig. 7. *Drag coefficient as a function of spacing. Configuration: Car-Truck-Van*

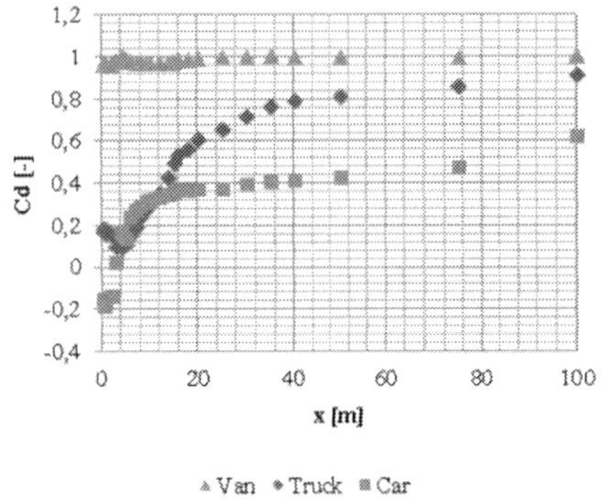

Fig. 8. *Drag coefficient as a function of spacing. Configuration: Van-Truck-Car*

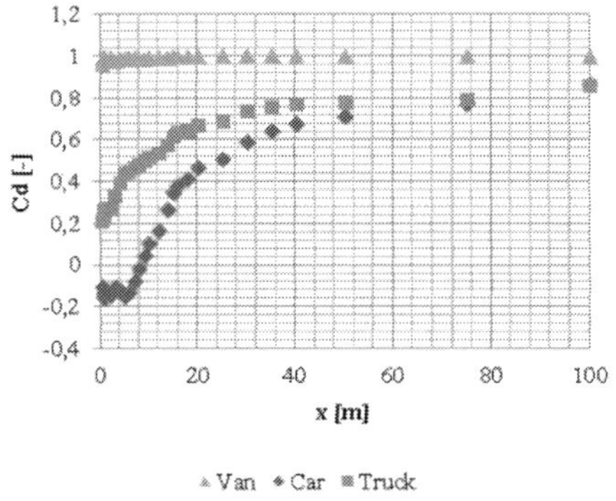

Fig. 9. *Drag coefficient as a function of spacing. Configuration: Van-Car-Truck*

Fig. 10. Velocity distribution at symmetry plane. Car - Van - Truck. Spacing between vehicles - 6m

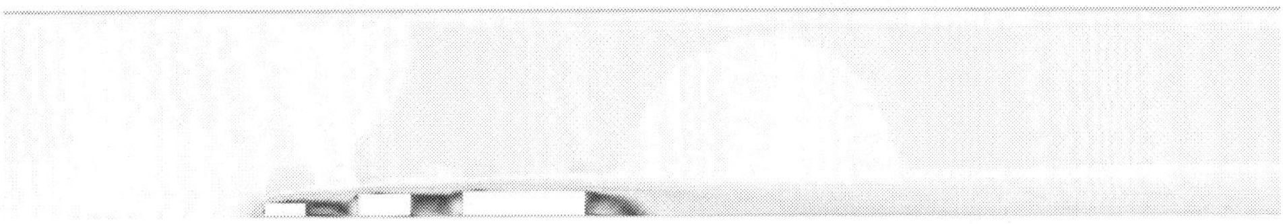

Fig. 11. Velocity distribution at symmetry plane. Car - Truck - Van. Spacing between vehicles - 6m

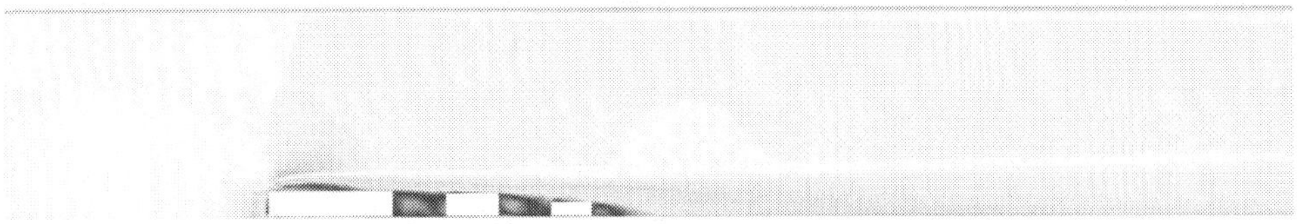

Fig. 12. Velocity distribution at symmetry plane. Truck - Van - Car. Spacing between vehicles - 6m

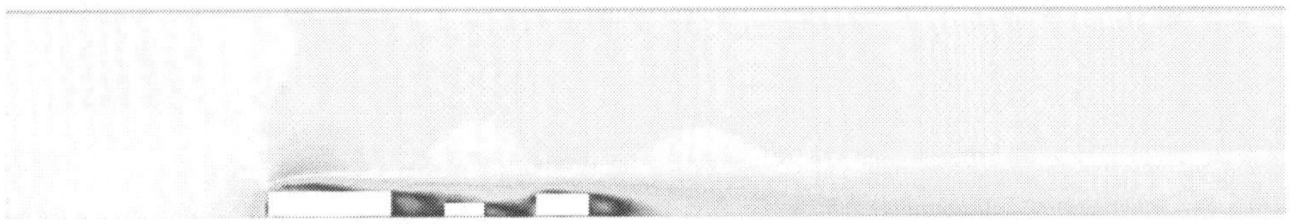

Fig. 13. Velocity distribution at symmetry plane. Truck - Car - Van. Spacing between vehicles - 6m

Fig. 14. Velocity distribution at symmetry plane. Van -Truck - Car. Spacing between vehicles - 6m

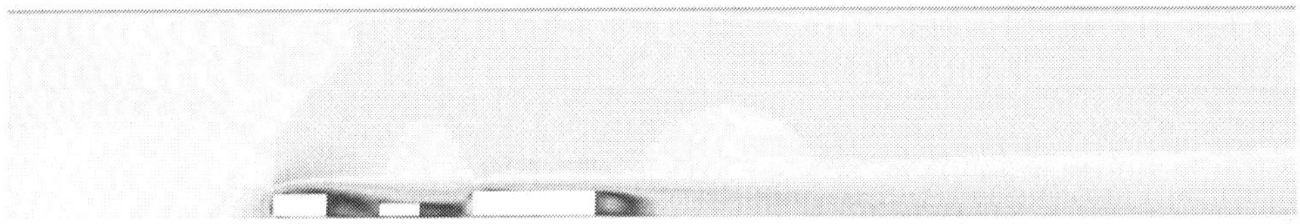

Fig. 15. Velocity distribution at symmetry plane. Van - Car - Truck. Spacing between vehicles - 6m

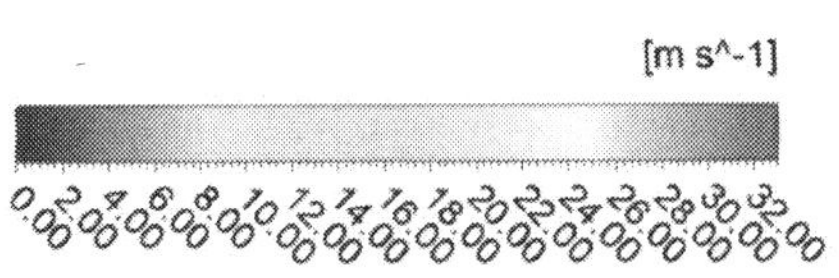

Fig. 16. Velocity distribution - legend

The CFD simulations of singular vehicles were also conducted. For increasing distance between vehicles, drag force tends to the value of singular vehicle. The value of drag coefficient going to coincide (Fig.4 - Fig.9).

With objective function to reduce total drag force of whole column, the most beneficial configuration is Car-Truck-Van (Fig.3). The shape of column, in this configuration, approximately, is similar to water drop.

Velocity distribution on symmetry plane is presented for distance between vehicles equals 6 meters (Fig.10 - Fig. 15). Comparison of velocity is shown for all configurations. Velocity distribution legend for Fig.10 - Fig. 15 is presented in Fig. 16. Wake behind column of vehicles is observable in all configurations and it is spread for whole domain.

4 Conclusion

In the paper was presented CFD study of airflow over the column of vehicles using ANSYS package. Drag coefficient, as a function of spacing, has been investigated for six different configuration. The validation process was executed on Ahmed body with backlight angle 25 degree and no stilts. Difference between calculated drag coefficient from simulation, and drag coefficient from literature, did not exceed 5%. Dependence of drag force and drag coefficient as a function of spacing was demonstrated on charts. Drag coefficient for small spacing (up to 1 meter) was significant reduced in all configurations, but only on second and third vehicle in the row.

Further study on aerodynamic parameters of convoy could be realize with more vehicles. Geometry of vehicles should be improve to achieve more accurate value of drag forces and drag coefficients. To reduce time of simulation, geometry in this work, was simplified. Presented numerical model will be use for investigation of acoustic emission, in aspect of its influence to the environment. The results presented in this work have a universal character and may be used to build intelligent acoustic environment management systems for boroughs, districts, cities and urban conurbations [2], [3].

References

1. Bello-Millan F. J., Makela T., Parras L., Pino C., Ferrera C.: Experimental study on Ahmed's body drag coefficient for different yaw angles, Journal of Wind Engineering and Industrial Aerodynamics, **157**, pp. 140-144, (2016)
2. Ciesielka W.: Management of environmental noise – Cracow example, Polish Journal of Environmental Studies, **16**, 3B, pp. 53–58, (2007)
3. Ciesielka W.: Management of environmental noise – the Cracow example, Archives of Acoustics, **32**, 4, pp. 983–994, (2007)
4. Guilmineau E.: Computational study of flow around a simplified car body, Journal of Wind Engineering and Industrial Aerodynamics, **96**, pp. 1207-1217, (2008)
5. Hesham M. Ebrahim, Robert G. Dominy, Pak S. Leung: Evaluation of Vehicle Platooning Aerodynamics using Bluff Body Wake Generators and CFD, Students on Applied Engineering (ICSAE), pp. 218-223, (2017)
6. Manimaran R.: Aerodynamic Study of the Ahmed Body in Road-Situations using Computational Fluid Dynamics, Fluids, Heat and Mass Transfer, Mechanical and Civil Engineering, pp. 140-147, (2015)
7. Schito P., Braghin F.: Numerical and Experimental Investigation on Vehicles in Platoon, SAE International Journal of Commercial Vehicles, **5**, 1, pp. 63-71, (2012)
8. Schoon, R. E.: On-road Evaluation of Devices to Reduce Heavy Truck Aerodynamic Drag, SAE Publication, SAE Paper No. 2007-01-4294, (2007)

9. Simmonds N., Pitman J., Tsoutsanis P., Jenkins K., Gaylard A., Jansen W.: Complete body aerodynamic study of three vehicles, *Proceedings of WCX17: SAE World Congress Experience, 4 - 6 April 2017, Detroit, MI*, Paper number 2017-01-1529, (2017)
10. Watkins S., Vino G.: The effect of vehicle spacing on aerodynamics of representative car shape, Journal of Wind Engineering and Industrial Aerodynamics, **96**, pp. 1232-1239, (2008)

Sensitivity of the Hawt noise level predicted based on Amiet's theory on the value of a turbulence intensity coefficient determined numerically

Katarzyna Suder-Dębska[1,*], *Dawid* Romik[1], and *Ireneusz* Czajka[1]

[1]AGH University of Science and Technology, Department of Power Systems and Environmental Protection Facilities, Mickiewicza 30 Av. 30-065Cracow, Poland

Abstract. In the paper the authors presented and compared two methods of the HAWT noise predicting. The priority, however, was to test the possibility of using Amiet's theory to determine the noise value in the far field. In this theory it is necessary to know the value of the turbulence intensity coefficient. The value of this coefficient was determined based on numerical modeling. The NACA 0012 profile was used for the airfoil shape. The ANSYS/Fluent program was used for numerical calculations, where the k–ω SST turbulence model was used to simulate the flow, and Ffocs-Williams and Hawkings model was used to determine the noise level. The turbulence intensity coefficient estimated in this way was then used to determine the noise value from the wind turbine airfoils using Amiet's theory.

1 Introduction

Sustainability requires that along with technological development providing adequate comfort of life, while at the same time carrying for the natural environment. It seems as close as possible to obtain energy from renewable sources as far as possible. Wind turbines, for example, can be such a solution. Often, however, there are situations in which residents of a given area have a negative feeling in relation to such investments. Their position is based, for example, on the impression of too loud sound generated by wind turbines. Because noise is a subjective impression, the solution to such a situation may be to present the obtained sound pressure level from a given object. For existing objects, these results can be obtained based on measurements in the environment. For designed object seem to be necessary the forecasts obtained, for example, on a numerical way. It is necessary to have full information about the sound source to correctly carry out the numerical simulations. It this paper the authors presented and compared two methods of the HAWT noise predicting. It also presents how the value of the sound level changes depending on the wind parameters.

* Corresponding author: suder@agh.edu.pl

1.1 Research object

For this example a model of a wind turbine consisting of three 45 m long blades was adopted. The chord length is 3.5 m at the root of the blade and 0.8 m at the tip. The blades geometries were chosen so that the angle of attack (AoA) over the entire blade length was 4°. However, according to [1] turbulent inflow noise is almost independent of AoA for symmetric airfoils such as NACA 0012 (described above).

2 Numerical simulation

The Ffowcs Williams Hawkings [2] acoustic model (FW-H) available in the ANSYS/Fluent program was used to determine the sound pressure level of the model rotor. The FW-H model takes the general form of the Lighthill's acoustic analogue [3] and is able to predict the sound generated by equivalent acoustic sources such as monopoles, dipoles and quadrupole. The ANSYS/Fluent program adopts an equation in which the acoustic pressure and acoustic signal at designated points of the receiver is calculated by determining several integrals on the surface. The FW-H equation is a non-homogeneous wave equation that can be determined by the coupling of continuity equations and Navier-Stokes equations. The FW-H equation can be written as:

$$\frac{1}{a_0^2}\cdot\frac{\partial^2 p'}{\partial t^2}-\nabla^2 p'=\frac{\partial^2}{\partial x_i \partial x_j}\{T_{ij}H(f)\}-\frac{\partial}{\partial x_i}\{\left|P_{ij}n_j+\rho u_i(u_n-\upsilon_n)\right|\delta(f)\}+ \\ +\frac{\partial}{\partial t}\{\left|\rho_0\upsilon_n+\rho(u_n-\upsilon_n)\right|\delta(f)\}=0 \tag{1}$$

where :

- u_i – fluid velocity component in the x_i direction,
- u_n – fluid velocity component normal to the surface $f = 0$
- v_i – surface velocity component normal to the surface,
- $\delta(f)$ – Dirac delta function,
- $H(f)$ – Heaviside function,
- p' – sound pressure at the far field $(p'-p_0)$
- n_i – unit normal vector pointing toward the exterior region $(f > 0)$,
- a_0 – far-field sound speed,
- T_{ij} – Lighthill stress tensor,
- P_{ij} – compressive stress tensor.

The surface $f = 0$ corresponds to the source (emission) surface, and can be made coincident with a body (impermeable) surface or a permeable surface off the body surface.

$$T_{ij}=\rho u_i(u_n-\upsilon_n)+P_{ij}-a_0^2(\rho-\rho_0)\delta_{ij} \tag{2}$$

$$P_{ij}=p\delta_{ij}-\mu\left[\frac{\partial u_i}{\partial x_j}+\frac{\partial u_j}{\partial x_i}-\frac{2}{3}\frac{\partial u_k}{\partial x_k}\delta_{ij}\right] \tag{3}$$

The solution of equation (1) is obtained using the free-space Green function $\delta(g)/(4\pi r)$. The complete solution consists of surface integrals and volume integrals. The surface integrals represent the contributions from monopole and dipole acoustic sources and partially from quadrupole sources, whereas the volume integrals represent quadrupole (volume) sources in the region outside the source surface. The contribution of the volume integrals becomes small when the flow is low subsonic and the source surface encloses the source region. In ANSYS/Fluent, the volume integrals are dropped. Thus, one has

$$p' = (x,t) = p_T'(x,t) + p_L'(x,t) \tag{4}$$

$$4\pi p_T'(x,t) = \int_{f=0} \left[\frac{\rho_0 \left(\dot{U}_n + U_{\dot{n}} \right)}{r(1-M_r)^2} \right] dS + \int_{f=0} \left[\frac{\rho_0 U_n \left\{ r\dot{M}_r + a_0 \left(M_r + M^2 \right) \right\}}{r^2 (1-M_r)^3} \right] dS \tag{5}$$

$$4\pi p_L'(x,t) = \frac{1}{a_0} \int_{f=0} \left[\frac{\dot{L}_r}{r(1-M_r)^2} \right] dS + \int_{f=0} \left[\frac{L_r - L_M}{r^2 (1-M_r)^3} \right] dS + \frac{1}{a_0} \int_{f=0} \left[\frac{L_r \left\{ r\dot{M}_r + a_0 \left(M_r + M^2 \right) \right\}}{r^2 (1-M_r)^3} \right] dS \tag{6}$$

where:

$$U_i = v_i + \frac{\rho}{\rho_0} (u_n - v_i) \tag{7}$$

$$L_i = P_{ij} \hat{n}_j + \rho u_i (u_n - v_n) \tag{8}$$

where :

n, r – unit vectors in the radiation and wall-normal directions,

M – Mach number of a source.

Taking into account the observer time t and the distance to the observer *r*, the integral of the above equation is computed in time defined by the equation :

$$\tau = t + \frac{r}{a_0} \tag{9}$$

2.1 Numerical model

All numerical calculations were done in ANSYS/Fluent. On the rotor blades and the walls of the enclosure, a Wall boundary condition is provided which prevents air from flowing through these surfaces. At the inlet, the boundary condition Velocity inlet U=8 m/s. At the outlet the pressure outlet was determined so that it corresponded to the barometric pressure $p = 0$. The rotational speed of the rotor is 14 rpm and the simulation uses the Sliding Mesh approach. The time step $\Delta t = 1 \cdot 10^{-4}$ s was adopted. The equation of continuity and Navier-Stokes with Reynolds averaging (RANS) were used for the calculations. The two-mode Menter's turbulence model *k–ω* SST was adopted [4][5]. For the above input data, the model quickly reached convergence and the results matched the expected results.

The *k–ω* SST model is a model that combines the advantages of the *k–ε* and *k–ω* model and introduces an additional limiting factor for the over-production of turbulent kinetic energy in the areas of strong positive pressure gradients (accumulation points, separation of the boundary layer regions). By examining the *k–ε* and *k–ω* models, Menter observed that the first one models well the turbulence in free flow and shear layers and is characterized by low sensitivity to inlet conditions for turbulence volumes. This is a desirable feature due to the fact that often in practical calculations these quantities are not exactly known. The *k–ω* model is much better at modeling the turbulent flow in the boundary layer, but it is very sensitive to turbulent values in free flow [4].

As an acoustic model Ffowcs-Williams and Hawkings was adopted. The sound pressure level were measured at 100 m distance from the rotor in rotor axis direction behind the turbine.

As a result of the simulations described above, the average values of turbulence intensity on the blade of approx. 14% (Fig.1) was obtained. Fig.2 shows the turbulence intensity values obtained along the length of the blade. It is easy to notice that the highest values of turbulence intensity are obtained on the tip of the blade, while the lowest in the root area. Between the root and tip areas, turbulence intensity values are approximately the same, i.e. they show differences of no more than 10 percentage points.

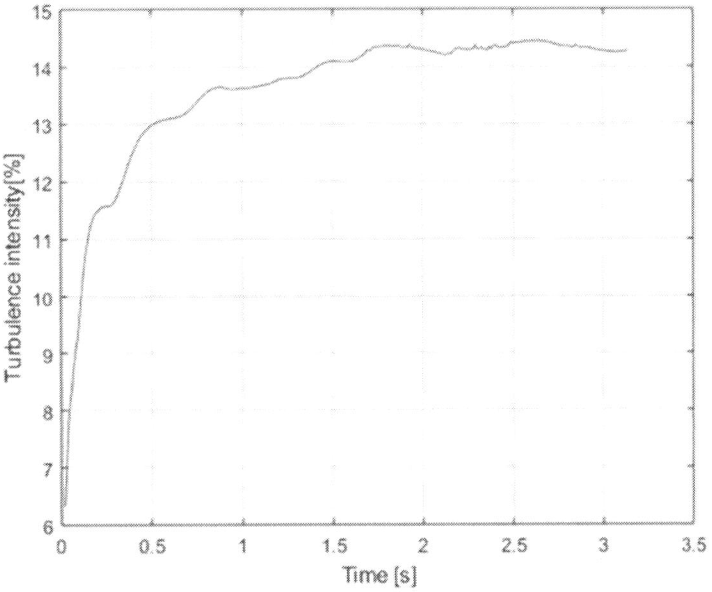

Fig. 1. Average values of turbulence intensity on the airfoil

Fig. 2. Values of turbulence intensity along the blade length

3 Main assumptions of Amiet's theory

Amiet's theory concerns the generation of noise through a thin plate under the influence of turbulent flow. The plate is an approximation of the airfoil. The eddies appearing on its surface, pressure fluctuations are treated as dipole sound sources.

In the classic Amiet's theory the airfoil is reduced to a flat plate with zero thickness and a zero angle of attack. The basic assumptions are as follows [6-11]:

- the flow is uniform with velocity U,
- the turbulence fluctuation is small in relation to mean flow velocity,
- the interactions between the airfoil and the turbulent flow can be described as inviscid, so the problem is reduced to the solution of the linearized Euler equation,
- the turbulence is frozen - turbulent gust properties are unchanged while it is convected by the mean flow,
- velocity fluctuation of turbulent gust is represented in terms of chordwise (k_x)and spanwise (k_y) wavenumbers.

In Amiet's theory is used the following equation to calculate the one-third octave sound pressure level (in all parameters, elements related to length are expressed in feet)

$$SPL_{1/3} = 10\log_{10}\left[\frac{L \cdot d/2}{z^2} M^5 \frac{\overline{u}^2}{U^2} \frac{\hat{k}_x^3}{\left(1+\hat{k}_x^2\right)^{7/3}}\right] + 181.3 \tag{10}$$

where :

L	–	integral length scale of turbulence,
d	–	airfoil span,
$\overline{u}^2$	–	turbulence intensity,
k_x	–	chordwise turbulence wavenumber,
$\hat{k}_x$	=	$\frac{k_x}{k_e}$,
k_e	=	$\frac{\sqrt{\pi}}{L}\frac{\Gamma(5/6)}{\Gamma(1/3)}$.

4 Results

Fig.3 presents the results of calculations of the sound pressure level in one-third octave bands for different (changing) values of the intensity coefficient of turbulence obtained by simulation. These values have changed in the range of 6.5-14.5% at one percentage point.

The highest value of turbulence intensity determined on the simulation way was 109%. For this value, the SPL in one-third octave bands generated by the wind turbine was determined at a distance of 100 m from the turbine. Also for this value of the turbulence intensity, the SPL in the one-third octave bands was determined based on the Amiet's theory. The obtained results have been presented in the graph (Fig.4).

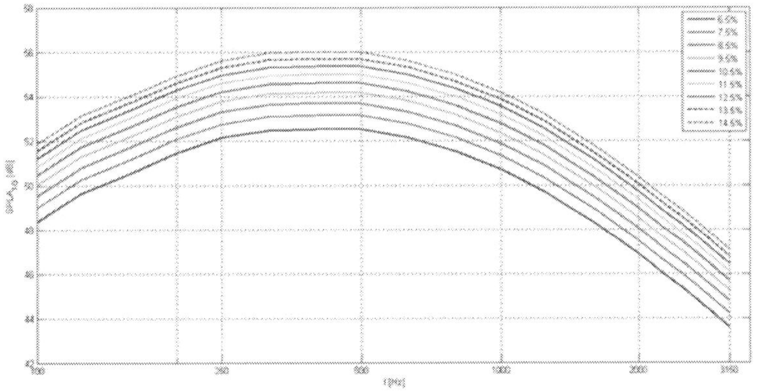

Fig. 3. SPL values depending on the turbulence intensity value

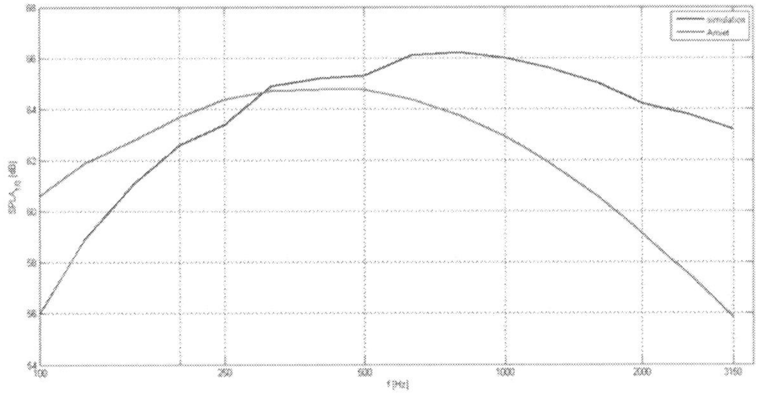

Fig. 4. SPL for turbulence intensity equal to 109% calculated based on simulation and Amiet's theory

The number of numerical experiments were reduced by using DOE. An extended central composite face centered (CCF) design in which factors adopt values on five levels were used. For four factors x_1, x_2, x_3 and x_4 there is a need to conduct 49 numerical experiments. The calculations were performed using "ZEUS" supercomputer at ACK Cyfronet Krakow. The table 3 presents the plan of the conducted simulations with the input and output parameters.

5 Conclusion

The article has presented the considerations on the influence of turbulence intensity values on the obtained SPL. In detail, this impact has been analyzed for the turbulence intensity in the range of 6.5-14.5%. As might be expected, along with the increase of the turbulence intensity, the obtained SPL also grows. For values within the range of up to 10%, a change by one percentage point may lead to a change in SPL values even by 1 dB. For the higher turbulence intensity values, these changes will of course also appear, but they are not so clear.

Since the greater the intensity of turbulence, the higher the SPL values, it can be assumed that for a turbulence intensity of 109%, the obtained SPL value will be the highest possible. Both in simulations and in calculations based on Amiet's theory, the SPLA values in the one-third octave bands at the level of approx. 65 dB were obtained. The SPLA values determined in both ways show differences in the lowest and highest frequencies analyzed. However, this is still a value of several decibels. The results for frequencies in the range of 200-1000Hz are promising - the differences in obtained values is much less than 3 dB.

Because the obtained SPLA values are similar, one can expect that similar values will also be obtained under real conditions. It seems important to verify the correctness of the obtained values of turbulence intensity, because as one can see it significantly affects the value of the noise generated by wind turbines. Fig

References

1. Devenport W., Staubs J., Glegg S., Sound radiation from real airfoils in turbulence, Journal of Sound and Vibration, **329**, pp. 3470–3483, (2010)
2. Ffowcs-Williams J. E., Hawkings D. L., Sound Generation by Turbulence and Surfaces in Arbitrary Motion, Proc. Roy. Soc., London, pp. A264:321–342, (1969)
3. Lighthill M.J., On Sound Generated Aerodynamically, Proc. Roy. Soc., London, pp. A211:564-587, (1952)
4. Menter F.R., Zonal Two Equation Turbulence Models for Aerodynamic Flows, AAIA Paper, pp. 93-296, (1993)
5. Ansys Fluent.: Theory Guide. Ansys Inc. (2015)
6. Amiet R.K. Acoustic Radiation from an Airfoil in a Turbulent Stream, Journal of Sound and Vibration, **41**, 4, pp. 407-420, (1975)
7. Amiet R. K. Noise due to Turbulent Flow past a Trailing Edge, Journal of Sound and Vibration, **47**, 3, pp. 387-393, (1976)
8. Paterson R.W., Amiet R.K. Acoustic radiation and surface pressure characteristics of an airfoil due to incident turbulence, National Aeronautics and Space Administration: Washington, USA, (1976)
9. Schlinker R.H., Amiet R.K. Helicopter Rotor Trailing Edge Noise, National Aeronautics and Space Administration: Washington, USA, (1981)
10. Roger M., Moreau S. Extensions and Limitations of Analytical Airfoil Broadband Noise Models, International Journal of Aeroacoustics, **9**, 273-305, (2010)
11. Tian Y., Cotte B. Wind Turbine Noise Modeling Based on Amiet's Theory: Effects of Wind Shear and Atmospheric Turbulence, Acta Acustica united with Acustica, **102**, 102, pp. 626-639, (2016)

E3S Web of Conferences **46**, 00027 (2018) https://doi.org/10.1051/e3sconf/20184600027
3rd International Conference on Energy and Environmental Protection

Smog – Poland's pressing problem. Anti-smog technologies

Jan Kiciński[1,*]

[1]Polish Academy of Sciences, Institute of Fluid Flow Machinery, Generała Józefa Fiszera 14, 80-231 Gdańsk, Poland

Abstract. The article also presents the situation of the Polish power engineering against the background of global trends, as well as proposals for solving the smog problem using anti-smog technologies developed at the Institute of Fluid-Flow Machinery of the Polish Academy of Sciences in Gdańsk (IMP PAN) and by launching a pilot project that will cover 1,000 households equipped with these technologies. The author and his closest co-workers recommend equipping boilers, including both older and newer ones, with low-power electrostatic precipitators. In this way, one can burn low-quality fuels while keeping the emission of airborne dust at a low level. It is a quick and not pricey solution to the smog problem.

1 Introduction

Scale of the problem. There are about 13.4 million households in Poland. Most of them (55.5%) are located in multi-family houses, and 44.5% in single-family houses. Currently, approximately 5-6 million households utilise coal and biomass for heating buildings and utility water. The vast majority of them use older generation boilers, which have low-efficiency ratings and high emission levels of pollutants. The oldest devices include solid fuel stoves (whose average age exceeds 24 years) and boilers (the average age of which is 10 years). On the other hand, the share of boilers and solid fuel stoves as sources of pollution to the atmosphere in Poland is estimated at **over 90%**.

Smog originates from energy poverty, we burn everything that has a calorific value!

So, how to solve the problem of smog? Do we have anti-smog technologies suitable for rapid implementation in our country?

An example of anti-smog technologies with a high degree level of technological readiness are the installations developed at the Institute of Fluid-Flow Machinery of the Polish Academy of Sciences in Gdańsk. At this point, it is worth mentioning the results of the works of research teams managed by Prof. D. Kardaś (low-emission boilers [3,4]), Prof. M. Dors, Dr. J. Podliński, Prof. A. Jaworek (electrostatic precipitators [5-9]), Prof. J. Kiciński and Dr. G. Żywica (micro power plants [1,2]), Dr. M. Lackowski (energy storage) as well as Prof. A. Cenian and Prof. P. Lampart (gasifiers, biogas power plants and ORC systems [10]).

The work of these teams led to the development of technologies which Poland so desperately needs.

* Corresponding author: jan.kicinski@imp.gda.pl

2 Anti-smog technologies elaborated in IMP PAN

There were proposed four stages of the implementation of these innovative technologies:

- equipping older generation domestic boilers with low-power electrostatic precipitators (IMP PAN patent No. P. 422507). Fast and relatively cheap implementation;
- development of technologically advanced coal-fired boilers with very low levels of dust emissions, equipped with electrostatic precipitators;
- introduction into the market of domestic micro power plants that enable the production of heat and electricity and can replace central heating boilers;
- introduction into the market of a series of micro-cogeneration power plants that can meet the energy needs of small companies, public buildings and multi-family houses.

2.1 Low-power electrostatic precipitator for use in older generation boilers operating in domestic installations (IMP PAN patent No. P. 422507)

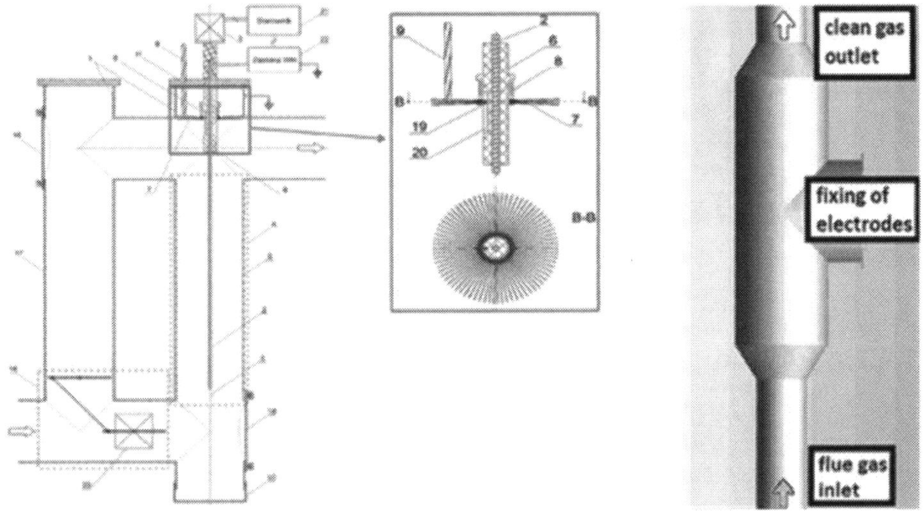

Fig. 1. Low-power electrostatic precipitator and its schematic diagram (for use in domestic boilers with thermal power capacities ranging from 20kW to 30kW) [19-25]

Device characteristics:

- quick implementation is possible, high degree of technological readiness (the device already passed laboratory tests and operational tests of the pilot installation),
- dust removal efficiency reached 90%,
- the price of one electrostatic precipitator would not be higher than PLN 2,000 (assuming the mass production).

Remark: Even though it is fast and cheap, it is not the ultimate solution to the problem.

2.2 Ultra-low emission coal boiler (highly efficient device plus an electrostatic precipitator)

The boiler enables combustion of various types of low-quality fuels (e.g. coal-mine waste that is much cheaper than eco-pea coal), and thanks to the use of the electrostatic precipitator the concentration of airborne dust can be kept at a low level.
This is a very attractive solution.

Characteristics of the installation:

- the electrostatic precipitator for a lower class boiler is a temporary solution. An ultra-low emission coal boiler equipped with an electrostatic precipitator is a more forward-thinking solution;
- the total manufacturing cost of an ultra-low emission coal boiler is PLN 10,600. Having analysed the market of central heating boilers, it can be said that the selling prices of coal boilers with a capacity of approx. 25kW (class V according to the PN EN 303-5:2012 standard) range from PLN 11,600 (Draco Versa Tekla) to PLN 12,600 (EGRO PLUS Witkowski);
- the cost of an ultra-low emission boiler with a capacity of 20 kW will be paid back after 2.5 years, and the profit from its use will exceed PLN 40,000 within 10 years;
- high degree of technological readiness (after laboratory tests).

An ultra-low emission domestic boiler will be an original solution based on a highly efficient combustion process at elevated temperatures, with a two-stage heating and will be equipped with an electrostatic separator of solid particles. In addition to increased energy efficiency, this solution will enable the use of low-quality solid fuels, i.e. coal dust containing a higher percentage of minerals, which will significantly reduce exploitation expenses. A product of this kind is not available either on the Polish or European market.

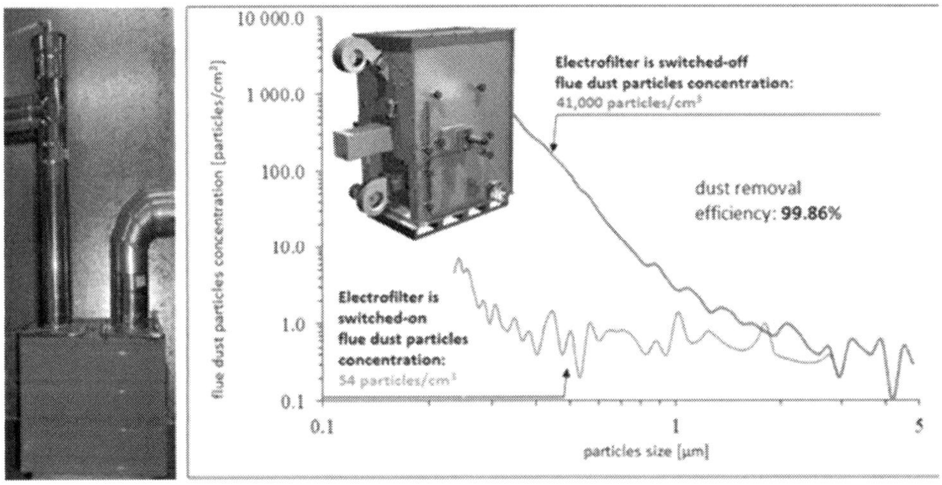

Fig. 2. Electrostatic precipitator behind the flue and the graph that shows the dust emission reduction. The dust removal efficiency is 99.86% [19-25].

2.3 Domestic cogeneration power plant (low-emission boiler, microturbine, electrostatic precipitator)

Characteristics of the installation:

- it is the most forward-thinking solution that allows producing heat and electricity in households while, at the same time, providing low emission levels of pollutants.
- The installation passed laboratory tests, a pilot installation is currently under construction.
- A coal-fired or wood-fired domestic micro power plant will be profitable mainly due to the synergy between the production of heat and electricity (cogeneration). Achieving profits from the production and sale of electricity can lead to a situation in which the technology demonstrated herein will become a technology of the future from the point of view of civic power engineering.

- It is estimated that a micro power plant with a thermal power capacity of 30kW and an electric power capacity of 4.5kW can give an income of PLN 1,000 a month.

Fig. 3. Micro power plant with a thermal power capacity of 25kW and an electric power capacity of 2.5kW. On the left-hand side: microturbines operating in the ORC system [18, 19 -25].

2.4 Scaling. Series of power plants with higher power capacities

The technologies listed in sections I, II and III relate to installations for use in small households, which mainly emit particulate pollutants. However, there are still municipal boiler rooms, boiler rooms for public utility buildings, etc., and all of them have a significant share in air pollution. For this market segment, it is worth developing anti-smog technologies for devices with higher thermal (several hundred kilowatts) and electric (20 – 30 kW) power capacity.

Mostly owners of large farms will be buyers of micro power plants with electric power capacities exceeding 20kW. There are over 30,000 large farms in Poland (with an area of land above 50 ha) and their owners are potential recipients of micro power plants. Another very important group of buyers of micro power plants will be municipal enterprises from small towns, which use hard coal as a basic source of thermal energy. In Poland there are 2,478 municipalities (of which 1,563 are rural municipalities). Assuming that each rural municipality has a school and a public building (e.g. a municipal office) that need heat and electricity, this gives a total of more than 3,000 potential users of the device.

In Poland, there are at least 50,000 potential buyers of solid fuel fuelled gas turbine systems with an electric power capacity of approximately 30 kW. That number covers large farms, small and medium-sized production and service enterprises from the agricultural and forestry sectors as well as municipal entities managed by local authorities.

3 Fight for clean air

The most advanced and most profitable method of fighting for clean air is to completely transform the existing farm heating system in Poland into the distributed power engineering based on small installation capable of generating electricity and heat. These installations would use cheap coal and locally available biomass as fuel. A transformation of domestic boilers and local heating-generating plants into small heat and power plants will additionally result in a drastic reduction of dust emissions, which will take place due to the need to maintain high operational efficiency of all components of micro and mini power plants.

Poland has a chance to become a vast repository of knowledge about air pollution, its formation, impact on people and the environment, but also a centre for technologies for the ecologically-friendly use of solid fuels as well as prevention and elimination of smog. The goal could be to make Poland the global leader in the field of eco-friendly utilisation of solid fuels for heating households and for the cogeneration of heat and electricity.

Domestic cogeneration power plants can be perceived as the last stage of the fight for clean air. In the first stage, it is worth implementing low-power electrostatic precipitators in both older and newer generation boilers.

The use of low-power electrostatic precipitators for both old and new generation boilers can help a lot in this fight. According to the estimates, one electrostatic precipitator will cost no more than PLN 2,500 (assuming the mass production), which seems to be a reasonable price. The eco-pea coal, the calorific value of which is at least 24 MJ/kg, is currently sold for PLN 775 per tonne whilst the small coal from the Wujek mine with the same calorific value is sold for PLN 509 per tonne. Also in other places in the country, there are big differences in the price of these two types of coal. For example, in the Carbon company from the city of Kielce, the cheapest coal of the eco-pea type is sold for PLN 787 per tonne and the small coal is sold for PLN 537 per tonne. In other companies, the difference between the price of eco-pea coal and small coal is usually around PLN 250 per tonne. **The old generation boiler with an electrostatic precipitator can be fuelled with cheaper coal and the emission of pollutants will be the same as in modern V class boilers!**

4 National Centre for clean air

Considering the above facts, it seems justified and urgent to establish instruments for financing innovative anti-smog technologies and, in particular, state funds for launching a pilot series of electrostatic precipitators or a set of boilers with electrostatic precipitators, e.g. in 1,000 households. Such activities seem to be a very good idea due to the reluctance of homeowners as well as local authorities to incur additional costs in the name of environmental protection.

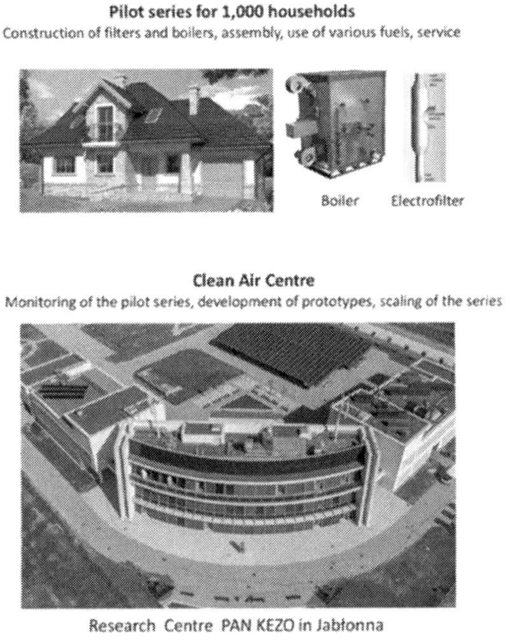

Fig. 4. Postulate for launching a pilot series of installations equipped with boilers and electrostatic precipitators (e.g. for 1,000 households), and the proposal of its monitoring and developing new prototypes at the Research Centre PAN in Jabłonna

This is a task for the state. And it is feasible because the total estimated costs of such a pilot series (manufacturing and assembling electrostatic precipitators and boilers, testing of various fuels and boilers and their service) should not exceed PLN 10 million. The author of this article finds this amount acceptable, considering the importance and scale of the problem.

Activities related to the development of these technologies, including scaling, monitoring and training, could take place at the Research Centre PAN in Jabłonna. It would become a national centre for clean air technologies – ***Clean Air Centre*** (Fig. 4). This choice is not accidental. It is the most modern centre in the country and third in Europe in the field of energy conversion and renewable sources of energy (with well-equipped laboratories).

The KEZO centre is considered to be the Centre of Excellence in the field of small-scale clean energy technologies, mainly due to:

- membership in international organisations:
  - EERA – European Energy Research Alliance (175 research entities and universities)
  - ESEIA – European Sustainable Energy Innovation Alliance.
- coordination of the Horizon 2020 project entitled "SuPREME: *Twinning for a Sustainable, Proactive Research partnership in distributed Energy systems planning, Modelling and management*".

5 Conclusion

Smog originates from energy poverty, we burn everything that has a calorific value.
New technologies: a low-power electrostatic precipitator installed in the boiler that enables combustion of various types of low-quality fuels is an attractive offer for households.

Launching a pilot series of installations equipped with boilers and electrostatic precipitators (e.g. for 1,000 households), which would be financed by the state is a proposal that can overcome existing difficulties in implementing new anti-smog technologies.

I hereby would like to thank my closest co-workers from the Institute of Fluid-Flow Machinery of the Polish Academy of Sciences in Gdańsk, and especially Prof. D. Kardaś, Prof. M. Dors, Prof. A. Jaworek, Dr. J. Podliński, Dr. M. Lackowski, Dr. G. Żywica, Prof. A. Cenian, Prof. P. Lampart and MSc S. Bykuć.

I also direct my words of gratitude to Msc J. Sawicki for the results achieved in the field of Industry 4.0 and the institute's position achieved due to it. Finally, I would like to give my special thanks to Mrs U. Kokosińska as well as to Mr K. Woźniak and Mr W. Czapla for the fruitful cooperation with Human World company.

References

1. Kiciński J., Do we have a chance for small-scale energy generation? The examples of technologies and devices for distributed energy systems in micro & small scale in Poland, Bulletin of the Polish Academy of Sciences, **61**, 4, (2013)
2. Kiciński J., Żywica G., Steam Microturbines in Distributed Cogeneration, Springer, ISBN 978-3-319-12017-1, (2014)
3. [3] D. Kardaś, G. Żywica, P. Klonowicz, Small domestic power engineering, "Nowa Energia" 2, (2017)
4. Kardaś D., Turzyński T., Ronewicz K., Combustion and heat transfer processes in biomass-fuelled boilers, IMP PAN Publishers, Gdańsk, Poland, (2015)
5. Podliński J., Niewulis A., Shapoval V., Mizeraczyk J., Electrohydrodynamic secondary flow and particle collection efficiency in a one-sided spike-plate type electrostatic precipitator, IEEE Transactions on Dielectrics and Electrical Insulation, 18, 5 (2011)

6. Podliński J., Niewulis A., Mizeraczyk J., Atten P., ESP performance for various dust densities, Journal of Electrostatics, **66**, 5-6, pp. 246–253, May (2008)
7. Podliński J., Niewulis A., Mizeraczyk J., Electrohydrodynamic flow and particle collection efficiency of a spike-plate type electrostatic precipitator, Journal of Electrostatics, **67**, 2-3, pp. 99–104, May (2009)
8. Jaworek A., Marchewicz A., Sobczyk A. T., Krupa A., Czech T., Two-stage electrostatic precipitator with co- and counter-flow particle prechargers, Journal of Electrostatics, **87**, pp. 180–190, June (2017)
9. Jaworek A., Marchewicz A., Sobczyk A. T., Krupa A., Czech T., Two-stage electrostatic precipitator with dual-corona particle precharger for PM2.5 particles removal, Journal of Cleaner Production, **164**, pp. 1645-1664 (2017)
10. Lackowski M., Karwacki J., Przybyliński T., Heda Ł., Kluska J., Cenian A., Lampart P., Measurement of characteristics of the cogeneration installation equipped with a biomass gasifier in Szepietowo (2016)

E3S Web of Conferences **46**, 00028 (2018) https://doi.org/10.1051/e3sconf/20184600028
3rd International Conference on Energy and Environmental Protection

Standard test methods applied for minimalisation of fire and explosion hazard during co-firing of biomass and coal

Marek Woliński[1,*]

[1]Szkoła Główna Służby Pożarniczej (Main School of Fire Service), Juliusza Słowackiego 52/54, 01-629 Warsaw, Poland

Abstract. The use of existing coal feeding system for the input of the crushed biomass to the energy boiler requires the least expenditures to start and run a multi-fuel combustion installation. And if losing popularity technology of co-firing biomass with coal is intended to allow meeting Polish promises to achieve the 15% share of renewable energy sources in the energy mix by the year 2020, such approach could be expected rather than construction of new, dedicated multi-fuel installations to produce electricity and heat. However, this involves the significant increase of fire and explosion hazard in the coaling system – designed for work with coal, a material with much better parameters characterizing this hazard, than biomass. The article presents the standardized research techniques (applied test stands and methodology) used to determine the data necessary for the design and the application of appropriate measures to reduce fire and explosion hazard related to operation of a mix of coal and biomass.

1 Introduction

The basic rules for the prevention of explosion and explosion protection include avoiding explosive atmospheres, avoiding any possible effective sources of ignition and limiting the effects of an explosion to an acceptable level [1, 2].

Correct assessment of the explosion hazard in accordance with [3] and the selection of appropriate (applicable in a given case) preventive and protective measures requires the determination of the properties of a mixture of a combustible substance with air, informing about the possibility of starting the explosion and behaviour of substances during combustion. The following indicators inform about these properties:

- explosion limits,
- limiting oxygen concentration,
- minimum ignition energy,
- minimum ignition temperature of dust – air mixture,
- minimum ignition temperature of dust layer,
- maximum explosion pressure,

* Corresponding author: mwolinski@sgsp.edu.pl

- maximum rate of pressure rise due to explosion.

2 Determination of flammable properties

The application of the principle of preventing the explosion by avoiding explosive atmospheres requires knowledge when the mixture has an explosive composition, and when not. This distinction can be made due to the content of flammable dust in the mixture or due to the oxygen content in the mixture. The critical parameter in the first case is the lower explosive limit LEL of combustible dust, while in the second - the limiting concentration of oxygen LOC in the mixture of dust - air - inert gas.

Determination of the lower explosive limit LEL is carried out in a laboratory test in accordance with PN-EN 14034-3 [4]. This standard provides for tests in a spherical or cylindrical chamber with a volume of 1 m^3 or - with a volume of 20 dm^3 (Fig. 1). The main components of the test equipment are: test chamber (capable of withstanding a minimum pressure of 20 bar, dust dispersion system, ignition source (for chamber 1 m^3: two pyrotechnical ignitors with energy 5 kJ each, for chamber 20 dm^3: two fuses with energy 1 kJ each), a control system, a system for collecting and processing measurement data. The experiment is based on the registration of the pressure course inside the test chamber during the explosion of the dispersed dust sample and subsequent analysis of this record. Research begins with the concentration of dust at which the explosion occurs, and then this concentration is reduced: 1000, 750, 500, 250, 125, 60, 30 g/m^3, until the explosion does not occur. In the case of a test chamber with a volume of 1 m^3, the explosion is considered to have occurred if the pressure measured during the experiment exceeds the initial value by at least 0.3 bar, and the concentration at which the explosion did not occur is considered to be the concentration of dust corresponding to the LEL value. In the case of a test chamber with a volume of 20 dm^3, the criterion is a pressure increase of at least 0.5 bar, and the concentration corresponding to the LEL value is the highest concentration of dust, at which the explosion did not occur in three subsequent tests.

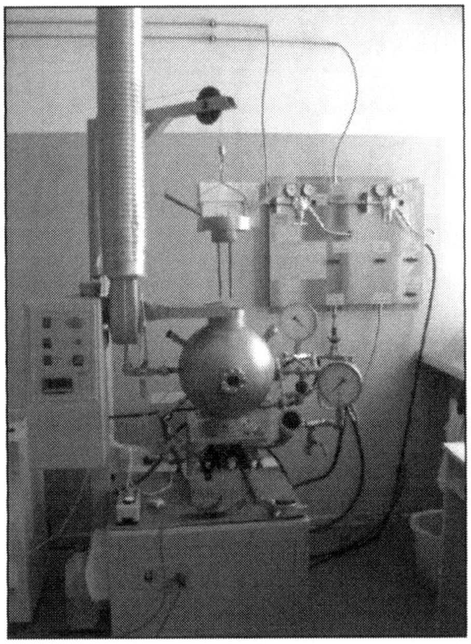

Fig. 1. Spherical test chamber of 20 dm^3 volume (laboratories of The Main School of Fire Service)

Determination of the limiting oxygen concentration LOC is carried out in a laboratory test in accordance with PN-EN 14034-4 [5]. The same test chamber is used as in the case of LEL (1 m^3 or 20 dm^3), additionally equipped with a system that allows the inert gas mixture to be supplied to the chamber (with the possibility of changing the oxygen content in this mixture every 1% by volume, measured with an accuracy of + 0.3% vol.). The same as for the LEL determination case are the explosion detection criteria for the tested dust concentration and oxygen content in the mixture. The test procedure consists in the step-by-step modification of the oxygen content in the mixture and the concentration of dust, with the same criteria as the LEL determination of the mixture ignition but the LOC value requires confirmation in three consecutive tests for each dust concentration that will not ignite.

3 Determination of ignition propertles

3.1 Minimum ignition energy MIE

The determination of the minimum ignition energy MIE of the air / dust mixture is carried out in a laboratory test in accordance with standard [6], using the test stand shown in Figure 2. The stand is designed for determining the minimum ignition energy of dust cloud at ambient temperature. The dust is lifted by a blast of air and then ignition attempt is performed by charging a high voltage spark with known energy value. The minimum energy used to ignite a dust cloud is considered to be the lowest energy determined for the current research.

Fig. 2. Test stand for MIE determination of dust - air mixtures (laboratories of The Main School of Fire Service)

The main part of the stand is a 1.2-liter Hartmann glass tube. The stand construction enables the energy of 1000 mJ, 300 mJ, 100 mJ, 30 mJ, 10 mJ, 3 mJ and 1 mJ of spark to be applied, with an ignition delay time in the range of 50 - 350 ms. During the test, the dust sample dispersed in the glass Hartmann tube is ignited with the high voltage spark. Adjusted energy value is stored in the capacitor and discharged during the ignition process. Dust ignition is evaluated visually by the operator. The test result is the lowest ignition energy obtained for a given dust (of characteristic particle size and moisture content), defined as:

$$E_1 < MIE < E_2 \tag{1}$$

where :

E_1 – the highest energy at which ignition does not occur,
E_2 – the lowest energy at which ignition occurs in one of the next 10 trials.

The energy value stored in the capacitor is calculated from the equation:

$$E = 0.5 \cdot C \cdot U^2 \tag{2}$$

where :

E – the stored energy, J,
C – the capitance, F,
U The voltage of the capacitor, V.

3.2 Minimum ignition temperature of dusty explosive mixture

Determination of the minimum ignition temperature of dusty explosive atmosphere is carried out according to the standard [7], method B: cloud of dust in a constant temperature furnace (test stand: Fig.3). This method is used to indicate the minimum temperature at which the cloud of a tested dust or other particulate solid is ignited. The test consists in observing whether an ignition of a sample of dust injected into the furnace at a specified temperature occurred. It is considered that the ignition occurred if the burst of flame is visible beyond the lower end of the furnace pipe. A certain ignition delay is allowed. Sparks without a flame do not constitute ignition. The minimum ignition temperature of the dust cloud should be the lowest temperature of the furnace where the ignition has been achieved using a standard measurement procedure with changing the dust sample mass and pressure of the dispersing air as well as furnace temperature, reduced by 20 K at furnace temperature above 300°C and 10K at furnace temperature equal to 300°C or lower. If at the temperature of the furnace even up to 1000°C ignition was not obtained, this fact should be noted in the test report.

Fig. 3. Test stand (oven) for determination of minimum ignition temperature of dust cloud (laboratories of The Main School of Fire Service)

3.3 Minimum ignition temperature of dust layer

Determination of the minimum ignition temperature of the dust layer is carried out according to the standard [7], method A (test stand: Fig. 4). This method is used to determine the minimum temperature at which the dust layer of a specific thickness placed on the hotplate reaches decomposition and /or ignition. This method applies in particular to industrial equipment with hot surfaces, where dust forms thin layers and is exposed to atmospheric air. The test consists in observing whether a sample of dust, formed and kept in a suitable way on the heating plate ignites or not. It should be considered that the ignition occurred if:

- glow or smoking has been observed, or
- the measured temperature of the dust sample has reached the value of 450°C, or
- the measured temperature of the dust sample has exceeded the temperature of the heating plate by 250K.

In order to determine the minimum ignition temperature of a dust layer of a certain thickness, a series of tests should be carried out using a new dust sample in each test. The temperature of the heating plate should be changed accordingly until the temperature that ignites the layer is found, higher not more than 10K from the temperature which did not lead to the ignition of the layer (it should be confirmed in at least three tests).

The highest temperature at which ignition has not occurred should be confirmed by continuing the test long enough to be able to conclude that the self-heating rate is decreasing, i.e. the temperature at the measurement point in the layer falls to a predetermined value, lower than the temperature of the heated surface. The tests should be interrupted if ignition of the dust layer does not occur at the temperature of the heating plate up to 400°C.

Fig. 4. Test stand for determination of minimum ignition temperature of dust layer (laboratories of The Main School of Fire Service)

3.4 Determination of the spontaneous ignition behaviour of dust accumulations

Determination of the spontaneous ignition behaviour of dust accumulations is carried out in accordance with standard [8] (standard [9] applies only to hard coal). This standard specifies testing and evaluation procedures for determining the self-ignition temperature of combustible dust as a function of volume by means of hot storage experiments in a constant temperature furnace (test stand: Fig. 5). If the test is carried out at temperatures well below the self-ignition temperature of the dust to be examined, the sample temperature reaches asymptotically the temperature of the furnace. At higher temperatures of the furnace reactions with oxygen are clearly visible in the sample material. For this reason, the sample temperature will be temporarily higher than the furnace temperature. This indicates that self-heating processes have started, without the spontaneous ignition of the sample. There are two ways to determine if the ignition occurs:

- when the temperature inside the sample rises by at least 60K above the furnace temperature,
- when the temperature course inside the sample as a function of time shows an inflection point, if this occurs above the furnace temperature.

Fig. 5. Test stand for determination of the spontaneous ignition behaviour of dust accumulations (laboratories of The Main School of Fire Service)

The choice of hot storage temperatures for the two final critical tests should be made in such a way that the temperature of the furnace causing the ignition and the temperature of the furnace not causing ignition differ by no more than 5 K. The result of the self-ignition temperature measurements is rounded down to the nearest integer value expressed in degrees.

4 Determination of properties characterizing the explosion course

The standard [10] indicates that the calculated size of the venting surface depends, among others, on two parameters describing an explosive dust/air mixture: the maximum explosion pressure p_{max} and the K_{St} coefficient (i.e. practically on the value of the maximum rate of explosion pressure rise $(dp/dt)_{max}$).

The value of the maximum explosion pressure p_{max} of the dust/air mixture is determined in laboratory tests in accordance with the standard [11], using the same test chamber as for the determination of the LEL and LOC values (Figure 1). The standard recommends tests for a number of concentrations of dust in the mixture: 60, 125, 250, 500, 750, 1000, 1250, 1500 g/m^3 and - for the 20 dm^3 chamber, it provides a method of calculating the p_{max} value and correcting this value due to cooling effect of chamber walls as well as impact of explosion pressure of pyrotechnical igniters.

The value of the maximum rate of explosion pressure rise $(dp/dt)_{max}$ is determined in accordance with standard [12], using a similar test procedure as for p_{max} (also for the same series of dust concentrations in mixture), and the K_{St} value is calculated as:

$$K_{St} = (dp/dt)_{max} * V^{1/3} \tag{3}$$

where :

V – the test chamber volume, m^3.

5 Conclusion

Correct assessment of the explosion hazard and the selection of appropriate (possible to apply in a given case) preventive and protective measures requires the determination of the properties of a mixture of a combustible substance with air, informing about the possibility of starting the explosion and the behavior of substances during combustion. Conducting normative tests described above (and systematized recently by [13]) allows to determine safety data - appropriate parameters describing these properties. However, it should be borne in mind that this type of safety data are not physical constants - they depend on the size and shape of the dust particles, the moisture content and the presence of additives even in trace amounts. Therefore, in order to identify hazards and to determine preventive and protective measures, it is necessary to test dusts collected in specific places and operating conditions of equipment in the technological process.

References

1. PN-EN 1127-1:2001 Atmosfery wybuchowe. Zapobieganie wybuchowi i ochrona przed wybuchem. Pojęcia podstawowe i metodologia.
2. Rozporządzenie Ministra Gospodarki z dnia 8 lipca 2010 r. w sprawie minimalnych wymagań, dotyczących bezpieczeństwa i higieny pracy, związanych z możliwością wystąpienia w miejscu pracy atmosfery wybuchowej. Dz. U. z 30 lipca 2010 Nr 138 poz. 931.
3. PN-EN 60079-10-2:2015 Atmosfery wybuchowe – Część 10-2: Klasyfikacja przestrzeni. Pyłowe atmosfery wybuchowe.
4. PN-EN 14034-3:2011 Oznaczanie charakterystyk wybuchowości obłoków pyłu – Część 3: Oznaczanie dolnej granicy wybuchowości DGW obłoków pyłu.
5. PN-EN 14034-4:2011 Oznaczanie charakterystyk wybuchowości obłoków pyłu – Część 4: Oznaczanie granicznego stężenia tlenu GST obłoków pyłu.
6. PN-EN 13821: 2004 Przestrzenie zagrożone wybuchem. Zapobieganie i ochrona przed wybuchem. Oznaczanie minimalnej energii zapłonu mieszanin pyłowo – powietrznych.
7. PN-EN 50281-2-1:2002 Urządzenia do stosowania w obecności pyłów palnych – Część 2-1: Metody badania. Metody oznaczania minimalnej temperatury zapłonu pyłu.
8. PN-EN 15188:2009 Oznaczanie skłonności nagromadzeń pyłu do samozapalenia.
9. PN-G-04558:1993 Węgiel kamienny – Oznaczanie wskaźnika samozapalności.
10. PN-EN 14491:2013 Systemy ochronne odciążające wybuchy pyłów.
11. PN-EN 14034-1:2011 Oznaczanie charakterystyk wybuchowości obłoków pyłu – Część 1: Oznaczanie maksymalnego ciśnienia wybuchu pmax obłoków pyłu.
12. PN-EN 14034-2:2011 Oznaczanie charakterystyk wybuchowości obłoków pyłu – Część 2: Oznaczanie maksymalnej szybkości narastania ciśnienia wybuchu (dp/dt)max obłoków pyłu.
13. PN-EN ISO/IEC 80079-20-2:2016 Atmosfery wybuchowe – Część 20-2: Właściwości materiałowe. Metody badań pyłów palnych.

E3S Web of Conferences **46**, 00029 (2018) https://doi.org/10.1051/e3sconf/20184600029
3rd International Conference on Energy and Environmental Protection

The analysis of "more electric engine" technology to improve the environmental performance of aircraft jet engine

Maciej Henzel[1,*], *Krzysztof* Falkowski[1], and *Aleksander* Olejnik[1]

[1]Military University of Technology, Faculty of Mechatronics and Aerospace, gen. Witolda Urbanowicza 2, 00-908 Warsaw, Poland

Abstract. In aviation, there is now a dynamic development of aircraft equipment related to the implementation of "more electric aircraft" technology. This concept offers the ability to improve the use of on-board systems, e.g. environmental operating conditions of aircraft jet engine. This technology is named "more electric engine". It allows the use of magnetic levitation technology at engine turbine shaft bearing. The development of this technology relates to the dynamic change of electronic power systems for civilian transport aircraft, the use of adaptive control methods and new materials in aviation technology. All technologies are improved the environmental operating conditions of the on-board system, e.g. operational flexibility, technological potential growth. [1] In the paper will be presented the TS-21 aircraft jet engine. This engine is modernized in the Jet Engine Laboratory of the Military University of Technology. The paper is presented a digital engine control system, the operating parameters acquisition system and magnetic bearing system. It is described the concept of active magnetic suspension of the turbine engine shaft support. The magnetic suspension technology allows eliminate mechanical bearing arrangements with an oil installation, friction forces and classical, mechanical bearings. The paper contains the simulation and experimental results of a modernized jet engine TS-21.

1 Introduction

In the last years, in aviation there is developed new trend in aircraft equipment. This concept is called "More Electric Aircraft" (MEA) technology. It is based on the use of electric equipment, instead of pneumatic and hydraulic systems. The expansion of electrical machines, power systems and other electric devices are caused increase of demand for on-board electric energy. Thanks to this technology, new aircraft are produced less noise, pollution. This idea influences increase of maintenance susceptibility, ensure systems flexibility during modification. It reduces operation costs and system weight, too. More Electric Aircraft concept is used in passenger aircrafts Boeing B787, Airbus A380. There are

* Corresponding author: maciej.henzel@wat.edu.pl

applied e.g. environmental control systems, aircraft engines, electrical power and actuation systems. [1,2]

Jet engine is the main aircraft element produced in MEA technology. It works in hard and changeable conditions, e.g. high and low temperature, low pressure, high and changeable loading, high rotational speed, high lateral and longitudinal overloads. So, jet engine roll bearings are worked in very hard conditions, too. New concept of engine bearing system is based on magnetic suspension phenomenon. It eliminates mechanical contacts between stator and rotor, and removes negative operating effects such as friction forces, heat and wear, lubrication, heat dissipation, noise, etc.

The active magnetic bearing system (AMB) is implemented in some commercial engines. These bearings are perfectly suitable to support fan in "hot" engine section. Active magnetic bearings are devices, which using attraction forces to provide stable shaft engine levitation at work point. New system can work properly in rugged conditions. In aircraft engine domain, the new concept is named "all electric engine". In laboratory of the Military University of Technology, this idea is tested on the TS-21 jet engine. This engine is presented in the Figure 1 [4].

Fig. 1. The TS-21 jet engine on the lab stand

2 New concept of jet engine shaft bearing

The main elements of classic jet engine are rotor fan, low- and high- pressure compressors and turbines, two or one drive shafts, combustion chamber and bearings. The turbine is a drive source for all other components. Loads acting on these elements are created by the generation of thrust or power. Rotary systems are also loaded by centrifugal forces. These forces are changed in wide range, and they can be disturbed by residual unbalance influences. Moreover, compressive and stretching loads are affected engine rotary systems depending on thrust-support bearing location.

In classical engine constructions, roll bearings are worked in high rotor speed (over 40 000 [rpm]), high longitudinal forces, high temperature (over 1200 [$^{\circ}$C]) and difficult lubrication conditions. Axial and radial forces are growth during engine acceleration, hard landings and during curvilinear aircraft motion. These bearings support turbine rotors and compressor. So, they are exposed to a continuous heat flow from the heated rotors. For that reason, they have to be cooling by oil system (Figure 2a). The described conditions eliminate the use of sliding bearings. But they indicate the advantageous features of the roll bearing, that have the small friction resistance. [4,5]

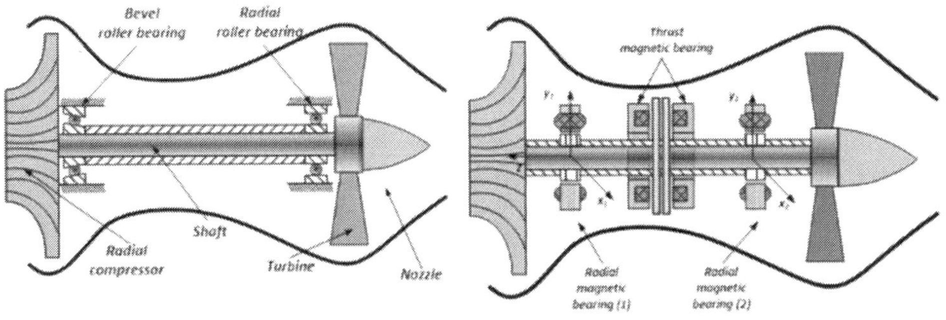

Fig. 2. Diagram of jet engine construction: a) classic construction, b) construction with magnetic suspension system[4]

New bearing concept bases on the use of active magnetic bearings for shaft jet engine (Figure 2b). These bearings consist of electro-mechanical actuator with stator and rotor. New construction of bearing system eliminates friction forces between co-operating kinematic pairs and makes possible to monitoring and diagnosing aircraft engine. Figure 2b is shown the aircraft engine structure with magnetic bearings. Two radial magnetic bearings generate radial forces in two perpendicular radial axes. The first magnetic bearing (1) is controlled in two radial axis x_1 and y_1. And the second radial magnetic bearing (2) is controlled in radial axis x_2 and y_2. The axial position in the z-axis is controlled by a thrust magnetic bearing. [2,3,6]

3 Finite element method analysis of the radial active magnetic bearings

The further analyses described the radial magnetic bearings of the TS -21 aircraft jet engine. These bearings are placed on the left and right plane of the shaft turbine. The magnetic bearing construction is modelled in COMSOL Multiphysics software based on the finite element method analysis. This analysis is enabled the phenomena in proposed solution. There are modelled the geometric dimensions, mass and inertia parameters of the engine. Magnetic bearings are designed for a maximum load about 100 N. The tests allowed to determine the magnetic circuits parameters and limits. The analysis are made possible to optimize the bearing weight and geometrical parameters. The established parameters of the magnetic bearing are described in the Table 1.

The results of FEM analysis of passive electrodynamic bearing are presented in Figure 3 and Figure 4. There are shown the distributions of magnetic fields and the values of magnetic forces in the magnetic bearing. There are presented the magnetic induction distributions, the induced eddy currents, and the stiffness and damping parameters of the bearing. Stiffness parameter is estimated by the determination of the relationship between the magnetic force and the position of the shaft in the air gap. However, the damping parameter is estimated as the relationship between the magnetic force and the velocity of the shaft.

Table 1. Vehicles dimensions

Parameter	Value	Unit
Maximum force	100	N
Saturation magnetic induction	1,6	T
Rotor diameter	25	mm
Air-gap	0,3	mm
Maximum current	10	A
Current density	8	A/mm^2

a)

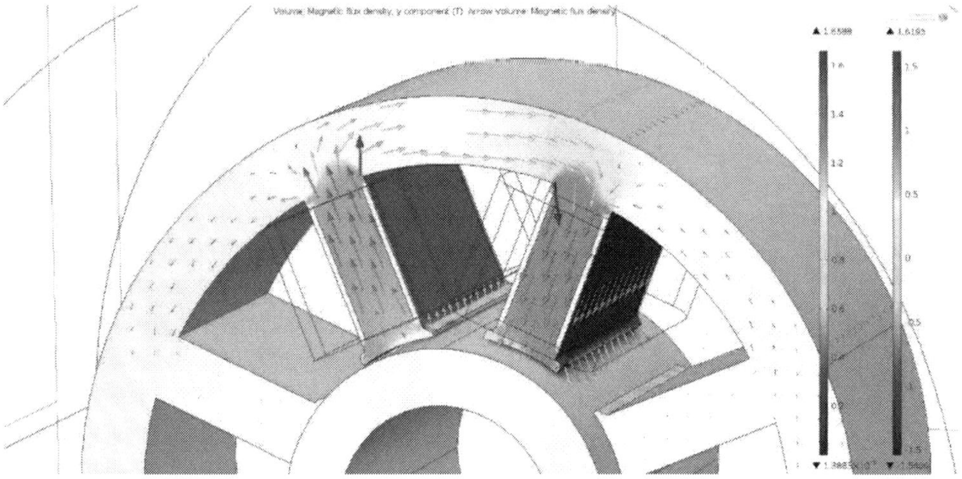

b)

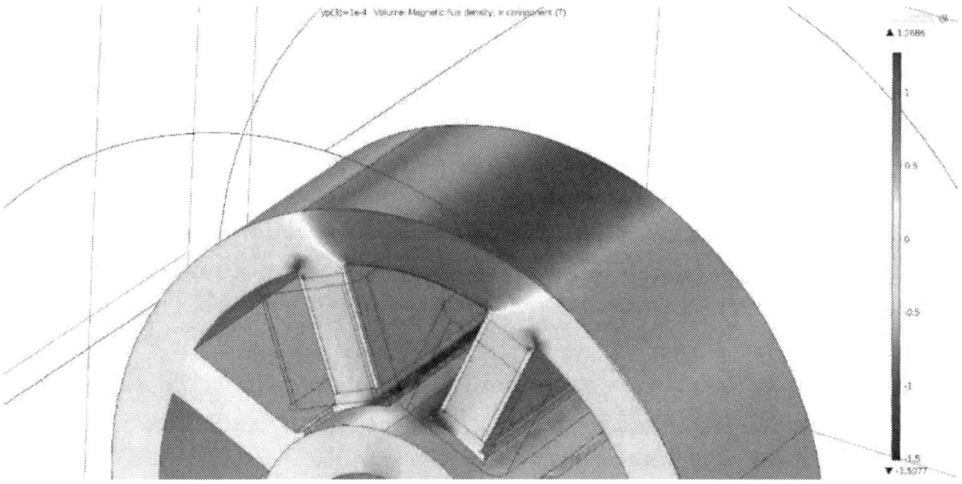

Fig. 3. Magnetic flux density: a) nominal position of the shaft; b) the 0,1 mm displacement of the shaft

a)

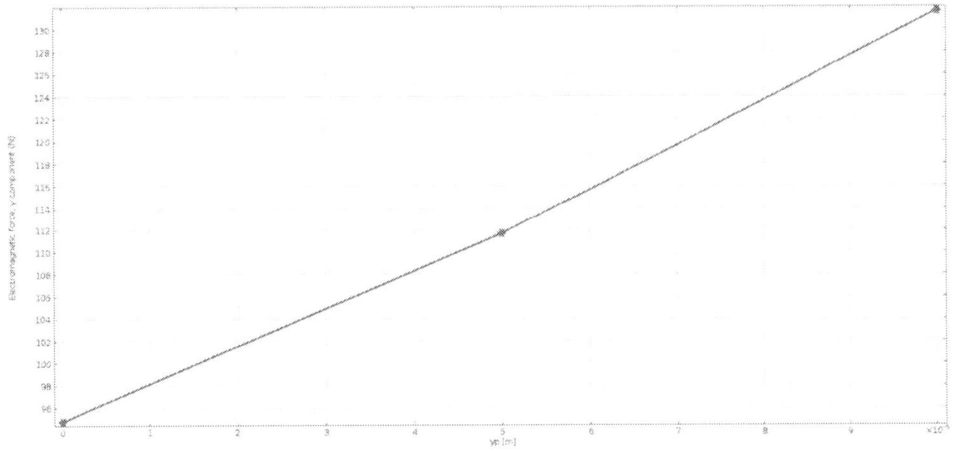

b)

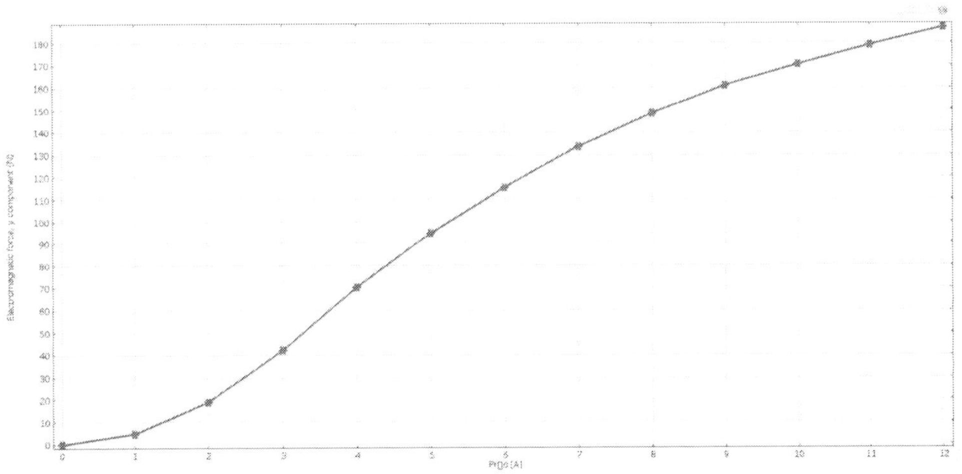

Fig. 4. The diagram of electromagnetic force: a) for the differ value of air gap; b) for the change of the operation current values

4 Laboratory analysis

The described active magnetic bearing system is elaborated and investigated in Jet Engine Laboratory. The laboratory stand is presented in Figure 1. While, in Figure 5 is shown the AMB laboratory model.

The experimental tests are consisted of the parameters measurement of the stator electrical windings. The RLC bridge is used during the tests. The measurements allow to verify the designed and made of active magnetic bearing components. The knowledge of these parameters is important for the selection of the power system parameters and to determine the controller parameters. These results are implemented in mathematical and simulation engine models. There are implemented in Comsol Multiphysics and Matlab software.

During the laboratory tests are measured the static characteristics of the magnetic bearings with various shaft position in stator (Figure 6 and 7), too. The position of rotor changes the air gap in magnetic bearing. During the tests, the work point is changed from -0.4 mm to +4 mm. The air gap is changed from 0.2 mm to 0 [mm] with the rotor moving in the vertical direction (blue characteristic) and from 0 to 0.2 [mm] (red characteristic). The move is relative to nominal position. The position measurements are proceed with use of two dial sensors located to the left and right rotor side.

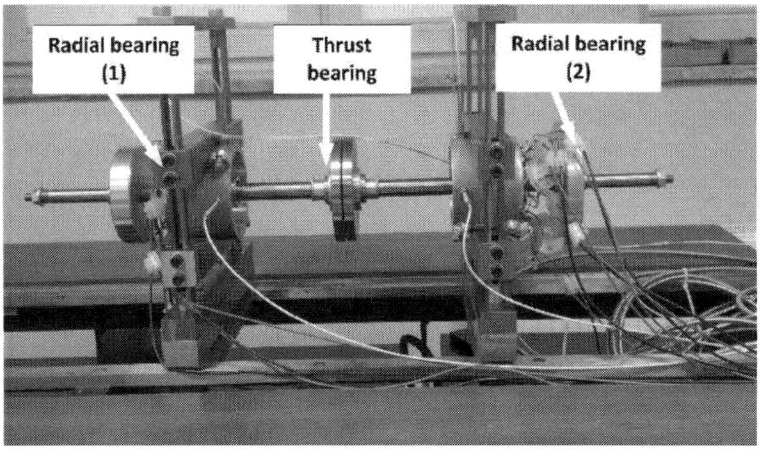

Fig. 5. The laboratory model of active magnetic bearing of jet engine

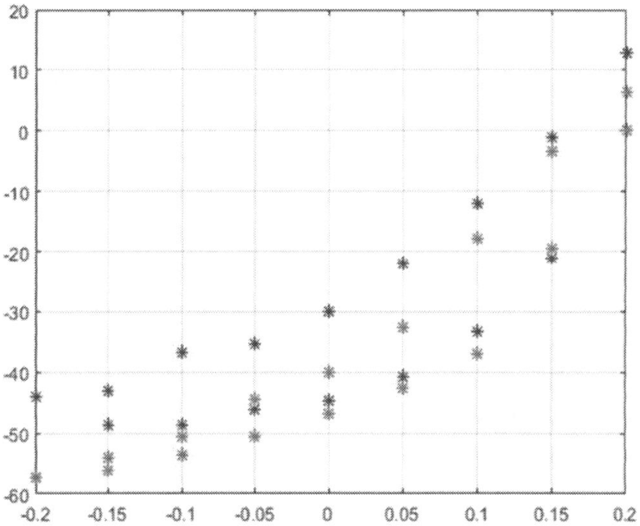

Fig. 6. The measure of air gap changes for the work point 2 A

During next test are measured the rotor attraction force foe differ value of work points. It is changed from 2 to 4 A with a step 0.2 A. Due to the non-linearity of these characteristics, the parameters are measured two times to determine the characteristic trends. Figure 7 is shown the current values for differ work point. These characteristics have been plotted to show the measurement accuracy for a circuit current in two directions (rotor movement).

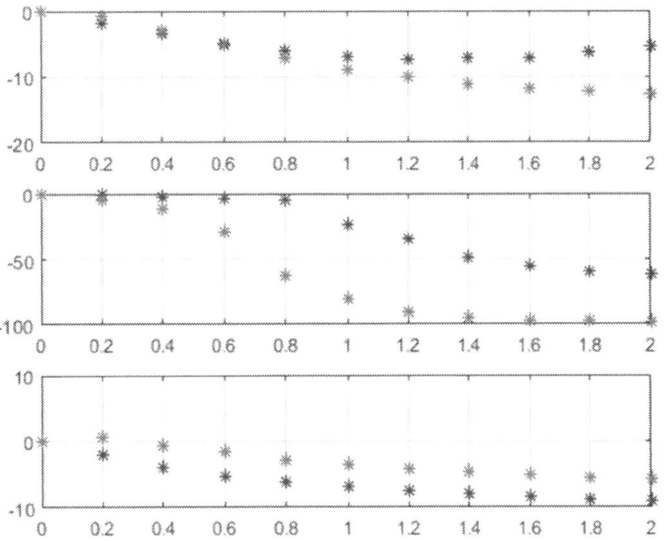

Fig. 7. The force characteristics in function of the work point current

5 Conclusion

In the paper are presented results of the simulation and laboratory analyses of the modernized TS-21 jet engine. These analyses are consisted in examining the properties of active magnetic suspension in engine technology. This construction has got a lot of advantages then classical electromechanical bearings. The finite element method is used to verify the construction parameters of designed magnetic bearing system. This method allows estimate parameters of the AMB in aircraft engine.

The active magnetic bearing technology are used in modernized jet engine to improve the environmental operating conditions.

References

1. Moir I., Seabridge A.: *"Aircraft systems - Mechanical, electrical and avionics subsystem integration"*, John Wiley & Sons, pp. 227-238, (2008)
2. Lempke, T.A., Design and analysis of a novel low loss homopolar electrodynamic bearing, Doctoral Thesis, KTH Electrical Engineering, Stockholm., (2005)
3. Falkowski K., Gosiewski Z.: Analytical Method of the Magnetic Forces Estimation in passive Magnetic Bearings, *The 8th International Conference on Motion and Vibration Control MOVIC 2006, Korea, September* (2006)
4. Falkowski K., Henzel M., Mazurek P., Magnetic suspension system for aircraft engine, *10th International Conference on Mechatronic Systems and Materials „MSM 2014", July 7÷10, Opole, Poland*, (2014)
5. Henzel M., Mazurek P.: „The analysis of the control system of the active magnetic bearing", Electrodynamic and mechatronic systems, *3rd International Students Conference on Electrodynamics and Mechatronics (SCE III), IEEE, Opole* (2011)

6. Henzel M., Mazurek P. The rapid prototyping of active magnetic bearing, Advances in Intelligent Systems and Computing, Recent Advances in Automation, Robotics and Measuring Techniques, Springer, **267**, pp. 155-166, (2014)

E3S Web of Conferences **46**, 00030 (2018) https://doi.org/10.1051/e3sconf/20184600030
3rd International Conference on Energy and Environmental Protection

The efficiency of the process of coal gasification in the presence of hydrogen

Gracjana Woźniak[1,*], *Rafał* Longwic[1], *Kamil* Szydło[1], *Adam* Kryłowicz[2], *Jarosław* Kryłowicz[2], and *Robert* Juszczak[2]

[1]Lublin University of Technology, Department of Vehilces, Nadbystrzycka 36, 20-618 Lublin, Poland
[2]Bicarbo Sp. z o.o., Grzybowska 2/29, 00-131 Warsaw, Poland

Abstract. The results of the research on the process of gasification of coal mixture in a closed system was evaluated in this article. The method presented includes oxygen-free gasification of coal at high pressure and its thermal degradation. Hydrogen is a gasifying factor. Gas containing, among others, methane, carbon dioxide, carbon monoxide, hydrogen is obtained as a result of reaction. Fossil coal as a chemical and energetic raw material is very significant in the process of demand for energy. Due to more and more difficult access to petroleum and natural gas, as well as growing prices, fossil fuels are becoming economically attractive. However, the works on more effective and environmentally friendly methods of energetic use of coal - including its gassing - should be continued. The use of hydrogen during coal gasification allows to get emission-free fuel gas, having better energetic properties than primal raw material.The laboratory site was built and consisted of: 4,5dm^3 reactor with a heating system, system of supply reactor with hydrogen and rinsing with argon and cooling system. The results of hydrogasification of coal with specific parameters, for selected time intervals were presented in this article. The efficiency of hydrogasification depending on the type of applied coal was assessed. Powdered charcoal, meeting the norm of German Committee for Standardization DIN EN1860 2 with a number 3H020, hard coal and lignite coal, as well as their mixtures were applied during the research.

1 Introduction

Limited resources of petroleum and natural gas, instability of the prices of petroleum-derived fuels, the necessity of controlling the emission of greenhouse gasses, development of new technologies are the factors affecting changes in global structure of the use of primary fuels. According to the forecasts of International Energy Agency (IEA), coal will remain basic chemical raw material and energetic fuel (Fig. 1.1). Therefore, coal gasification as a technology of energy production is becoming more popular due to global availability of this raw material and economic conditions. Global coal resources are estimated to last for about 200 years. Polish resources of hard coal amount to about 43,3 billion tons, whereas,

* Corresponding author: g.wozniak@pollub.pl

lignite coal – 13,7 billion tons. Global petroleum resources are estimated to last for about 45 years, and natural gas for about 60 years.

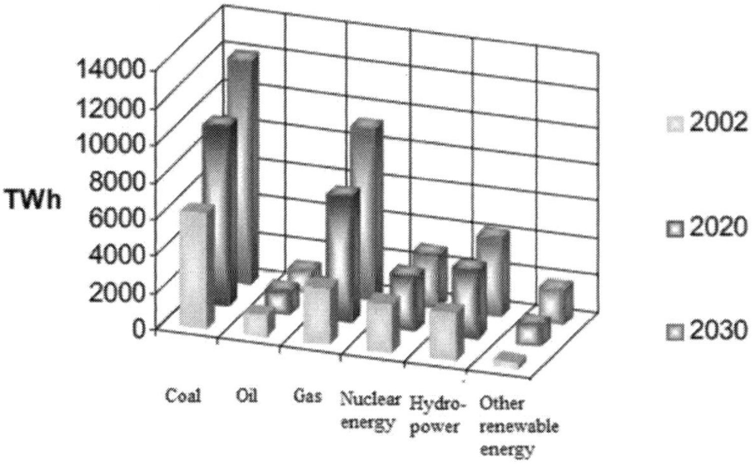

Fig. 1. IEA forecasts – perspective within the scope of amount of energy that can be obtained from various primal sources

Coal gasification is a complex of multidirectional changes, both chemical and thermal, occurring in high temperature between coal and gasifying factor. Air, oxygen, steam, carbon dioxide, rarely hydrogen, are usually applied as a gasifying factors. Gasifying factor is usually transmitted through heated coal to ensure close molecular contact for chemical reaction. The goal of the process of gasification is a complete conversion of coal to combustible gas that can be used as fuel or synthesis gas. The efficiency of gasification depends on the type of coal and gasifier. Technology of coal gasification allows to use generated gas in many ways [26]. The majority of the process of gasification is endothermic reactions, part of the element of coal is oxidized to carbon dioxide and applied in the process. Unfortunately, raw material such as coal is not only hydrogen and coal as the element, it is also the so-called heteroatoms, which generate such harmful substances in the process of coal gasification as: hydrogen sulphide, ammonia, carbon oxysulfide in the amount not exceeding 1% of generated gases.

Technology of coal gasification has been applied around the world for many years. It dates back to the 19th century, when Lurgi gasifier with slidable deposit of coal (1887) was patented. There are many processes of gasification. They can be divided due to the way of flow of raw material inside the reactor, therefore, the reactors can be divided into three categories: with permanent or slidable deposit, with fluidal deposit and dispersive reactors. The process of coal gasification can be conducted in two ways. Firstly, in the gasification reactors, the so-called gas generators, located on the surface, secondly as the so-called ground-based gasification or in the georeactor located in the coal, as underground gasification (gasification in situ). The results of research on coal gasification in a prototype of ground-based reactor were presented in the article. Hydrogen was applied as a gasifying factor Hydrogen is a coal carrier affecting improvement of efficiency of the process. The research results showed the positive impact of application of H_2 as a gasifying factor. Technologies based on hydrogen may use energy in a more efficient way, provide ultra clean fuels, eliminate emission of pollutants and considerably limit emission of greenhouse gasses to atmosphere, particularly carbon dioxide.

2 Empirical research

2.1 Research site

Figure 2 shows a scheme of the research site used for coal gasification. The system consists of the following main elements: about 4,5 l reactor with a heating system, system of supply reactor with hydrogen and rinsing with argon and cooling system.

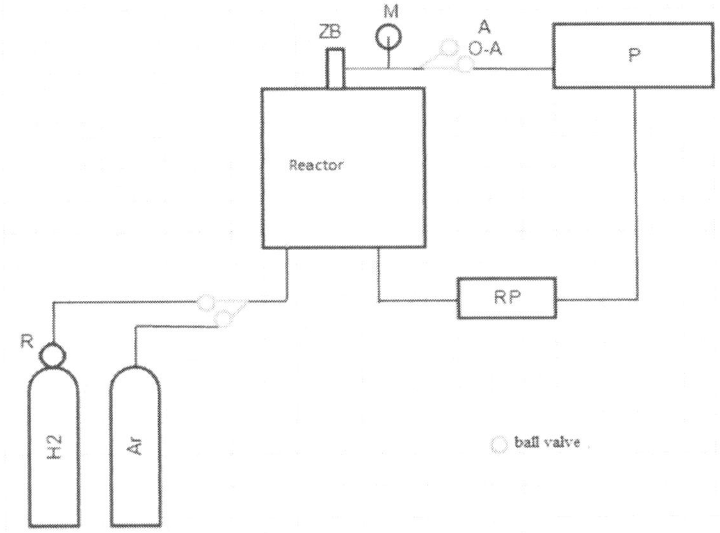

Fig. 2. Scheme of the measuring site used for coal gasification: R- reducer, H2 - hydrogen cylinder, Ar- argon cylinder, ZB – cylinder valve, M – manometer, A O-A- receiving-analysing apparatus, P – pump, RP – flow regulator

The casing of the reactor (Fig.2, Fig. 3) consisted of a cover made of heat resisting steel with the stub pipes charging and discharging gas and coolant (water). Double walls of the cylinder form an external chamber. Temperature of the charge was measured with the use of a sensor containing thermoelement PTTN TKb-30.

Fig. 3. Reactor

Fig. 4. Thermal shield

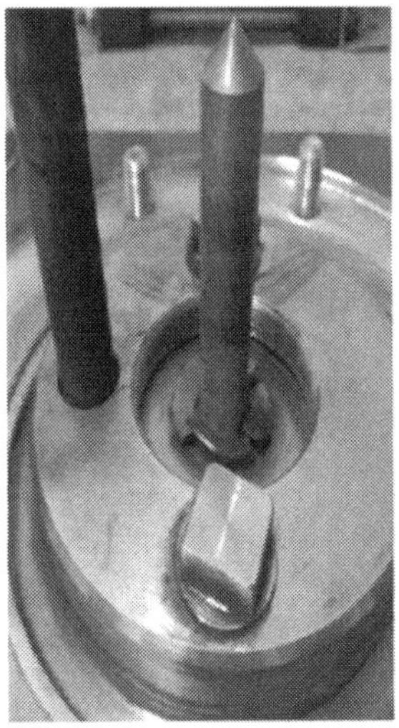

Fig. 5. Diffuser

Fig. 6. Site used for coal gasification

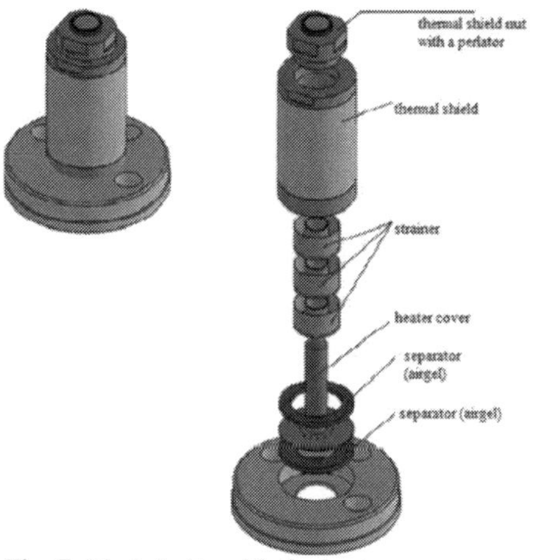

Fig. 7. Mody1 site with the heater cover

Hydrogen supply system – technical compressed hydrogen 5,0 and rinsing with argon – compressed technical argon 5,0 consists of a pressure cylinder, manometer of reducing and discharge valves. Post-reaction gas flows to the radiator. The main element of the reactor is a furnace. The research was conducted on MODY1 site (Fig. 4). The furnace consisted of a thermal shield with a nut containing perlator (Fig. 5), distance and diffuser (Fig. 6). Figure 7 shows a site used for coal gasification.

2.2 Methodology of the research

The process of gasification was conducted in repeatable initial conditions. Sample of coal in the ground form of the weight about 9-19 g was heated. At first, the samples of coal for the research were prepared and weighed, then charge in the form of two filling sieves with coal in the furnace, was placed in the bottom part of the reactor, and thermal and cylindrical shield placed. When tightness was checked, the reactor was screwed together. Argon under the pressure of 1,2 Mpa was raised and dropped three times to rinse out every sample. The same action was repeated when the samples were rinsed out two times with hydrogen under the pressure of 1 MPa. Upon rinsing with gases, the reactor was filled with hydrogen up to 1,2 MPa. Heater and cooling of the bottom cover of the reactor were turned on. The measurements at every attempt were started from 800°C. Temperature in the oxidation zone should be maintained at the level of 700-900°C (this temperature would not exceed 1000°C due to the structure of the reactor). When temperature in the oxidation zone is lower than recommended, carbon dioxide is not completely reduced to oxide. If the temperature is above recommended value, ash is melting and slag is forming, causing suspension of the charge and, as a result, limitation of flow of gas [6]. The pressure inside the reactor was controlled by elastic pressure gauge on the outlet of the reactor and initially set with the use of a pressure regulator. In order to increase fluidization in the reactor, constant blow through was performed with the use of a piston pump forced pressure 0,1 MPa. Aerogel was applied as insulation.

3 The results of the research

Coal gasification for its various types, at constant parameters specified in point 2.2 were analysed on the laboratory scale during experimental research. The research was conducted for various times of hydrogen gasification. Three measurements for various time intervals were taken, that is, 15 minutes, 30 minutes and 60 minutes. The tests were performed for charcoal, hard coal and lignite coal and their mixtures, with constant blow through, temperature was between 800 °C and 950 °C, pressure was 1,2-1,3 MPa. The results of the research are presented in Tab. 3 and Fig. 8. Relative amount of gasified coal was calculated using dependencies 1 and 2.

$$\Delta mk = mp - mk \tag{1}$$

$$G\% = \frac{\Delta mk}{mp} \tag{2}$$

where :

Δmk – absolute amount of gasified coal, kg,

mp – initial mass of coal (before gasification), kg,

mk – final mass of coal (after gasification), kg,

$G\%$ – relative amount of gasified coal.

Charcoal, meeting the norm of German Committee for Standardization DIN EN1860 2 with a number 3H020, as well as hard coal and lignite coal were applied during the research. Description and content of elements in charcoal are presented in table 1 and table 2. The coals for research gasification were dried and grinded. Technical hydrogen produced by Linde of 5.0 (99,999%) purity were used as a gasifying factor.

Table 1. Technical analysis of the sample of charcoal

Element	Contents [% by weight]
Mg	0.018 ± 0.002
Al	0.0026 ±0.0006
Si	0.0167 ± 0.0020
P	0.0136 ± 0.0015
S	0.0139 ± 0..0007
K	0.176 ± 0.005
Ca	0.133 ± 0.004
Ti	0.016 ± 0.002
Mn	0.0083 ± 0.0015
Fe	0.0197 ± 0.0025
Cu	0.00064 ± 0.00020
Zn	0.00057 ± 0.00015
Rb	0.00042 ± 0.00013
Sr	0.0011 ± 0.0003

Table 2. Elemental analysis CHN

Sample name	%C	%H	%N
P-LA-79/2016/Z – charcoal - A	average: 77.435	average: 3.45	average: 0.49

During gasification of hard coal, slight relation between gasification and process duration has been observed. The difference between the measurements for various times was about 2%. After 15 minutes of the process, gasification process at the average level of about 48% was obtained. The lowest standard deviation was obtained for the process lasting 60 minutes, the process was repeatable for this value. In the event of mixture of hard coal and charcoal in the proportion 50% to 50%, moderate relation between gasification and process duration has been observed. Between 15-minute and 30-minute and 60-minute measurements, there was an absolute difference at the level of about 8-9%. In this case, the lowest standard deviation was obtained for the process duration of 15 minutes, which proves that the process was most repeatable then. The relation between gasification and process is small in the case of lignite coal. After 15 minutes of the process, gassing process at the average level of about 59% was obtained. Standard deviation for all three times was constant and slight, which may show high repeatability of the process. In the case of mixture of lignite coal and hard coal, gassing process at the average level of about 55% was obtained after 15 minutes. The highest results for carbon mixture were achieved for 15 minutes, and then for 30 minutes, however, the most repeatable process with the lowest standard deviation was achieved for the process that lasted 60 minutes. In the case of charcoal and its mixture with lignite coal, slight relation between gasification and process duration has been observed. After 15 minutes of the process, gassing process at the average level of about 39,18% was obtained. The lowest standard deviation was obtained for the process that lasted 30 minutes.

Table 3. Averaged percentage values of gasified samples, minimum, maximum and standard deviation for all types of coal

Hard coal	**MIN**	**MAX**	**AVERAGE**	**STANDARD DEVIATION**
15 minutes	42.92 %	53.92 %	47.89 %	0.055760829
30 minutes	43.46 %	53.14 %	48.23 %	0.048416664
60 minutes	44.64 %	53. 04%	50.06 %	0.046990247
			48.72 %	
Hard coal and charcoal	**MIN**	**MAX**	**AVERAGE**	**STANDARD DEVIATION**
15 minutes	34.01 %	39.13 %	35.76 %	0.029165
30 minutes	42.02 %	49.47 %	44.81 %	0.040618
60 minutes	39.84 %	46.75 %	43.00 %	0.034917
			41.19 %	
Lignite coal	**MIN**	**MAX**	**AVERAGE**	**STANDARD DEVIATION**
15 minutes	58.19%	60.90 %	59.26%	0.0144411
30 minutes	56.57 %	60.83%	58.91%	0.0216083
60 minutes	60.81 %	62.85 %	61.52%	0.0115539
			59.89 %	
Lignite coal and hard coal	**MIN**	**MAX**	**AVERAGE**	**STANDARD DEVIATION**
15 minutes	51.69 %	58.45 %	55.03%	0.0338083
30 minutes	50.49 %	56.97 %	52.75%	0.0365503
60 minutes	47.30 %	53.70%	50.65%	0.0299377
			52.81%	
Lignite coal and charcoal	**MIN**	**MAX**	**AVERAGE**	**STANDARD DEVIATION**
15 minutes	47.62%	52.43%	50.03 %	0.0340118
30 minutes	48.65 %	52.26%	50.46%	0.0255266
60 minutes	58.99%	58.99%	58.99%	0.0210526
			53.16%	
Charcoal	**MIN**	**MAX**	**AVERAGE**	**STANDARD DEVIATION**
15 minutes	35.83 %	49.27 %	39.81 %	0.064108
30 minutes	38.93 %	41.70 %	40.10 %	0.012858
60 minutes	37.91 %	49.60%	41.91 %	0.052563
			40.61 %	

The research presented above shows that the coal that is gasified to the largest extent is lignite coal, which may result from its properties and chemical composition. This type of coal has the highest repeatability of the measurements for each time interval. The mixtures with lignite coal also achieved high degree of gasification above 50%. Hard coal is also a good raw material for gasification. It has been observed that coke was formed in the filling sieves after every measurement, which may have impact on low losses during the research.

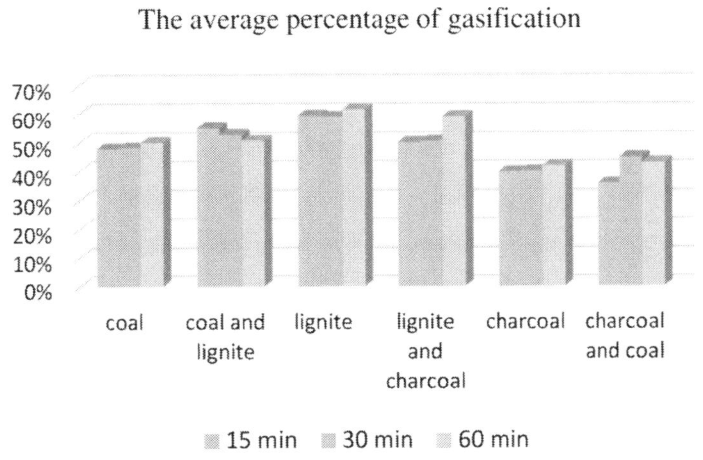

Fig. 8. Relative amount of gasified coal for various types of coal, process duration 15 min, 30 min, 60 min

The results were repeatable at constant level of about 48%. Apart from repeatability of the results, the mixture of hard coal and lignite coal was characterized by high temperature during the measurement, which results from fusion of coals of similar physical properties (temperature exceeded 950 degrees two times). Charcoal is characterized by the lowest loss of initial mass of coal at the level of about 40%. During the research on charcoal and its mixtures, it was difficult to achieve the temperature of 800 degrees. Time of conducting the research only slightly affects the value of gasification mixtures - at the level of a few per cent. Gas formed during coal gasification contains H_2, CO, CH_4, NO, NO_2, CO_2, CH_4, as well as side tarry substances.

The average value of gasification of carbon mixtures was obtained in conducted research, understood as relative loss of initial mass of coal (Fig. 9):

- lignite coal 59,89 %,
- the mixtures of lignite coal and hard coal 52,81 %,
- the mixtures of lignite coal and charcoal 53,16%,
- hard coal 48,72%,
- the mixtures of hard coal and charcoal 41,19 %,
- charcoal 40,61%.

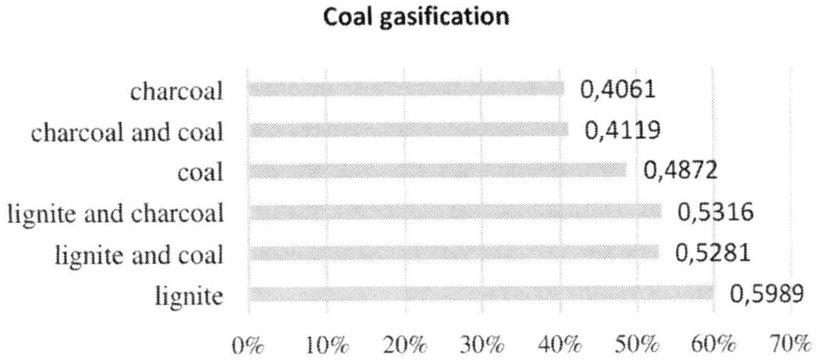

Fig. 9. Relative loss of initial mass of coal during gasification

4 Conclusion

Coal gasification is a controlled process, characterized by high efficiency. Its advantage is the fact that it reduces emission of pollutants from combustion process, which makes this process environmentally friendly. One of the products of coal gasification in the presence of hydrogen is methane, depending on the source of origin, it can be biomethane. It can be applied for production of electric and thermal energy, but also as a renewable motor fuel, which is characterized by high efficiency. Conducted research showed that temperature slightly affects the process of gasification, however, in order to increase the share of burnable components in the reactions, we must accept minimal temperature of 800 degrees. Lignite coal is characterized by the highest reactivity in comparison with hard coal or charcoal, achieving coal gasification at the level of nearly 60% after 15 minutes of experimental research. The use of hydrogen as a gasifying factor seems to be a prospective direction of development of the process of coal gasification. The works on coal gasification in the presence of hydrogen and the works on higher deposits of coal should be continued. It will cause many problems, mainly rinsing the deposit with hydrogen.

References

1. Bigda J., Burchart-Korol D., Porada S., Chart of technology solutions for coal gasification processes. Przegląd górniczy, pp. 86 – 96, (2014)
2. Borkiewicz M. Zgazowanie węgla wodór nośnikiem energii. Biuletyn Górniczy, **5-6**, pp. 155-156, (2008)
3. Chmielniak T., Sobolewski A., Tomaszewicz G., Zgazowanie węgla przy wykorzystaniu CO2 jako czynnika zgazowującego. Doświadczenia IChPW. Przemysł Chemiczny **94,** 4, pp. 442-448, (2015)
4. Chmielniak T., Bigda J., Czardybon A., Popowicz J., Tomaszewicz G., Technologie oczyszczania gazu ze zgazowania węgla. Przemysł Chemiczny, **97**, 2, pp. 1000÷1010, (2014)
5. Chmielniak T., Stelmach S., Współczesne technologie zgazowania węgla, **13**, 2 p. 69-76, (2009)
6. Gawlik L., Mokrzycki W., Paliwa kopalne w krajowej energetyce – problemy i wyzwania. Energy Policy Journal, **20**, 4, pp. 6-26, (2017)
7. IEA Publications 2014. World Outlook EIA
8. Kamble AD., Saxenaa VD., Chavan PD., Mendhe VA., Co-gasification of coal and biomass an emerging clean energy technology: Status and prospects of development in Indian context. International Journal of Mining Science and Technology, (2018)
9. Kezhong L., Zhangand R., B and J., Experimental study on syngas production by co-gasification of coal and biomass in a fluidized bed. Int J Hydrogen Energy, **35**, 7, pp. 2722–2726, (2010)
10. Kumabe K., Hanaoka T., Fujimoto S., Minowa T., Sakanish K., Co-gasification of woody biomass and coal with air and steam. Fuel, **86**, 5-6, pp. 684–689, (2007)
11. Minchener A. J., Coal gasification for advanced power generation. Fuel, **84**, pp. 2222-2235, (2005)
12. Porada S., Otrzymywanie substytutu gazu ziemnego na drodze hydrozgazowania węgla. Karbo Energochemia Ekologia, **5**, 186-188, (1996)
13. Smoliński A., Howaniec N., Badania procesu zgazowania węgla kamiennego w obecności K_2CO_3 Research reports mining and environment, **1**, pp. 81- 92, (2012)

E3S Web of Conferences **46**, 00031 (2018) https://doi.org/10.1051/e3sconf/20184600031
3rd International Conference on Energy and Environmental Protection

The impact of the condensation process on the degree of cleaning of flue gases from acidic compounds

Piotr Szulc[1,*], *Tomasz* Tietze[1], and *Daniel* Smykowski[1]

[1]Wroclaw University of Science and Technology, Wybrzeże Wyspiańskiego 27, 50-370 Wroclaw, Poland

Abstract. The paper presents studies on the impact of the process of condensation of water vapour on the process of cleaning of flue gases from acidic compounds. The measurements were carried out on a pilot-scale plant for waste heat recovery from flue gases, taking into account the process of condensation of the water vapour contained in them. The plant was connected to a lignite-fired power unit with a capacity of 360 MW located at PGE GiEK S.A., Bełchatów Power Plant Branch. The impact of the condensation of water vapour on the reduction of sulphur, chlorine and fluorine forming acidic compounds was examined. The studies show that the condensation process is conducive to removal of acidic compounds from flue gases.

1 Introduction

In the majority of coal-fired power plants, flue gases, after cleaning in a desulphurization plant, are directed to a chimney. The temperature of flue gases is higher than the temperature of condensation of water vapour, because the condensing water in combination with sulphur compounds causes, inter alia, the corrosion of the flue gas duct walls. However, there are plants, in which flue gases are cooled below the temperature of condensation of water vapour. Such solutions enable the recovery of waste heat from flue gases and a reduction in pollution emissions. The most interesting plants are those used in Scandinavia, in which waste heat recovery systems are integrated with flue gas desulphurization systems [5,6,7]. In 1998, a flue gas cleaning system with condensation of the water vapour contained in flue gases was launched in Gothenburg [5]. In this system, the temperature of flue gases is gradually reduced from 220°C to 140°C upstream of the flue gas cleaning reactor. In the reactor, flue gases are subjected to preliminary cleaning and in addition they cooled down to the temperature of 60°C and then directed to the condensing part, where their temperature is reduced to 40°C. The process of condensation increases the reduction in the emission of pollutants. After cleaning, the flue gases are heated to the temperature of 150°C in order to avoid the corrosive action on the duct and chimney walls. A similar solution was implemented in the Holstebro power plant [7]. In this case the flux of heat recovered from flue gases is approximately

* Corresponding author: piotr.szulc@pwr.edu.pl

14 MW. This is not a large value, as the main objective of the system is to reduce the emission of pollutants. Like in the power plant in Gothenburg, flue gases must be heated upstream of the chimney duct. In the largest power plant in Denmark in Copenhagen, the maximum flux of the recovered heat, which is used for district heating purposes, is 17 MW. Another interesting solution is the heat and water recovery system with flue gas condensation implemented in the power plant in Sønderborg in 2007 [6]. The condensate formed as a result of condensation of water vapour is returned to the flue gas cleaning process and thereby no fresh water is used during normal operation. The recovered waste heat with the flux of 6 MW is directed to a district heating network. However, in all these cases the flux of the recovered heat is low, while the primary purpose of these plants is to clean flue gases.

Large streams of water are used for the needs of cleaning of flue gases and process gases. In order to limit the consumption of fresh water, it is advantageous to use condensation systems for cooling down the flue gases coming from the combustion of lignite due to high moisture content. As shown in [1,2,3,4], the use of condensing heat exchangers enables recovery of a large flux of waste heat and reduces the emission of the pollutants. In Poland, the first condensing heat exchanger for recovery of waste heat from flue gases was built in the Białystok S.A. combined heat and power plant in 2015. The flue gases coming from a biomass-fired boiler are subjected to the processes of cooling and cleaning by sprinkling them with the condensate. As a result of this process, significant amounts of chlorides, heavy metals and dust are removed from flue gases. Then, the flue gases get to the condenser, in which water vapour is condensed. The recovered heat of condensation is directed to a municipal district heating network. Contaminated condensate coming from flue gases is subjected to the process of cleaning through microfiltration, ultrafiltration and reverse osmosis filter. The condensate obtained in this way is reused for industrial processes in the combined heat and power plant. The use of the heat recovery system significantly reduces the emission of pollutants into the atmosphere, as well as increases the efficiency of the combined heat and power plant through recovery of heat and water from flue gases. The paper presents the results of studies on the impact of condensation of water vapour from flue gases coming from the combustion of lignite on the reduction of emissions of pollutants, i.e. acidic compounds.

2 Test rig

The tests have been carried out in a pilot-scale plant for recovery of waste heat from flue gases, which was built at PGE Górnictwo i Energetyka Konwencjonalna S.A., Bełchatów Power Plant Branch. The pilot-scale plant was supplied with flue gases from the combustion of lignite in a power unit with a capacity of 360 MW [2]. Diagram of the plant is shown in Fig. 1, while its view is presented in Fig. 2. The main element of the pilot-scale plant for waste heat recovery was a shell and tube condensing heat exchanger with a capacity of 312 kW. The heat exchanger was installed in the flue gas duct downstream of the main induced draft fan and upstream of the flue gas desulphurization system. In order to adjust the mass flow rate of flue gases passing through the condensing heat exchanger, a supporting fan was installed upstream of the heat exchanger. Flue gases flowing through the condensing heat exchanger were cooled with water using a system of PTFE pipes. As a result of cooling down the flue gases to a temperature below the dew point, condensation of the water vapour contained in the flue gases and liquefaction of the condensate took place. The condensate collected by the tray was directed to the sewage system. The pilot-scale plant was equipped with a measurement recording system as well as systems for controlling the main thermal and flow parameters. In order to determine the composition of the flue gases, there were made holes, through which samples were collected at the inlet and at the outlet of the condensing heat exchanger. The measurements were carried out continuously for 220 minutes.

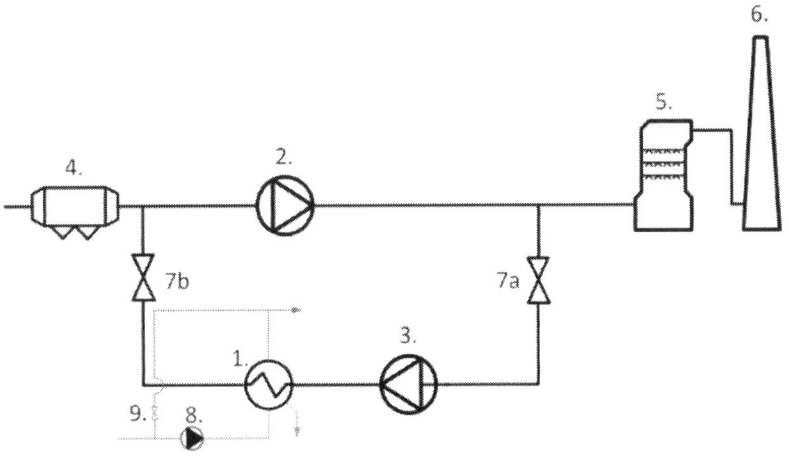

Fig. 1. Diagram of the pilot-scale plant for investigating waste heat recovery from flue gas. 1-condensing heat exchanger, 2 – main induced draft fan in the flue gas duct, 3 – supporting fan, 4 - system for removal of ash from flue gas, 5 – flue gas desulphurization system, 6 – chimney, 7 – shut-off gate valve, 8 – cooling water pump, 9 – control solenoid valve

Fig. 2. View of the test rig

3 Research methodology

The measurements of the composition of flue gases were carried out on a continuous basis. Their aim was to determine the degree of reduction in content of pollutants present in flue gases as a result of the condensation of water vapour. The scope of the studies presented in the paper covered measurements of concentrations of NOx, SO_2, HCl, and HF, as well as measurements of the volumetric flow rate of flue gases. Sampling of the values took place at a frequency of 1 Hz and the recorded values were averaged every 60 seconds. The values averaged over the period of 600 seconds were taken for the analysis. In order to draw in a sample of flue gases, a heated probe, heated hoses and a particulate filter were used. Dust-free flue gases were transported through heated hoses to a diaphragm pump with a heated head. A part of the flue gas stream was directed to a two stage cooler that removed the water vapour contained in the flue gases, while dry and clean gas was directed to the analyser. The main measurements of the concentration of gas components in the flue gases were performed in the analyser. The uncertainty of the determination of the content and concentration was as follows: volumetric flow – approx. 3.3%, particulate matter content – approx. 5.7%, sulphur dioxide content – approx. 2.0%, nitrogen oxides content – approx. 7.0%, carbon monoxide content – approx. 3.0 %, hydrogen chloride content – approx. 4.0%, hydrogen fluoride content – approx. 12%, moisture content – approx. 7%, oxygen content – approx. 2%.

4 Analysis of the chemical composition of flue gases

In the condensing heat exchanger, to which the flue gases containing water-soluble and/or water-reactive compounds are supplied, there occur both physical and chemical processes. Water in the liquid phase, which occurs as a result of condensation, absorbs a certain part of these compounds, while some of them not only dissolve in water, but also react with it, which leads to the formation of aggressive compounds. Acidic compounds and those reacting with water to form acids are of particular importance in this case. This group includes, inter alia, nitrogen and sulphur oxides, as well as hydrogen chloride and hydrogen fluoride. These gases form acids as a result of the reaction with water. In the case of nitrogen oxides (NO_x), the following reactions are of greatest importance:

$$2NO + O_2 \leftrightarrow 2NO_2$$

$$2NO_2 \leftrightarrow N_2O_4$$

$$2NO_2 + H_2O \rightarrow HNO_2 + HNO_3$$

$$3NO_2 + H_2O \leftrightarrow 2HNO_3 + NO$$

$$N_2O_4 + H_2O \leftrightarrow HNO_3 + HNO_2$$

$$2HNO_2 \leftrightarrow NO + 1/2N_2O_4 + H_2O$$

$$3HNO_2 \leftrightarrow 2NO + HNO_3 + H_2O$$

$$NO + 3/4O_2 + 1/2H_2O \leftrightarrow HNO3$$

which result in formation of nitric acid (V) – HNO_3. Sulphur oxides (SO_x) also react with water to form an acid, in particular SO_2, which usually is the main sulphur compound contained in flue gases. Depending on the conditions, e.g. presence of water and oxygen, sulphur dioxide may oxidise to sulphur trioxide

$$SO_2 + 1/2O_2 \leftrightarrow SO_3,$$

which in turn reacts with water, which results in formation of sulphuric acid (VI)

$$SO_3 + H_2O \leftrightarrow H_2SO_4.$$

Two other gases present in the composition of the flue gases: HCl and HF, after the dissolution in water, form hydrochloric and hydrofluoric acid, respectively. As a result, the condensate discharged from the heat exchanger is a mixture of four strong acids (this is confirmed by measurements of pH of the condensate), which shows an aggressive action in relation to the materials (mainly metals), of which the plant is built. This is proved by advanced corrosion processes leading to the destruction of the elements that are in direct contact with the condensate.

The analysis of the content of individual components of flue gases identified at the inlet and at the outlet of the heat exchanger in combination with the analysis of the composition of the condensate allows determining the predominant chemical processes accompanying the phenomenon of condensation.

The content of the flue gas components analysed at the inlet and at outlet of the condensing heat exchanger is shown in Figs. 3–6, while their reduction – in Fig. 7.

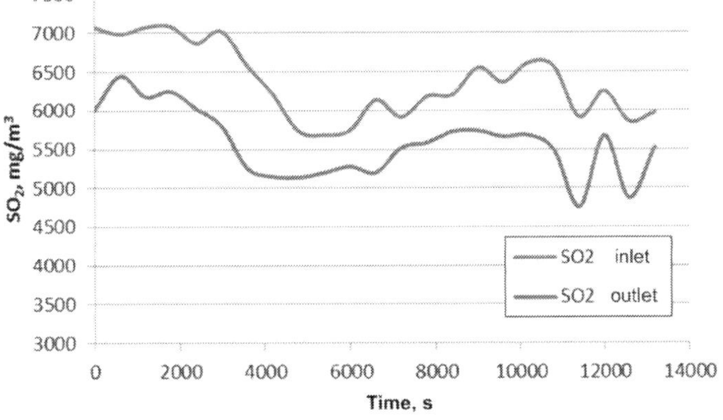

Fig. 3. *Content of SO_2 in flue gases at the inlet and at the outlet as a function of time*

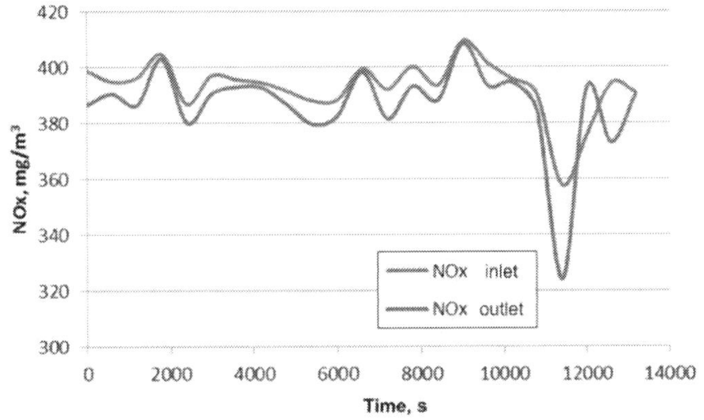

Fig. 4. *Concentration of NO_x at the inlet and at outlet of the condensing heat exchanger as a function of time*

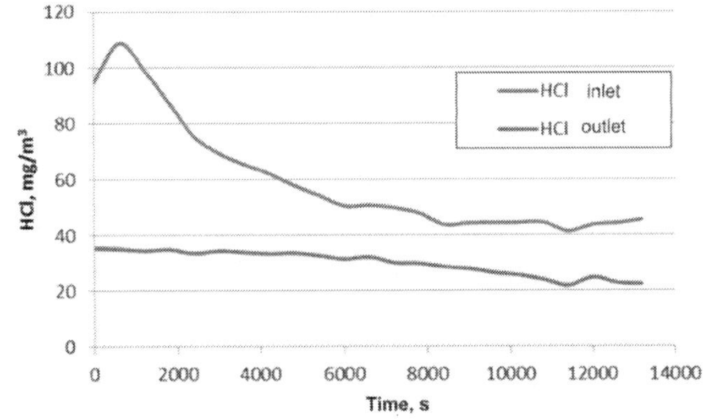

Fig. 5. *Content of HCl in the flue gases upstream and downstream of the condensing heat exchanger as a function of time*

Among the acidic compounds analysed, sulphur dioxide has the largest share in the flue gases – both at the inlet and at the outlet (Fig. 3). The components with the second largest content are nitrogen oxides (Fig. 4), while other components occur in much smaller amounts. A comparison of the content of individual gases in the flue gases allows determining which of them are absorbed to the highest degree by the condensate formed in the heat exchanger. The largest difference in the concentration between the inlet and outlet can be noticed for hydrogen fluoride (HF) and for hydrogen chloride (HCl). In turn, the concentration of nitrogen oxides (NOx) changes only slightly. A significant difference also occurs in the case of sulphur dioxide (SO_2), which is shown in Fig. 3.

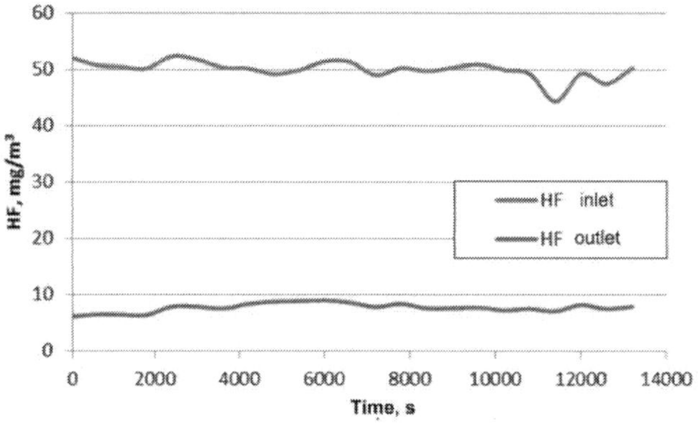

Fig. 6. *Content of HF in the flue gases upstream and downstream of the condensing heat exchanger as a function of time*

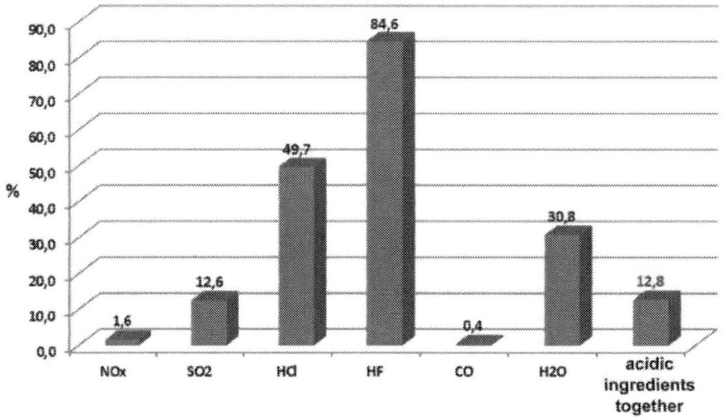

Fig. 7. *Reduction in the content of NOx, SO_2, HCl, HF, CO, H_2O and the total reduction of acidic components in the condensing heat exchanger, %*

The difference in the concentration of individual components in the inlet and outlet flow of flue gases was characterized by the reduction, % (Fig. 7), defined as:

$$R = \frac{(c_1 - c_0)}{c_1} \cdot 100\% \tag{1}$$

where :

- R – reduction, %,
- c_I – content (concentration) of the component at the inlet of the heat exchanger, mg/m^3,
- c_O – content (concentration) of the component at the outlet of the heat exchanger, mg/m^3.

Fig. 7 shows that the content of hydrogen fluoride changes to the highest degree as a result of using a condensing heat exchanger. The flue gases leaving the heat exchanger contain less than 20% of the HF contained in the flue gases entering the heat exchanger, while in the case of HCl – it is approximately half. After passing the flue gases through the heat exchanger, the content of SO_2 drops by only 12.6%, however it should be emphasized that there is much more sulphur dioxide in the flue gases than other components. As a result, this allows reducing the emission of sulphur dioxide to a significant degree. The total percentage of the reduction in the content of acidic components, i.e. HF, HCl, NOx and SO_2 is 12.8%, which is almost the same as in the case of the reduction in the content of SO_2. This is due to the fact that sulphur dioxide has a predominant share in the flue gases among these compounds.

When comparing the level of reduction in the content of individual components in the flue gases, it can be observed that it shows a certain correlation to the solubility of these components in water. It happens like that because the time of contact between the flue gases and the condensate is the same for each component.

5 Analysis of the chemical composition of the condensate

Fig. 8 shows the content of the selected ions in the condensate. As expected, sulphate ions (VI) have the largest share in the condensate. They are formed as a result of the reaction of sulphur dioxide with water. Although only slightly more than 12% of SO_2 is absorbed in the condensate (assuming that the entire loss of the component in the flue gases gets to the condensate), a large amount of this gas in the flue gases causes that ions of SO_4^{2-} are a predominant component of the condensate. Among the ions that are acid radicals, the fluoride ion is in the second place in terms of the content in the condensate and is followed by the chloride ion. This means that the presence of three acids (sulphuric (VI), hydrofluoric and hydrochloric) is responsible for the aggressive properties of the condensate. The presence of acids in the condensate is confirmed by its pH value of 2.6.

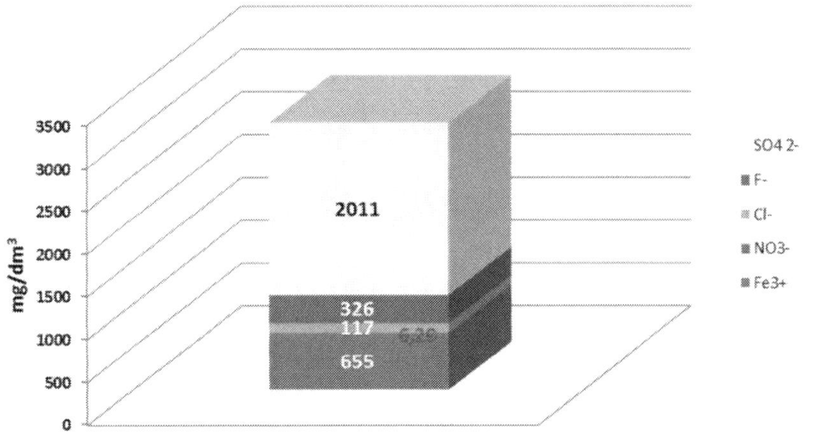

Fig. 8. *Content of SO_4^{2-}, F^-, Cl^-, NO_3^-, Fe^{3+} in the condensate, mg/dm^3*

It is noteworthy that the condensate contains a significant amount of iron – its content in the condensate is the largest among all metal ions. This is most likely caused by the corrosive processes consisting in an aggressive action of acidic condensate on steel, from which the plant is built.

6 Conclusion

The paper presents studies on the impact of the process of condensation of water vapour on the process of cleaning of flue gases from acidic compounds, such as sulphur, nitrogen, chlorine and fluorine compounds. The measurements were carried out on a pilot-scale plant for waste heat recovery from flue gases, taking into account the process of condensation of the water vapour contained in them. The plant was connected to a lignite-fired power unit with a capacity of 360 MW located at PGE GiEK S.A., Bełchatów Power Plant Branch. The studies were carried out by measuring the content of acidic compounds upstream and downstream of the heat exchanger. The content of these compounds in the condensate was also examined. Based on the research results presented, it can be concluded that the condensation of water vapour from flue gases has a decisive impact on the composition of flue gases downstream of the heat exchanger. The concentration of hydrogen fluoride is reduced to the highest degree – on average by 85%, while hydrogen chloride is reduced by 47% and sulphur dioxide by 12.5%. The concentration of nitrogen oxides is reduced only to a small extent – by 1.6%. From the analyses of the condensate it appears that SO_4^{2-} ions constitute a predominant component of the condensate. Among the ions that are acid radicals, the fluoride ion is in the second place in terms of the content in the condensate and is followed by the chloride ion. The conclusion is that the condensation process is conducive to removal of acidic compounds from flue gases.

References

1. Szulc P., Tietze T., Wójs K. Numerical analysis of a waste heat recovery process with account of condensation of steam from flue gases, Archives of Civil and Mechanical Engineering. **15**, **4**,:pp 1017–1023, (2015)
2. Wójs K., Szulc P., Tietze T. *Odzysk niskotemperaturowego ciepła odpadowego ze spalin wylotowych*, scientific editor: Wójs K., Warszawa, Polish Scientific Publishers PWN,pp:1–111, (2015)
3. Szulc P., Tietze T., Wójs K. *Wpływ strumienia objętości spalin na strumień niskotemperaturowego ciepła odpadowego odzyskanego w kondensacyjnym wymienniku ciepła. Analiza systemów energetycznych: collective work*, edited by Bohdan Węglowski, Piotr Duda. Kraków: Publishing House of the Cracow University of Technology, pp. 317–330 (in Polish), (2013)
4. Mikielewicz D., Wajs J., Ziółkowski P., Mikielewicz J., Utilisation of waste heat from the power plant by use of the ORC aided with bleed steam and extra source of heat, Energy, **97**, pp. 11-19, (2016)
5. Condensing flue gas cleaning system with electric power and heat production at the waste incineration heating plant in Gothenburg, Götaverken Miljö AB, www.gmab.se.
6. Extended energy recovery using flue gas condensation with condensate treatment, Götaverken Miljö AB. www.gmab.se .
7. Multi-functional wet ADIOX® scrubbers with integrated energy recovery, Götaverken Miljö AB. www.gmab.se .

E3S Web of Conferences **46**, 00032 (2018) https://doi.org/10.1051/e3sconf/20184600032
3rd International Conference on Energy and Environmental Protection

Utilization of rigid polyurethane foam waste in removal of uncontrolled releases of hazardous substances

Tomasz Węsierski[1,*]

[1]Szkoła Główna Służby Pożarniczej (Main School of Fire Service), Juliusza Słowackiego 52/54, 01-629 Warsaw, Poland

Abstract. In 2017 in Poland 12522 incidents involving hazardous substances were recorded, of which 23 have been classified as large and very large. Apart from the share of carbon monoxide, most of them concerned uncontrolled petroleum product releases. In the same year over 427 m^3 of foaming agents were used. Consequently, an extremely important issue at the intervention level is the efficient and safe removal of the resulting pollution. Tests of maximal absorption proprieties with the use of the modified Westinghouse method executed on a wide range of industrially used diverse industrial toxic substances and various classes of waste of firefighting foam agents have demonstrated substantial usability of rigid polyurethane foam waste. The obtained maximal theoretical sorptivity were within the range of 1.87-5.11 $kg{\cdot}kg^{-1}$. It has also been proven that absorption proprieties in relation to waste firefighting foaming agents were reduced. However, the maximum absorption proprieties of the tested rigid polyurethane foam waste are much higher than those for standard mineral sorbents used in rescue operations, which indicates a high potential for their practical use for emergency services.

1 Introduction

Data of the Polish Headquarter of State Fire Service indicate that only in 2017 there were 12522 incidents involving uncontrolled releases of hazardous substances in Poland, of which 23 were large and very large scale [1].

Apart from the share of carbon monoxide, most of them concerned uncontrolled petroleum products releases. In the same year over 427 m^3 foaming agents were used. A significant number of accidents involving dangerous substances make it necessary to search for new and more ecological solutions that enable the removal of environmental hazards generated during the incidents. An ever increasing attention is paid to problems related to post-incident waste of foaming agents. These agents, due to the surfactants contained in them, may have an adverse effect on the environment [2-10]. One of the solutions enabling an easy and safe removal of hazardous substances is the use of sorbents, which, depending on their bulk density, can be used both in land and water actions [11]. Their

* Corresponding author: wesierskitomas@poczta.onet.pl

non-reactivity to the absorbed substances, the ease and intuitive use, as well as the possibility of subsequent controlled utilization allows their use as a cheap and safe means of removal smaller quantities of dangerous substances on the spot. Given the characteristics of sorbents during actions carried out only by the State Fire Service in 2017 almost 797 tons were used.

2 Research methodology

Maximal theoretical sorptivity measurements were carried out using the modified Westinghouse method. The presented modification is an original research and interpretation concept. Water, water solutions of Fomtec AFFF 3% S and BIO FOR C foam concentrate (conc. 0.1-3%) and selected dangerous substances and their solutions used industrially were subjected to testing (Table 1). Pumice powder from PUMICE-I.M.P.A S.C being a waste of rigid polyurethane foam was used as sorbent. Before proceeding with the specific tests, the material was sieved to obtain a uniform powder with a grain size of Ø ≤1 mm. The microscopic picture of rigid polyurethane foam waste used in research is shown in Figure 1.

Table 1. Substances and solutions used in experiments

Measured substance and their content	Producer
1-Propanol (99.5%), Toluene (99.5%), Benzyl alcohol (99.8%), Ethylene glycol (98.5%), Cyclohexanol (99%), 2-Butanol (99.5%)	CHEMPUR
Glycerol (99.5%), n-Hexane (995%), Cyclohexane (99.5%), Propylene glycol (99.5%), Methanol (99.8%), 1-Butanol (99.5%) 3-Methylbutan-1-ol (99.5%), 1-Octanol (99%), Acetone (99.5%), 2.4-Dihydroxy-2-Methylpentane (99%), 2,2`-oxydiethanol (99%), chloroform (99.5%), Phosphoric acid (85% water solution), Ammonia (25% water solution), sodium hydroxide (40% water solution), sulphuric acid (pure and 30% water solution)	POCH
Acetic acid (99,5%), 2-Ethoxyethanol (99%), 1,4-Dioxane (99,8%), Cyclohexanone (99,5%), 4-Nitrotoluene (98%), Methyl acetate (98%), 2-Methyl-2-hydroxy-2-pentanone (99%), Ethyl benzoate (98%), 2-Methoxy-2-methylbutane (min 97%), Nitrobenzene (99%)	Sigma-Aldrich
4-methyl-2-pentanone (98%), 1-Decanol (98%)	Fluka
n-Octane (99,5%)	Merck
1-Undecanol	SERVA
Gasoline, Diesel oil	-
Formtec AFFF 3% S (solutions 0,1%, 0,5%, 3%). Fomtec AFFF 3% S - an aqueous film forming foam concentrate consisting of fluorocarbon and hydrocarbon surfactants	FORMTEC
BIO FOR C (solutions 0,1%, 0,5%, 3%). BIO FOR C - a synthetic class A foaming concentrate.	BIO-ex

10 grams (m_o) of the previously sieved sorbent were placed in a conical stainless steel strainer with a diameter of 70 cm and a height of 7.5 cm. The test sample prepared in this way was immersed in the sorbed liquid (sorbate). The sorbent was immersed in the sorbed liquid until it was completely wetted. Then the sample was pulled from the liquid and hung on the previously tared balance and the time measurement was started. During the measurement, the mass of rigid polyurethane foam together with sorbed liquid as a function of time for 40 minutes was recorded. The mass of sorbed liquid was calculated by subtracting the mass of the sieve and sorbent. Dividing this value by the sorbent mass, absorbency was obtained. On the basis of the obtained results, diagrams of temporary sorptivity as a function of residence time in the strainer were made. The value of the maximal theoretical sorptivity

(A_{max}) of the sorbent relative to a given liquid was obtained by extrapolating the evaporation straight to the OY axis.

An evaporating straight is defined as a straight related to the mass loss of sorbed liquid from the sorbent space due to the evaporation process. Extrapolation to the OY axis allowed taking into account the loss of sorptivity resulting from the evaporation of the sorbed liquid from the sorbent space, whereby the maximal theoretical sorptivity (Amax) was obtained at time t = 0. The method for determining A_{max} is shown in Figure 2.

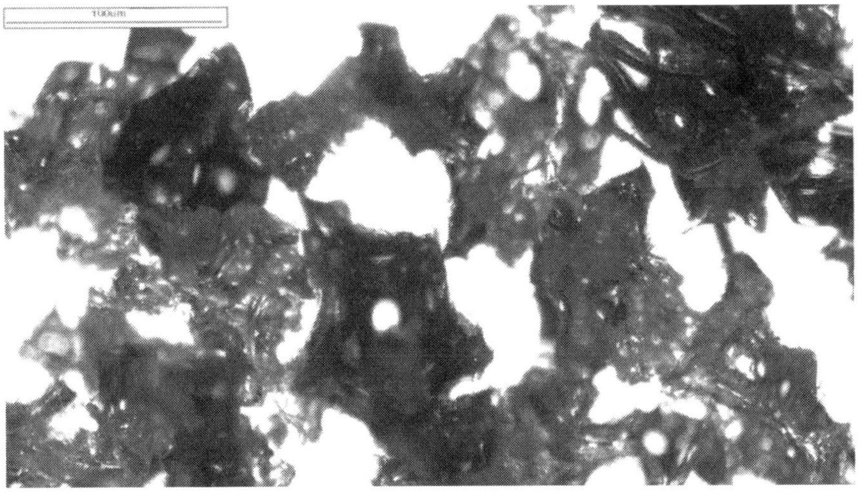

Fig. 1. Microscopic picture of rigid polyurethane foam used in investigations

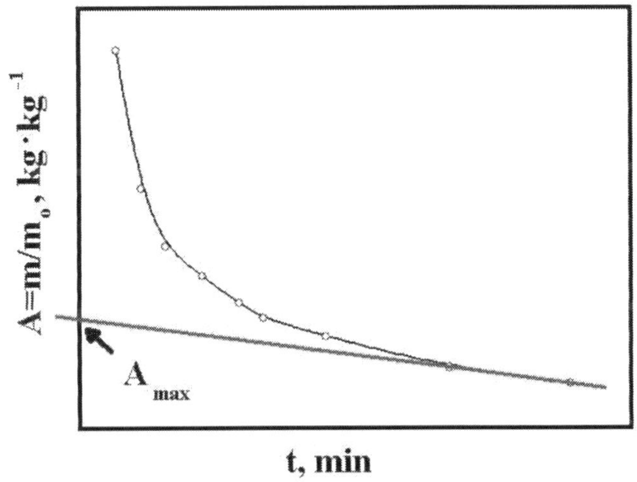

Fig. 2. Determination of the theoretical maximal sorptivity (Amax) obtained by extrapolation of evaporation straight line to the OY axis

In the initial stage of suspension in the strainer there is a leak and then dripping takes place of the sorbed liquid from the sorbent space. At a later stage, the dripping process ceases. Mass loss occurs as a result of evaporation of the sorbed liquid. The standard Westinghouse method in the calculation included only the sorbed mass of the sorbate after 40 min. of dripping without taking into account measurement and sorbate mass analysis as a function of time.

3 Results and discussion

Exemplary curves of dependence of temporary sorptivity as a function of residence time in sieve A = f (t) for selected sorbates being hazardous substances are shown in Fig. 3. Fig. 4 shows curves A = f (t) of selected solutions of foaming agents with different concentrations, and a comparison was then performed with the curve of water sorption. Extrapolation of the curve points associated with the evaporation stage to point t = 0 allowed the determination of the maximal theoretical sorptivity of sorbent A_{max} in relation to the sorbate, taking into account the mass which has evaporated. The results obtained are summarized in Table 2 and 3. The plot of A_{max} versus density of sorbed liquids is presented in Fig. 5.

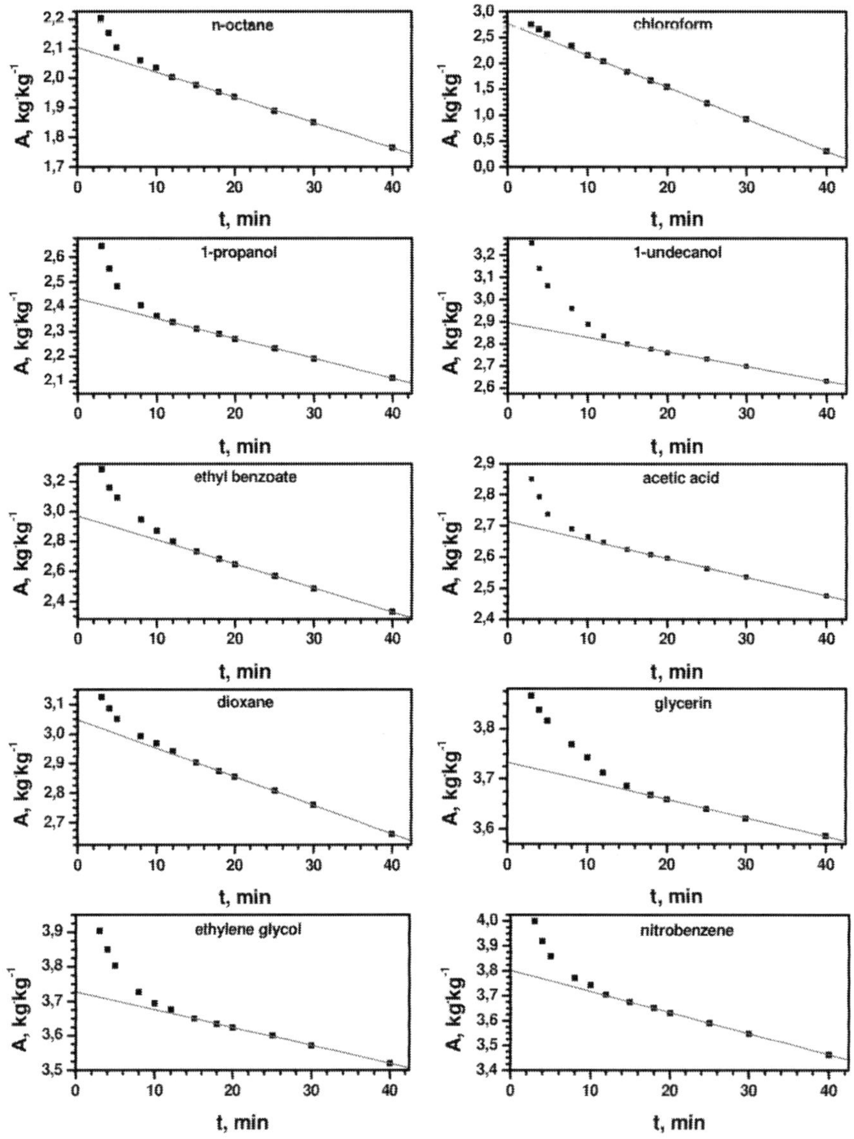

Fig. 3. Exemplary curves of dependence of temporary sorptivity as function of residence time in sieve A = f (t) for selected sorbates.

The obtained results indicate that the maximal theoretical sorptivity ranges from 1.87 to 5.11 kg·kg^{-1}, reaching the lowest value for n-Hexane and gasoline and the highest one for 85% phosphoric acid solution. A linear relationship between Amax and the density of sorbed liquid was ascertained. This is due to the fact that the available sorption space of the sorbent is the same, and any differences that may be observed could arise from excessively high surface tension or excessively high density that hinders the maintenance of liquid in the open pores. A surface tension that is too low facilitates leakage of the substance, similarly as an excessive specific gravity. This can be found by analyzing the curves A = f (t) for solutions of foam agents (Figure 4).

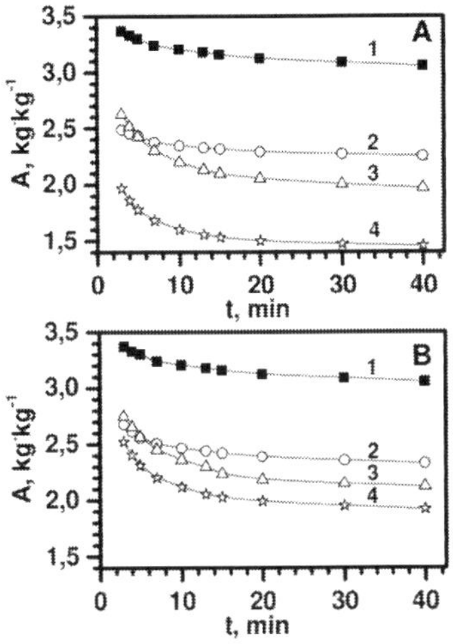

Fig. 4. Dependence of temporary sorptivity as a function of residence time in sieve A = f (t) for water and selected foam agent solutions of different concentrations. Fig. A: 1) water; 2) 0.1% Formtec AFFF 3% S; 3) 0,5% Formtec AFFF 3% S; 4) 3% Formtec AFFF 3% S; Fig. B: 1) water; 2) 0.1% BIO FOR C; 3) 0,5% BIO FOR C; 4) 3% BIO FOR C.

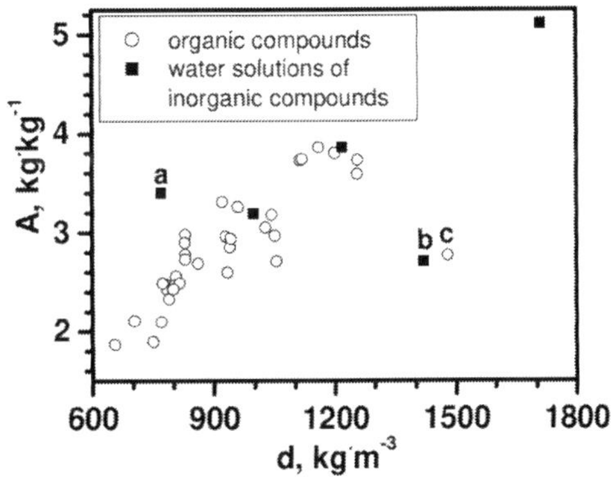

Fig. 5. Dependence of maximal theoretical sorptivity as a function of sorbate density with a breakdown into studied organic substances and aqueous solutions of inorganic substances and water. As points a, b and c are marked Amax of solutions that stand out from the trend line where: a) 25% ammonia water solution b) 40% NaOH water solution c) chloroform.

Their presence causes a decrease in surface tension and as the concentration increases, the decrease of the sorptivity is observed. On the other hand, the effect of specific gravity may be responsible for low chloroform sorption. The capillary forces would be unable to support the weight of sorbed liquid, which may be clearly seen on the basis of the very low A_{max} determined for this substance. Rigid polyurethane foam reaction with the investigated sorbates has not been ascertained except for concentrated H_2SO_4.

Table 2. Determined values of maximal theoretical sorptivity A_{max} and relative porosity AV.

Substance	d, kg·m^{-3}	A_{max}, kg·kg^{-1}	Av, m^3·m^{-3}
Metanol	787	2,47	0,590
1-Propanol	785	2,43	0,582
1-Butanol	806	2,56	0,597
2-Butanol	802	2,47	0,579
3-Methylbutan-1-ol	815	2,49	0,574
Cyclohexanol	960	3,26	0,638
1-Octanol	830	2,78	0,630
1-Decanol	830	2,98	0,675
1-Undecanol	828	2,90	0,658
2-Ethoxyethanol	930	2,96	0,598
Benzyl alkohol	1042	3,18	0,574
2-Methyl-2-hydroxy-2-pentanone	940	2,85	0,570
Ethylene glycol	1113	3,73	0,630
Propylene glycol	1258	3,59	0,537
2,2`-oxydiethanol	1118	3,74	0,629
2,4-Dihydroxy-2-Methylpentane	922	3,31	0,675
Glicerol	1258	3,73	0,557
n-Hexane	655	1,87	0,537
Gasoline	750	1,90	0,476
Cyclohexane	774	2,49	0,605
n-Octane	703	2,11	0,564
Toluene	862	2,69	0,587
Diesel oil	830	2,73	0,618
Acetone	790	2,33	0,554
4-methyl-2-pentanone	800	2,43	0,571
Cyclohexanone	942	2,94	0,587
Nitrobenzene	1200	3,80	0,595
4-Nitrotoluene	1160	3,86	0,626
Methyl acetale	934	2,60	0,523
Ethyl benzoate	1050	2,97	0,532
2-Methoxy-2-methylbutane	770	2,10	0,513

Substance	d, kg·m-3	A_{max}, kg·kg-1	Av, m3·m-3
Chloroform	1480	2,77	0,352
1,4-Dioxane	1027	3,05	0,558
Acetic acid	1055	2,71	0,483
85 % Phosphoric acid	1710	5,11	0,562
30% Sulphuric acid	1,22	3,86	0,595
25 % Ammonia	770	3,40	0,830
40% Sodium hydroxide	1420	2,71	0,359
Water	997	3,19	0,602

4 Conclusion

The non swimming sorbents most frequently used in rescue operations have sorptivity within the range of 0.5-1.2 kg·kg^{-1}. Therefore, due to its sorptivity the waste of rigid polyurethane foam may be a very good starting material for the generation of a high performance sorbent. The sorptivity of the sorbent is reduced in the presence of foam agent solutions. Apart from concentrated sulphuric acid, the reaction of rigid polyurethane foam with the investigated sorbates has not been ascertained. Thanks to the low bulk density the sorbent could be used as filling of sorption mats designated for sorbing oil-based materials from the water surface, however, this requires additional surface hydrophobization. Low bulk density would hinder the use of sorbent in wind conditions.

The research was carried out as part of the DOB-BIO6 / 06/11/2014 project financed by NCBiR "Mobile turbine rescue and firefighting system."

References

1. http://www.straz.gov.pl/panstwowa_straz_pozarna/interwencje_psp(access 07.05.2018)
2. Król B., Prochaska K., Chrzanowski Ł., Biodegradability of Firefighting Foams, Fire Technology, **48**, pp. 173-181 , (2012)
3. Bourgeois A., Bergendahl J., Rangwala A., Biodegradability of fluorinated fire-fighting foams in water, Chemosphere, **131**, pp. 104-109, (2015)
4. Colville S., McCarron N., Environmental issues associated with defence use of aqueous film forming foam (AFFF), Environmental Stewardship, Environment, Heritage and Risk Branch, May (2003)
5. Rakowska J., Porycka B., Zagrodzka M., Ekologiczne kierunki zmian bazy surowcowej w koncentratach gaśniczych, Bezpieczeństwo i Technika Pożarnicza, **19**, pp. 107-116, (2010)
6. Hamilton S., Larson D., Finger S., Poulton B., Vyas N., Hill E., Ecological effects of fire retardant chemicals and fire suppressant foams, "Northern Prairie Wildlife Research Center", Jamestown (1998)
7. Caracciolo A. B, Cardoni M., Pescatore T., Patrolecco L., Characteristics and environmental fate of the anionic surfactant sodium lauryl ether sulphate (SLES) used as the main component in foaming agents for mechanized tunneling, Environmental Pollution, **226**, pp. 94-103, (2017)

8. Lechuga M., Fernández-Serrano M., Jurado E., Núñez-Olea J., Rios F., Acute toxicity of anionic and non-ionic surfactants to aquatic organisms, Ecotoxicology and Environmental Safety, **125**, pp. 1-8, (2016)
9. Teresa Garcia M., Kaczerewska O., Ribosa I., Brycki B., Materna P., Drgas M., Biodegradability and aquatic toxicity of quaternary ammonium-based gemini surfactants: Effect of the spacer on their ecological properties, Chemosphere, **154**, pp. 155-160, (2016)
10. Garcia M.T., Campos E., Sanchez-Leal J., Ribosa I., Effect of the alkyl chain length on the anaerobic biodegradability and toxicity of quaternary ammonium based surfactant, Chemosphere, **38**, pp. 3473-3483, (1999)
11. Gancarczyk D, Jakubiec J., Sobolewski M, Effect of Sorbent Type on Efficiency of Oil Pollution Removal from Road Surface, Zeszyty Naukowe SGSP, **61**, pp. 53-63, (2017)

AUTHOR INDEX